# Springer Series in Synergetics

Editor: Hermann Haken

Synergetics, an interdisciplinary field of research, is concerned with the cooperation of individual parts of a system that produces macroscopic spatial, temporal or functional structures. It deals with deterministic as well as stochastic processes.

Volume 40 **Information and Self-Organization**
A Macroscopic Approach to Complex Systems By H. Haken

Volume 41 **Propagation in Systems Far from Equilibrium**
Editors: J.E. Wesfreid, H.R. Brand, P. Manneville, G. Albinet, N. Boccara

Volume 42 **Neural and Synergetic Computers** Editor: H. Haken

Volume 43 **Cooperative Dynamics in Complex Physical Systems**
Editor: H. Takayama

Volume 44 **Optimal Structures in Heterogeneous Reaction Systems**
Editor: P.J. Plath

Volume 45 **Synergetics of Cognition** Editors: H. Haken, M. Stadler

Volume 46 **Theories of Immune Networks** Editors: H. Atlan, I.R. Cohen

Volume 47 **Relative Information** Theories and Applications
By G. Jumarie

Volume 48 **Dissipative Structures in Transport Processes and Combustion**
Editor: D. Meinköhn

Volume 49 **Neuronal Cooperativity** Editor: J. Krüger

Volume 50 **Synergetic Computers and Cognition**
A Top-Down Approach to Neural Nets By H. Haken

Volume 51 **Foundations of Synergetics I** Distributed Active Systems
By A.S. Mikhailov

Volume 52 **Foundations of Synergetics II** Complex Patterns
By A. Yu. Loskutov, A. S. Mikhailov

Volume 53 **Synergetic Economics** Time and Change in Nonlinear Economics
By W.-B. Zhang

Volume 54 **Quantum Signatures of Chaos** By F. Haake

Volume 55 **Synergetics of Rythms**
Editors: H. Haken, P. Koepchen

Volumes 1–39 are listed on the back inside cover

Wei-Bin Zhang

# Synergetic Economics

## Time and Change in Nonlinear Economics

With 92 Figures

Springer-Verlag

Berlin Heidelberg New York London
Paris Tokyo Hong Kong Barcelona

Dr. Wei-Bin Zhang
Institute for Futures Studies, Hagagatan 23A 3 tr., S-11347 Stockholm, Sweden

*Series Editor:*
Professor Dr. Dr. h. c. Hermann Haken
Institut für Theoretische Physik und Synergetik der Universität Stuttgart,
Pfaffenwaldring 57/IV,
D-7000 Stuttgart 80, Fed. Rep. of Germany and
Center for Complex Systems, Florida Atlantic University,
Boca Raton, FL 33431, USA

ISBN-13: 978-3-642-75911-6 e-ISBN-13: 978-3-642-75909-3
DOI: 10.1007/978-3-642-75909-3



2154/3154-543210 – Printed on acid-free paper

*To my Parents*

*who have suffered much from my having been away from home for so long*

. . . if orthodox economics is at fault, the error is to be found not in the superstructure, which has been erected with great care for logical consistency, but in a lack of clearness and of generality in the premises.

*J. M. Keynes* (1936)

# Foreword

This is a book about the dynamics of economic and other social systems. It was written at the Swedish Institute for Futures Studies and is oriented towards the problem of understanding economic evolution and rapid structural change.

The analysis is very closely related to synergetics. This implies that Dr. Zhang has focused considerable attention on the fact that economic and other social variables can be subdivided into subsets regulated by *slow* and *fast* processes, respectively. Some of the slow variables turn out to be of collective importance, i.e. they tend to act as order parameters of economic and social systems.

More or less mathematically, such a distinction has also been made in earlier analyses of dynamic economics. Alfred Marshall made the distinction in his nineteenth century textbook and Paul Samuelson did likewise in his *Foundations of Economic Analysis* in the 1940s. But they did not see the possibility of explicit solutions to the important dynamic economic problems implicit in such an approach to economics. Dr. Zhang not only points in the same direction, he also shows how the synergetic approach works in long term dynamic analyses of important development problems. One of the most important conclusions is that, with a proper subdivision into fast and slow interactive subsystems, predictability can be achieved in a system which would otherwise be unpredictable, i.e. chaotic. Furthermore, the analysis shows that certain order-enforcing variables are suitable as strategic policy instruments available for policy making. Most of these variables are slowly changing and can thus be seen as order parameters at the level of the economic system. This characteristic automatically implies that they become part of strategic decision making, i.e., instruments of future-oriented policies.

Studies of the future are clearly of importance; however, they can easily become utopian daydreaming if not based upon a stringent methodological foundation. With this book Dr. Zhang has provided one of the pillars of such a foundation.

Åke E. Andersson
Professor of Economics at Umeå University
and Director of the Swedish Institute for Futures Studies

# Preface

With the passage of time, not only does economic life change, but there is even a shift in the dominant economic ideas. While the classical economists, such as Smith, Ricardo, Malthus, Marx, Mill, Walras, and Marshall, have had and will have their days in different cultures at different times, it is still too early to judge the historical significance of contributions to economics by modern economists. Time is the arbiter of truth. Only time can make us wise enough to recognize the superficiality of ideas which initially sounded important and far reaching.

Not only the general public, but even many economists are tending to lose confidence in the applicability of economics to reality, although the level of knowledge of economics has increased greatly in recent years: there seems to be no simple relation between knowledge of and confidence in science.

One may conceive of different reasons why economics fails to explain reality. On the one hand, the complexity of the real world has increased dramatically in recent decades. For instance, technology, institutions, values and goals of human life, and morals, which were relatively slow to change in the past, are now subject to change in the short term as well. This character of modern society makes it difficult, if not impossible, for pure economics to accurately explain the complexity of economic life. On the other hand, traditional economics also has intrinsic limiations. For instance, traditional economics has been mainly limited to static or stabilized economic systems. Nonlinear unstable phenomena such as regular and irregular oscillations, which are the main concern of this study, have been considered to be temporary or insignificant in traditional economic analysis.

This book studies problems related to time and change in nonlinear unstable economic systems. We will concentrate on particular aspects of dynamic economic systems, for example nonlinearity, instability, bifurcations and chaos. We propose a new theory – "synergetic economics" – based on Haken's synergetics, for analyzing the characteristics of nonlinear dynamic economic systems. Fundamentally, synergetic economics emphasizes the interplay between linearity and nonlinearity, stability and instability, continuity and discontinuity, permanence and structural change, in contrast to pure linearity, stability, continuity and permanence in economic evolutionary processes. Synergetic economics treats nonlinearity and instability as sources of the variety and complexity of economic dynamics, rather than as a nuisance and as temporary phenomena as traditional economics did.

In a sense, this book aims to complete the task which was suggested by Paul A. Samuelson when he wrote his celebrated *Foundations of Economic Analysis*. He broadly classified the development of analytical economics into five steps. First, in

Walras we have the final culmination of the notion of determinacy of equilibrium and the static level. Pareto and others took a second step, which laid the basis of a theory of comparative statics. The third step, which is characterized by maximizing action within an economic unit, was mainly carried out by Johnson, Slutsky, Hicks and Allen. The fourth advance is due to the discovery of the correspondence principle. "A natural fifth step to take after we have investigated the response of a system to change in given parameters is to investigate its behavior as a result of the passage of time." Furthermore, Samuelson emphasized that "The usefulness of any theoretical structure lies in the light which it throws upon the way economic variables will change when there is a change in some datum or parameter. This commonplace holds as well in the realm of dynamics as in statics. It is a logical next step, therefore, to begin to create a theory of comparative dynamics. This will include the theory of comparative statics as a special case, and indeed all of the earlier five subjects, but it will cover a much richer terrain" (Samuelson 1946). The fifth step will be cultivated in this book.

This book is suitable for students and researchers in economics. It may also be useful to scholars interested in applications of nonlinear dynamic theory to economic problems.

Stockholm, July 1990 *W.B. Zhang*

# Acknowledgements

I have many intellectual debts to my teacher, Prof. Åke E. Andersson. His influence throughout this book is unmistakable, as it is, of course, throughout the profession. I am grateful to him for having written the Foreword to the book.

I would also like to express my deep thanks to Prof. Hermann Haken, Prof. Börje Johansson, and Prof. Tönu Puu for their important comments on the work. I am grateful to Dr. A.M. Lahee and Ms. I. Kaiser of Springer-Verlag for their effective cooperation.

I also want to acknowledge my debts to CERUM at the University of Umeå and the Institute for Futures Studies in Stockholm, for providing me with facilities and a stimulating intellectual environment for this research. I am grateful for the financial support of CERUM and the Institute for Futures Studies.

# Contents

# 1. Introduction

We are seekers for truth but we are not its possessors.

*Karl R. Popper* (1972)

What is time? How does change take place? Is there any universal law to govern change over time? Is it possible to find such a universal law if it exists? The human being has been confronted with these questions even since civilization began. All of these questions are essential in both western and eastern cultures, and not only confined to the sciences. Thinking about evolution is itself an evolutionary process due both to the complexity of the problem and to our limited capacity for understanding.

This book studies problems related to time and change in economic systems. Economic evolution has been investigated systematically even since Adam Smith, though no unique theory has been developed. Standing upon the shoulders of others, we try to see a little further than traditional economists.

Not all aspects of the economic evolutionary process will be investigated here. We will concentrate on some aspects such as nonlinearity, instability, bifurcations and chaos in dynamic economic systems. First we review some theories in traditional economics. Then we propose a new theory – "synergetic economics" – for analyzing the characteristics of nonlinear dynamic economic systems. Fundamentally, synergetic economics emphasizes the aspects of linearity versus nonlinearity, stability versus instability, continuity versus discontinuity, permanence versus structural change, in contrast to those of linearity, stability, continuity and permanence in the economic evolutionary process. Synergetic economics treats nonlinearity and instability as sources of the variety and complexity of economic dynamics, rather than as nuisances and temporary phenomena as traditional economics did.

"Synergetic economics" is developed from the science of synergetics created by Hermann Haken (1977, 1983). Synergetics is defined as the science of collective static or dynamic phenomena in closed or open multi-component systems with "cooperative" interactions occurring between the units of the system. In physics, chemistry, and biology, synergetics concentrates on the structural self-organizing space-time features of systems on a macroscopic level. It turns out that on this level there exist close analogies between various systems, though they are composed of different units with completely different elementary interactions. From this new scientific perspective, theories are beginning to emerge on how order gives way to chaos, order is discovered within chaos, and order is again created out of chaos. Synergetic economics also studies these characteristics of evolutionary systems.

Some aspects which synergetics emphasizes can be found in traditional economics. Traditional dynamic economic theories have been aware of interactions and cooperations among different parts of economic systems. There are few economists who will deny the existence of nonlinear interactions in economic systems. However, we will show that it is the role of instability in nonlinear systems which we have little knowledge of. Haken's synergetics and Prigogine's works on dissipative structures gave me a hint for a new way to systematically study the complexity of economic evolution.

Synergetic economics treats economic evolution as an irreversible process. Time and chaotic dynamics play an essential role in understanding irreversible processes. Irreversibility and evolution appear as illusions related to the complexity of collective behavior of intrinsically simple objects. This conception is considered as one of the driving forces of western science (Prigogine 1980, Prigogine and Stengers 1984). Having been strongly influenced by Newtonianism, economists have (implicitly) treated economic evolution as a reversible process. There are few economic models which can explicitly explain its irreversibility. The existence of chaos suggests a new way to reach an understanding of irreversible processes.

Synergetic economics is developed upon the basis of traditional economics. It refutes some ideas in traditional economics and treats the results from traditional economics as special cases rather than as general ones. The basic concepts of rational behavior and perfect competition in traditional economics still play a fundamental role in the development of synergetic economics. Our disagreement with traditional economics is that we treat instabilities in nonlinear systems as sources of the complexity of economic dynamics.

Although it cannot be denied that "the human mind has always struggled like a frightened bird to escape the chaos which caged it ... ", this cannot apply to our case. We try to examine how chaos can occur endogenously due to dynamic interactions of different forces in the evolutionary process. We show the way in which economic systems far from equilibrium evolve elaborate structures: cycles, aperiodic motion, chaos and well-organized time-dependent urban pattern formations. The study of all of these phenomena is what characterizes the difference between traditional economics and synergetic economics.

Many models from different economic theories and "schools" are used to explain "synergetic economics". As many schools are concerned, and the differences among them are very subtle, we will not explain these models in detail. Since we are mainly concerned with methodologies in economic analysis, this will not affect our purpose.

Each theory can only explain some aspects of the real world. An economic theory which tends to explain long-run economic evolution may be meaningless to explain short-run economic phenomena; and similarly the short-run Keynesian theory may be invalid for the long-run Schumpeterian social system. However, if economists are aware of the assumption that an economic theory explicitly and implicitly makes, and are sure of the economic phenomena (e. g., short-run or long-run) that the theory explains before they initiate debates, then misunderstandings among different schools could be greatly reduced. It must be noted that synergetic economics is not intended to follow any special school in current economics. It emphasizes aspects of advantage and disadvantage in each theory. It is important to understand approximately under what conditions the results from an economic theory can be applied.

Confronted with the failure of traditional economics in explaining actual phenomena, economists have tried to introduce assumptions such as imperfect competition, imperfect information, and irrationality into economics. Many theories have recently been proposed in this way. We have disequilibrium macroeconomics, family economics, the share economy and so on. Also, it is very popular to derive each theory of macro-economics upon the basis of a micro foundation. In modern times, both economic phenomena and economics have become "chaotic". We cannot simplify economic phenomena. But it is human nature to look for truth with simplicity and beauty. We always try to search for some simple and unique way to explain the complexity of the real world. It is rather natural for us to attempt to develop a theory which is able to explain complicated phenomena by using concepts and methods which are as simple as possible.

In the 1930s and 1940s, some economists were aware of the significance of instability in nonlinear dynamic systems. They were mainly the economists who studied business cycles. However, it is hard to say that they treated instabilities systematically and considered them as sources of the complexity of the real world. In fact, it had been almost impossible for economists to understand nonlinear phenomena comprehensively. Chaotic phenomena in unstable nonlinear dynamic systems can only be understood with the help of mathematics since they are beyond our intuitive senses. However, at that time even mathematicians had little knowledge about nonlinear unstable dynamic systems.

Since this book is written for and devoted to economic students, from an analytical point of view, it may appear to be too technical because almost all of the essential ideas are expressed via mathematical analysis. However, we try to make the analysis as simple as possible, although sometimes we have to use complicated analytical methods. It seems impossible to discuss difficult topics using simple tools. If mutual understanding (communication) between the reader and the author is to be realized, either partner should try his or her best.

The remainder of the book is organized as shown in Table 1.1.

We now describe the contents of each chapter in detail.

Chapter 2 is concerned with time and change in economies and economic analysis. First, we generally describe the complexity of economic development. Next, we review equilibrium theories and dynamic theories in economics. Then, we discuss the correspondence principle and its limitations. Finally, we emphasize the importance of nonlinearity and instability.

In Chap. 3, we provide some analytical methods for dynamic systems which are important in synergetic economics. Section 3.1 is concerned with the definitions of dynamic systems and the concepts of stabilities. We study the possible behavior of two-dimensional (linear) differential equations in Sect. 3.2. Section 3.3 presents an important theorem which gives the relations of stability between nonlinear and their corresponding linearized systems and then applies the theorem to the Tobin model. Section 3.4 discusses Lyapunov's direct method and its applications to economics. Section 3.5 defines the concepts of structural stability and instability and applies these concepts to the predator-prey system (and the Goodwin model). Section 3.6 defines the concepts of conservative and dissipative systems and studies some properties of conservative systems. Section 3.7 deals with bifurcation theory and its possible applications to dynamic economics. Section 3.8 defines some concepts in singularity theory and shows how these concepts can be applied to the pitchfork bifurcation

**Table 1.1.** The organization of the book

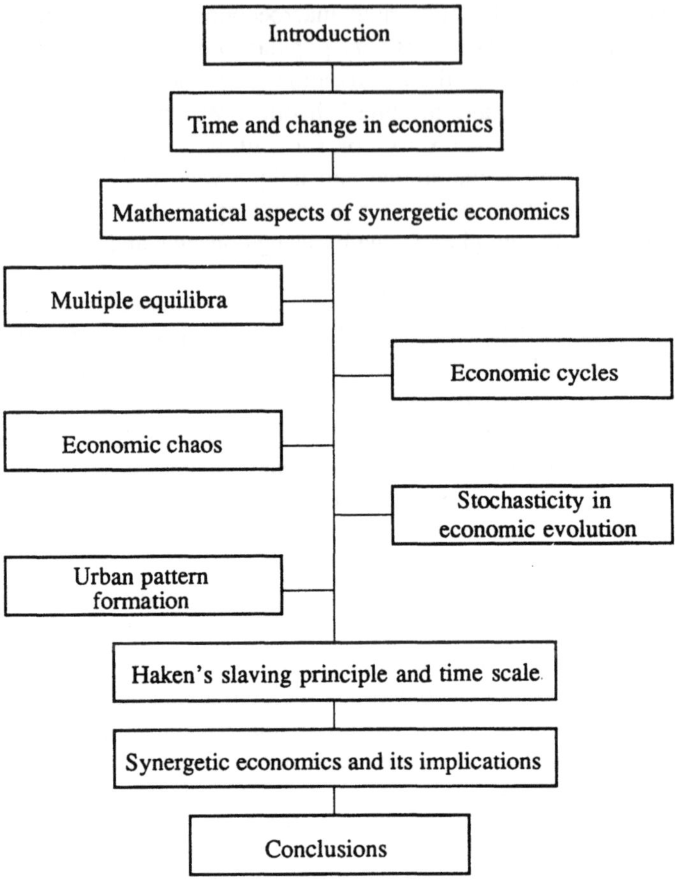


equation. In Sect. 3.9, we show how (elementary) catastrophe theory can be applied to analyze the behavior of dynamic systems. In the appendix to Chap. 3, we give some remarks on bifurcation theory.

Chapter 4 is concerned with the existence of multiple equilibria and structural changes in economic systems. In Sect. 4.1, we show the limitations of traditional comparative statics analysis and discuss how catastrophe theory can be applied to study structural changes. Section 4.2 examines the dynamic behavior of the Andersson regional development model. Section 4.3 deals with possible dynamic interactions between the inflation rate and the rate of interest. In Sect. 4.4, we provide some examples to illustrate the concept of structural change. First, we deal with the Kaldor business cycle model reformulated by Varian. Next, we show how a small change in the parameter can cause sudden movement in the variable in the fisheries management model proposed by Clark. Then we carry out the bifurcation analysis for the dynamic transportation modal choice model suggested by Deneubourg, Palma and Kahn. Finally, we show the existence of multiple equilibria in Wilson's retail model. In Sect. 4.5, we apply the bifurcation method of Iooss and Joseph to an economic growth model recently suggested by Zhang. Section 4.6 suggests possible applications of singularity theory to economic analysis. Section 4.7 concludes the chapter.

We study economic cycles in Chap. 5. Section 5.1 revises traditional business cycle theories. Section 5.2 presents the Poincaré-Bendixon theorem and the Hopf bifurcation theorem and discusses their applications to economic problems. Section 5.3 examines the existence of limit cycles in the simplified Keynesian business cycle model. In Sect. 5.4, we show how nonequilibrium can occur in a disequilibrium macro-economic model suggested by Eckalbar and Zhang. Section 5.5 proves the existence of monetary cycles in the generalized Tobin model. In Sect. 5.6, we show how structural change – from a stationary point to a limit cycle – occurs due to small shifts in the bifurcation parameter in van der Ploeg's hybrid growth model. Section 5.7 examines the existence of a periodic optimal employment policy in a micro-economic model which deals with the behavior of a firm in a perfectly informed environment. Section 5.8 studies endogenous fluctuations in the multiple sector growth models. Section 5.9 provides some analytical methods to identify further bifurcations to the Hopf bifurcations established in this chapter. Section 5.10 proves the existence of economic cycles in the overlapping generations model.

Chapter 6 studies economic chaos which occurs in deterministic dynamic systems. Section 6.1 defines the concept of chaos and discusses some possible ways to create chaotic behavior in deterministic systems. Section 6.2 defines some concepts for discrete maps and provides an example for the existence of chaos in the discrete one-sector economic model proposed by Stutzer. In Sect. 6.3, we guarantee the existence of aperiodic solutions in optimal multiple sector economic growth models. Section 6.4 shows that the Lorenz equations can be used to describe the dynamics of a small urban system. Section 6.5 shows that international trade among economies which exhibit limit cycles in the case of independence, may involve the occurrence of a strange attractor and hence of chaos. In Sect. 6.6, we prove the existence of economic chaos in a two-regional growth model suggested by Puu. Section 6.7 examines the implications of the existence of economic chaos. Section 6.8 provides some remarks on the chapter. In the appendix to Chap. 6, we prove some criteria to distinguish regular motion such as stationary points, limit cycles and aperiodic solutions from genuine chaos.

Chapter 7 examines effects of stochastic processes (with zero average values) upon economic evolution. Section 7.1 reviews some ideas in traditional economics about the effects of stochastic processes upon economic evolution. In Section 7.2, we define some basic concepts for the study of stochastic processes. First, we define some concepts in probability theory. Then, we define the concept of stochastic process and provide different examples of stochastic processes. Section 7.3 shows that small fluctuations can drive the system far away from its original paths, mainly by studying the birth-death process and the master equation. In Sect. 7.4, we illustrate the quantitative aproach to social dynamic systems recently proposed by Weidlich and Haag and provide an example to show how the approach can be used to explain the "Schumpeter clock". Section 7.5 further explains the effects of noise on the nonlinear stochastic systems close to critical points. In Sect. 7.6, we particularly examine the effects of a random environment on a two-dimensional deterministic system near critical points. Section 7.7 concludes this study.

In Chap. 8, we study various urban pattern formation processes. We are interested in the implications of structural stability for urban pattern formation processes. Section 8.1 describes different approaches to urban dynamic processes in regional science, urban economics and geography. In Sect. 8.2, we examine the implications of structural stability for urban pattern formation processes. Section 8.3 describes Puu's spatial multiplier-accelerator business cycle model to explain the complexity of urban dynamics. In Sect. 8.4, using Zhang's urban model we show that urban systems may be stabilized by the introduction of diffusional terms. Section 8.5 deals with the dynamic process of separation and coexistence of residents, which is described by a set of nonlinear partial differential equations. In Sect. 8.6, we examine an urban model which exhibits travelling-wave-like behavior when the system is close to critical points. Section 8.7 discusses the implications of instabilities for urban pattern formation processes. In the appendix to this chapter, we provide two examples of pattern formation models – a model for morphogenesis and the "Brusselator".

Chapter 9 presents some methods for dynamic economic analysis and discusses the implications of the adjustment speeds of economic variables and the time scale for economic analysis. Section 9.1 discusses the Haken slaving principle and its implications for economic analysis. Section 9.2 presents the centre manifold theorem. Section 9.3 presents some methods of singular perturbations. In Sect. 9.4, we show that economic variables such as money, wages, prices, output, capital, rate of interest, and technology have various adjustment speeds in different theories. Section 9.5 examines relations between the time scale and adjustment speeds. Section 9.6 proposes some implications of synergetic economics for understanding the dynamic behavior of human beings. In the appendix to Chap. 9, we show how the Haken slaving principle can be applied to stochastic differential equations.

Chapter 10 defines synergetic economics, discusses its relations to synergetics and traditional economics and examines its implications for different economic problems. Section 10.1 defines synergetic economics and examines relations between synergetics according to Haken and synergetic economics. Section 10.2 discusses the relations between synergetic economics and traditional dynamic economics. Section 10.3 examines the implications of synergetic economics for competitive and planned economies. Section 10.4 deals with the implications of synergetic economics for economic development. Section 10.5 studies relations between chance and necessity in economic life from a synergetic economics point of view. Section 10.6 examines the implications of synergetic economics for economic policy decisions. In Sect. 10.7, we study the relations between microeconomics and macroeconomics.

Chapter 11 concludes the study.

# 2. Time and Change in Economics

The difficulty lies, not in the new ideas, but in escaping from the old ones, which ramify, for those brought up as most of us have been, into every corner of our minds.

*J.M. Keynes* (1936)

## 2.1 Economic Evolution – An Introduction

The essence of life is change.

*Lao Tze*

After World War II, astonishing dynamic economic phenomena occurred. The two countries – West Germany and Japan – which were seriously destroyed (at least physically) during the war, have been rebuilt and developed at much faster speeds than the countries which won the war. Since the end of the war some countries have experienced strong economic growth, fast urbanization and numerous economic opportunities, although large (irregular) fluctuations sometimes characterize this development. Even in the same country different regions have shown great variations in economic development. The differences may be so large that one may think that people from different places are not living in the same country. It cannot be denied that time and space have played a deterministic role in forming and changing the characteristics of individuals and societies.

Economists have proposed many theories to explain economic evolution. Dynamic economics, however, is still concerned with simple behavior. The existence, uniqueness and stability conditions of stationary states have been the main concern of the literature. Unfortunately, these kinds of behavior are not identified by typical economies on record. Instead, actual economies exhibit complex dynamics: periodic cycles, irregular fluctuations and chaos. There are gaps between the actual economic development and economic theories. Such gaps have not been shortened as time passes. Whenever economists touch upon dynamic problems, they tend to have different points of view. This may result, on the one hand, from the complexity of economic systems and, on the other hand, from misunderstandings among economists. Moreover, it is interesting to note that the gaps are not completely filled by the innovation of computers and efforts in collecting data. The belief that improvements in capacity for calculation and simulation can solve all problems in economic development tends not to be borne out. The analysis of actual data can tell little, if the analysis is not firmly supported by some well built economic theory.

## 2.2 Equilibrium Theories in Economic Analysis

The beauty of physics reveals itself only after asking the right questions ...

*H.G. Schuster* (1988)

Before reviewing the main approaches in dynamic economics, we will explain the importance of static economics in understanding dynamic economies.

It may be argued that most fruitful results in economic analysis have been derived from equilibrium theories. Like some other concepts in economic analysis, "equilibrium" is borrowed from mathematical mechanics. The concept of equilibrium was familiar to mechanics long before the publication of the *Wealth of Nations* in 1776, though there is obvious evidence that Adam Smith drew his ideas from some analogy with mechanics. As no actual economy can be fixed in a stationary state, there are obviously limitations in equilibrium analysis. The question is whether the methods of equilibrium analysis can shed some light on the economic problems under consideration. The development of economics has proved that equilibrium analysis is quite useful.

The economists' interest in equilibrium situations may be justified by two kinds of arguments. An equilibrium has special claims on our attention because when we ask ourselves how to characterize a decentralized economy that is also efficient we find that it is often an economy in competitive equilibrium (steady state). Surely, this does not imply that a perfectly competitive system must have a strong tendency to a steady state. The simplest example of this is the "cobweb theorem". Furthermore, recent literature on business cycles and chaotic economics have shown that this argument can only be held for some limited cases. The second argument, which was originated by Marshall, is that there are forces at work in any actual economy that tend to drive the economy towards an equilibrium if it is not already in equilibrium. Plenty of examples in this book will show that this argument cannot be held in general.

These fundamental arguments can also be found in the concept of the "invisible hand". This notation means that a social system moved by independent actions in pursuit of different values is consistent with a final coherent state of balance. Thus, the outcomes of competition may be quite different from those intended by the agents. Smith perceived the most important implication of general equilibrium theory, the ability of a competitive system to achieve an allocation of resources that is efficient in some sense. Ricardo (1817), Mill (1848) and Marx (1867), whose work filled in some of Smith's logical gaps, can all be regarded as early expositors of general equilibrium theory. However, none of the classical economists had a true general equilibrium theory: none gave an explicit role to demand conditions.

Schumpeter (1934, 1975) had a similar point of view on the second argument about equilibrium. However, for Schumpeter, economic equilibrium in the capitalist system will never actually be obtained because there are always innovations which drive the system from its equilibrium. However, the study of equilibrium structures is significant because this can shed light on understanding the actual processes of economic evolution.

The full recognition of the general equilibrium concept can be attributed to Walras, though many elements of the system had been worked out independently by W.S. Jevons and C. Menger. His work has provided the foundation on which a good deal of theoretical economics is currently based. He developed the general equilibrium theory to encompass the fields of exchange, production, capital, and money in a unified formulation.

Walrasian equilibrium models gained some momentum in the early 1950s (see, e.g., Debreu 1959, Arrow and Hahn 1971). Their study, often referred to as the general equilibrium theory, was focused on the existence of competitive equilibria safeguarded by equilibrium prices. The associated dynamic analysis was therefore preoccupied with the stability of such equilibria. The dynamics in this tradition mainly consists of "artificial" price adjustment processes. It is essentially concerned with the stability of the equilibrium determined in a static framework (Arrow and Hahn 1971).

In his celebrated *Foundations of Economic Analysis*, Samuelson (1947) defended the use of the concept of equilibrium, arguing that many problems could be viewed as maximization and minimization problems. The theory of consumer behavior and the theory of the firm were, for Samuelson, simple applications of the theory of constrained maximization. Not only did this framework reveal a unified structure underlying apparently diverse problems, but it was a source of predictions. In this approach, one of the most popular methods is comparative statics analysis, which is concerned with the effects of shifts in parameters on economic variables. It provides information about how equilibrium is changed after exogenous shocks. The method has found wide applications to various problems in economics. In Sect. 4.1, we will discuss this method in particular.

## 2.3 Dynamic Theories in Economics

> The history of science is far from being a linear unfolding that corresponds to a series of successive approximations toward some intrinsic truth. It is full of contradictions, of unexpected turning points.
>
> *I. Prigogine* and *I. Stengers* (1984)

Time must be included in the description of any economic variables. Any dynamic economic theory cannot escape from being concerned with time. In equilibrium theories time is explicitly excluded because it is assumed that the system can be kept in a state where interactional relations among variables are invariant. This assumption is valid if the study period is very short, or if we are only concerned with stationary states. The role of time is thus fixed. In a sense, equilibrium analysis may be considered as a special case of dynamic analysis.

There is a large amount of literature on dynamic economics. The main works on economic evolution have been related to economic growth and business cycles. Since we will review business cycle theory in Chap. 5, we will also focus on development of economic growth theory.

Economic growth is one of the classical topics in economics. Adam Smith, D. Ricardo, T.R. Malthus, Karl Marx, J. Mill, and other classical economists made significant contributions to the theory.

During the period 1870 to 1920 the partial equilibrium approach of Marshall and the general equilibrium approach of Walras dominated the literature. The theories of capital and interest also made great advances with Bohm-Bawerk, Clark, Wicksell, and Fisher. However, their treatments are, at least in the light of present knowledge, frequently oversimplified in formulation and defective in reasoning. Within the framework of the classical tradition one can also find J.A. Schumpeter's *Socialism, Capitalism, and Democracy* (1975) and *The Theory of Economic Development* (1934), and W.A. Lewis' *The Theory of Economic Growth* (1955). All of these works are characterized by taking into account not only "purely economic variables", but also some other social factors such as morals, ethics, institutions and so on, in explaining economic development.

In Schumpeter's evolutionary system, the concept of innovation plays a significant role in developing the theory. This notion refers to several aspects of novelty such as the emergence of new needs and changing preferences as part of social learning processes, the development of new products satisfying the needs of consumers, the use of new products and equipment in increasing competitive advantage, the adoption of new organizational methods and the opening of new markets. Because of the existence of innovation, competitive capitalist economies cannot be stable. However, for Schumpeter such unstable evolution does not mean destruction of the systems. Each exogenous shock establishes a new equilibrium towards which the actual system moves. Recently inspired by Schumpeter's perspectives, some scholars have developed "evolutionary economics" to explain the dynamics of economic systems (e. g., Nelson and Winter 1982). Here, the term evolution mainly refers to processes of long-term and progressive change. "Out of equilibrium", rather than equilibrium, is a key term in this approach. Synergetic economics is rather similar to "evolutionary economics" in regard to this aspect.

The *General Theory* of Keynes revived the interest in macroeconomic growth theory, although the work is characterized by "peso-dynamics" rather than dynamics. The *General Theory* provides a picture of the interactions of the aggregated macroeconomic variables.

It should be noted that for Keynes and some of his followers, the evolution of a capitalist system is potentially unstable, while for neoclassical economists economic growth is characterized by stability. Keynes thought that a government can stabilize an economy by making appropriate policy decisions. Hence, how to stabilize economic systems becomes one of the main subjects in Keynesian economics.

The works by Samuelson, Solow, Morishima, Hicks, Leontief and others have played a significant role in the development of modern economic growth theory (see Zhang 1989, 1990b). These works focus on the process of capital accumulation intertwined with increases in production and consumption. But almost all of these approaches are carried out under presumed stabilities. Little has been said about what will happen if the system is placed in unstable states.

There are many differences between classical and modern economic growth theories. For instance, the empirical facts that modern growth economics attempts to explain are quite different from those which classical growth economics is confronted with. It appeared that some of the most important empirical predictions of, for example, the Malthus and Ricardo theories were not realized. The share of landowners did not seem to be increasing, and population was not growing faster than output. The importance of agriculture relative to manufacturing declined markedly. On the other hand, the primary object of modern growth economics is to explain the movements in the output, employment, and capital stock of a growing economy, and the interrelations among these variables, and to explain the movements in the distribution of income among the factors of production. The modern growth theory tries to provide a conceptual framework within which much more meaningful empirical research can take place. The economies which the modern theory attempts to describe are essentially advanced and industrialized. In such economies, capital and labour are the two inputs upon which attention is focused. Land, which is an important input in classical growth theory, is usually ignored. It is the consumption-investment decision, and not the allocation among alternative investment or alternative consumption goods, upon which the analysis has been primarily focused.

Although some models in this book are developed within the framework of modern growth economics, the emphases of our approach to economic development are different. We will mainly investigate the economic phenomena which occur when the presumed stability is relaxed.

## 2.4 Samuelson's Correspondence Principle and Its Limitations

In modeling an economic process, we usually introduce some parameters which are kept fixed during the study period. The parameters are used to describe exogenous factors such as environment, economic policies and the structure of the system under consideration. As these parameters are changeable, it is important to know the effects of changes in different parameters upon the behavior of the system. Comparative statics analysis and the correspondence principle are proposed to analyse such effects.

With regard to economic predictions, Samuelson found two sources of information. First, some results about comparative statics are derived from the assumption of maximizing (or rational) behavior of individuals. We will examine this topic in Sect. 4.1. Next, Samuelson argued that more important information can be derived from stability conditions. We will now show the usefulness and limitations of this method.

First, consider the Walrasian theory of explanations of the process by which supply and demand are equated. It is assumed that if, at any price, demand exceeds supply, then the price will rise; if supply exceeds demand, the price will fall. This can be precisely stated as follows

$$\frac{dp}{dt} = H\left[D(p, \alpha) - S(p)\right] , \tag{2.4.1}$$

where $H(0) = 0$, $H' > 0$, $p$ represents the price, $\alpha$ is a parameter measuring exogenous factors, and $D$ and $S$ represent demand and supply, respectively. For simplicity, let $H = 1$. In the neighborhood of the equilibrium point $p = p_0$, (2.4.1) can be written approximately as

$$\frac{dp}{dt} = (D_p - S_p)(p - p_0) + \dots , \tag{2.4.2}$$

where the terms involving higher powers of $(p - p_0)$ are omitted. If the initial price is denoted by $p(0)$, then the solution of (2.4.2) is given by

$$p(t) = p_0 + \left[p(0) - p_0\right] \exp\left[(D_p - S_p)t\right] . \tag{2.4.3}$$

If the equilibrium is stable,

$$p(t) = p_0 \text{ as } t \to +\infty .$$

This strictly holds if and only if

$$D_p - S_p < 0 . \tag{2.4.4}$$

If the supply curve is positively inclined, this will be realized. If it is negatively inclined, it must be less steep than the demand curve. If the stability conditions are realized, the problem originally proposed is answered. Price must rise when demand increases. Thus the comparative statics results can be deduced from stability conditions.

In contrast to Walrasian theory, the quantity supplied in Marshallian theory of normal price is assumed to adjust itself accordingly. If "demand price" exceeds "supply price", the quantity supplied will increase. Preserving the notations of (2.4.2) and neglecting higher-order terms, we have

$$\frac{dq}{dt} = \left(\frac{1}{D_p} - \frac{1}{S_p}\right)(q - q_0) , \tag{2.4.5}$$

whose solution is

$$q(t) = q_0 + \left[q(0) - q_0\right] \exp\left[\left(\frac{1}{D_p} - \frac{1}{S_p}\right)t\right] . \tag{2.4.6}$$

In order that the equilibrium is stable, we require that

$$\frac{1}{D_p} - \frac{1}{S_p} = \frac{S_p - D_p}{D_p S_p} < 0 , \tag{2.4.7}$$

i. e., the demand-curve slope referred to as the quantity axis is less than that of the supply curve. As $D_p < 0$, one has $(S_p - D_p)/S_p > 0$. Consequently, the Marshallian stability conditions require that quantity supplied increases when demand increases in any case, while the change in price is necessarily ambiguous depending on the sign of the slope of the supply curve. Thus, from given information on stability conditions, we can immediately deduce that a rise in demand will lead to an increase in output.

Such relationships between stability conditions and comparative statics results are named the "correspondence principle" by Samuelson. It was considered that if the principle is correct, then the method of comparing equilibria may be a legitimate way to predict the consequences of given parameter changes. If stability is presumed, small shifts in parameters can only result in smooth changes in the variables. No sudden development can be observed.

The validity of the correspondence principle depends on the presumed stability of economic systems. It is significant to investigate what will happen if this assumption is relaxed.

## 2.5 Instabilities in Economic Analysis

From the discussion above, we see that the hypothesis of stability is important since we can often obtain meaningful economic results. A number of successful applications of the correspondence principle to different economic problems have proved that this method is rather useful. It should be emphasized that the acceptance of stability does not mean that economists deny the existence of instabilities. It is the attitude toward instabilities in economic analysis that should be changed.

The attitude of most economists toward instabilities in economic analysis can be illustrated by Samuelson's recollection of his 1932–37 days as a classical theorist: "as an equilibrium theorist [I] naturally tended to think of models in which things settle down to a unique position independently of initial conditions ... technically speaking, we theorists hoped not to introduce hysteresis phenomena into our model, ... in our more realistic moods, we tacitly used models involving hysteresis: Spain would never be the same after Colombus ... obviously, in such models all real variables do not end up unchanged as a result of certain unbalanced introductions of new M into the system" (Samuelson 1972, pp. 540–1).

The acceptance of stability in economic analysis is strongly affected by the development of natural sciences in which dynamic systems are required to be stable in order to make the analysis meaningful. For experimental sciences, this implies that a descriptive model must lead to the same qualitative result if the experiment is repeated in a minimally changed environment. The attitude toward reality arising from this requirement is that reality is indeed stable in such a structural sense. And the conviction that small variations in the environment of a real system do not lead to drastically and qualitatively different kinds of behavior is a heritage of the mechanically oriented 19th century. Guided by the idea of deterministic mechanics, complex phenomena which could not be explained by the usual models led either to the postulate that such phenomena should be analytically neglected, or that the system is subject to purely stochastic influences. Consequently, chaotic phenomena in evolutionary systems are treated as temporal phenomena or mere disturbances to long-run equilibrium evolution.

This point of view regarding stabilities has been changed. Stabilities are not presumed in the sciences. It has been argued that small shifts in parameters may result in structural changes in dynamic systems. Such structural changes are not exceptional, but universal in evolutionary systems. Complex phenomena such as regular oscillations and chaos are characteristics of nonlinear unstable systems. The spontaneous formation of well-organized structures out of chaos can be observed in a relatively simple nonlinear dynamic system. It has also been found that well-organized spatial, temporal, or spatial-temporal structures arise out of chaotic states, and, in such self-organizing systems, instead of finding stability and harmony, we discover evolutionary processes leading to diversification and complexity (Nicolis and Prigogine 1977, Haken 1977, 1983).

We show that these ideas about evolution can be applied to economics as well. In modern economic systems sluggishness tempered with abrupt and sometimes unexpected changes appear to be the order of the day. Economic systems such as labor markets, monetary markets, urban systems, transportation and communication systems are characterized by chaos. All of these well-observed complicated phenomena have not been appropriately explained by the existing economic theories. A growing recognition of such discontinuities or structural changes and chaotic phenomena prompts a fundamental need for new theoretical frameworks and analytical tools, which can probe beyond the boundaries of traditional economics provided by optimization theory, stability analysis, and comparative statics. Synergetic economics is in a position to provide a new theoretical framework and some analytical methods to meet this need.

Inspired by the modern works on nonlinear dynamic systems by mathematicians and natural scientists, some economists have begun to explain complicated economic phenomena by introducing instabilites and nonlinearities into dynamic analysis. This study will follow this new direction for analyzing economic phenomena.

# 3. Mathematical Aspects of Dynamic Systems

The approach to a more profound knowledge of the basic principles of physics is tied up with the most intricate mathematical methods.

*Albert Einstein*

Mathematics is the handmaiden of modern sciences. Many of today's profound insights into nature could hardly be obtained from the sciences without the help of mathematics. On the other hand, mathematics has its own life. The works of Newton, Leibnitz and von Neumann provide good examples of interactions between mathematics and the sciences.

It may be said that modern economics is characterized by applying mathematics to various economic problems. Most deep insight into purely economic questions cannot be obtained without applying mathematics. The concepts of equilibrium versus non-equilibrium, stability versus instability, and steady states versus chaos in economics are difficult to explain without mathematics.

The history of applying mathematics to economics is as long as that of mathematics. Even since the beginning of the 19th century differential calculus has been applied to economics (Cournot 1838). It is by the use of this technique that Walras (1874) and Pareto (1908) created the theory of general economic equilibrium, which, around the time of World War II, culminated in Hicks' *Value and Capital* (1939), and Samuelson's *Foundations of Economic Analysis* (1947). Since World War II, advanced mathematics such as convexity theory and topology has found wide application in economics (Nikaido 1968, Arrow and Hahn 1971, Takayama 1985, Mas-Colell 1985, Anderson, Arrow and Pines 1988). Recently, catastrophe theory and bifurcation theory have been used to study economic evolutionary processes. It seems that the temporal lag between the discovery of mathematical results and their application to economics is getting shorter. It took about one and a half centuries after its discovery before calculus was applied to economics. However, it has taken only a few years for economists to apply catastrophe theory and bifurcation theory.

This chapter discusses some mathematical methods which are potentially useful in synergetic economics. The chapter only studies those dynamic systems which are described by deterministic ordinary differential equations. Other types of dynamic systems associated with space and stochasticity will be discussed later on.

## 3.1 Dynamics and Equilibrium

Ordinary differential equations have found wide applications in dynamic economics. Dynamic interactions among economic variables such as prices, wages and capital are generally described by a set of differential equations. Although some economists are interested in dynamic problems which are described by (parabolic) partial differential equations, these kinds of equation will be investigated when we study urban pattern formation problems. In general, dynamic systems may be written as

$$\frac{\partial x}{\partial t} = x_t = f(x) + D\nabla^2 x \ , \tag{3.1.1}$$

where $x = x(r, t)$ is the vector of dependent variables, $r$ is a spatial distance, $f(x)$ is a nonlinear vector-valued function of $x$, and $D$ is the diffusion matrix. For instance, in the simplified Keynesian system defined in Sect. 5.3, the vector $x$ (with $D = 0$) represents the output and the interest rate. In the urban model defined in Sect. 8.4, $x(r, t)$ represents the population density and land rent and $r$ is the distance from a point in the urban area to the CBD (the central business district). Thus, system (3.1.1) may be used to describe an urban pattern formation process which is reflected by the dynamics of the variable $x$ over space. In what follows, we neglect diffusional terms. We will especially study partial differential equations in Chap. 8.

Without the spatial dimension, system (3.1.1) may be rewritten as

$$\dot{x} = f(x) \ . \tag{3.1.2}$$

We will sketch some methods for analysis of these equations. For a fuller treatment the reader is referred, for example, to Coddington and Levinson (1955), or Chow and Hale (1982).

The primary fundamental truth related to the equations is known as the Picard-Cauchy-Lipschitz theorem, which is given as follows.

**Theorem 3.1.1.** Consider the system of equations

$$\dot{x} = f(x, t) \ .$$

Let the functions $f_i(x, t)$ satisfy the Lipschitz conditions in their arguments. Then there exists a unique solution $x = x(t)$ in the neighborhood of $t = t_0$ satisfying initial conditions $x(0)$. Moreover, this solution is a continuous function of the initial conditions. If

$$\dot{x} = f(x, t, r) \ ,$$

where $r$ is a parameter, and each $f_i$ also satisfies a Lipschitz condition uniformly in $r$ in the neighborhood of $r_0$, and is continuous in $r$, then the same conclusions hold in the neighborhood of $r_0$. Moreover, $x = x(t, r)$ is a continuous function of $r$ in this neighborhood.

The concept of stability is defined as follows.

**Definition 3.1.1.** (*Stability*). Consider the system $dx/dt = f(x, t)$. The solution $x = u(t)$ defined in $[t_0, \infty]$ is stable if, for any given $\varepsilon > 0$, there exists $\delta$ such that if $u^*(t_0)$ is any given vector satisfying

$$|u(t_0) - u^*(t_0)| < \delta \ ,$$

then the solution $x = u^*(t)$ with the initial conditions $x(t_0) = u^*(t_0)$ exists in $[t_0, \infty]$ and satisfies

$$|u(t) - u^*(t)| < \varepsilon ,$$

for all $t \geqslant t_0$.

**Definition 3.1.2.** (*Asymptotic Stability*). A solution $u(t)$ is asymptotically stable if (a) it is stable and (b) there exists $\mu > 0$ such that if

$$|u(t_0) - u^*(t_0)| < \mu ,$$

then

$$|u(t) - u^*(t)| \to 0 \quad as \quad t \to +\infty .$$

It is globally asymptotically stable if $\mu$ may be chosen arbitrarily large.

**Definition 3.1.3.** (*Instability*). A solution $u(t)$ is unstable if, for any given positive $\varepsilon$ sufficiently small and any $\delta > 0$, there is a solution $u^*(t)$ such that (a)

$$|u(t_0) - u^*(t_0)| < \delta ,$$

and (b)

$$|u(t) - u^*(t)| > \varepsilon ,$$

for some $t > t_0$.

Instability can be understood from an example in everyday life. A liquid in a quiescent state which starts a macroscopic oscillation is leaving an old state and entering a new one, and thus loses its stability. In physical experiments when we change certain conditions, e. g., power input, a system may run through a series of instabilities leading to quite different patterns of behavior. The complexity of instabilities may also be understood from the dynamics of exchange rates in economic markets.

**Definition 3.1.4.** (*Orbital Stability*). A solution $u(t)$ of an autonomous system $dx/dt = f(x)$ is orbitally stable if for any given $\varepsilon > 0$ there exists $\delta > 0$ such that if

$$|u(t_0) - u^*(t_0)| < \delta ,$$

then

$$\inf_{\Theta \geqslant t_0} |u(t) - u^*(\Theta)| < \varepsilon ,$$

for each $t \geqslant t_0$.

The concepts of stability and orbital stability should not be confused. Figure 3.1 provides an example. Assume that $C$ and $C'$ are two orbits of different periods.

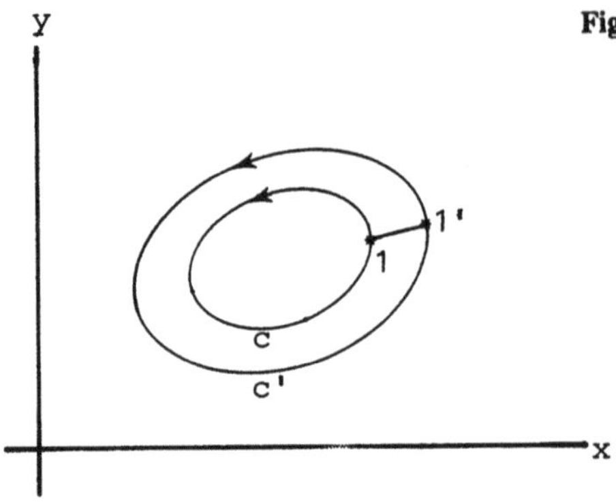


**Fig. 3.1.** Comparison between stability and orbital stability

Although the distance between them remains bounded for all times, the distance between two points 1 and 1′ on these orbits can increase in time owing to a phase shift induced by the difference between periods. Thus state 1 need not be stable, even if orbit $C$ is orbitally stable.

These definitions for $dx/dt = f(x,t)$ are applicable to an autonomous system: $f(x,t) = f(x)$.

An equilibrium of an autonomous system is a point $x = x_0$, where $x_0$ is defined by $f(x_0) = 0$. In this case, the equations have a solution $x = x_0$. The equilibrium is (asymptotically) stable if the solution $x = x_0$ is (asymptotically) stable.

To illustrate these concepts, consider the Solow economic growth model. This model plays a significant role in the development of the neoclassical economic growth theory. It may be said that the majority of the neoclassical growth models are extensions and generalizations of the pioneering papers of Solow (1956) and Swan (1956) (see also, Zhang 1989).

It is assumed that there is only one (durable) good; factor markets work well; factors are inelastically supplied; and the available factors are fully utilized at every moment. All savings volunteered by households are absorbed by firms for accumulation of capital. There are two productive factors: capital, $K$, and labor, $L$. There is no technical change. The production process is described by some sufficiently smooth function

$$Y = F(K,L)\ , \tag{3.1.3}$$

where $Y$ is the output flow attainable with given variables $K$ and $L$. The production function $F$ is neoclassical if it satisfies: (1) $F(K,L)$ is non-negative if $K$ and $L$ are non-negative; (2) $F(0,0) = 0$; (3) marginal products $F_K$ and $F_L$ are non-negative; (4) there are second partial derivatives of $F$ with respect to $K$ and $L$; (5) the function is homogeneous at degree one: $F(rK,rL) = rF(K,L)$, for all non-negative $r$; (6) the function is strictly quasi-concave.

It is assumed that the labor force $L$ exogenously grows at a fixed relative rate $n$:

$$L = L_0 \exp(nt)\ .$$

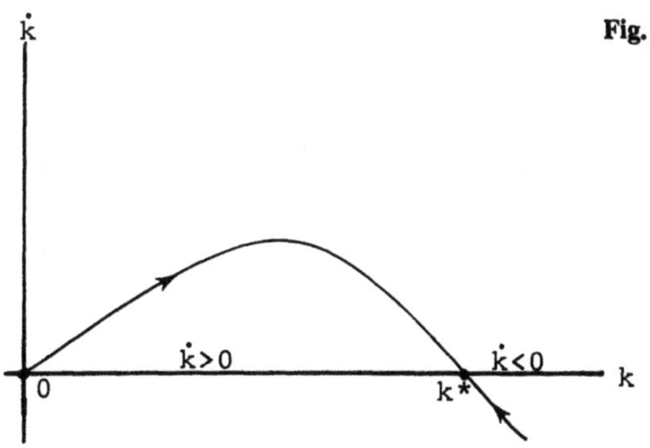


Fig. 3.2. Dynamics of the Solow model

It is assumed that a constant fraction of the total output flow $s$ is saved and set aside to be added to the capital stock. If we neglect depreciation of capital, then one has $dK/dt = sY$, $K(0) > 0$. As the production function is neoclassical, one has

$$\dot{k} = sf(k) - nk \,, \tag{3.1.4}$$

where $k = K/L$, $f(k) = F(K, L)/L = F(k, 1)$. The function $f(k)$ has the properties: $f(0) = 0$, $f'(k) > 0$ if $k \geqslant 0$, $f''(k) < 0$ if $k \geqslant 0$. The existence of solutions of the differential equation (3.1.4) can be guaranteed. It is well known that in the Solow model, once the capital per capita is determined, all of the variables – $K$, $Y$, consumption, savings, wages, rentals – can be calculated accordingly.

**Theorem 3.1.2.** (*The Existence of Equilibria*). If $n$ and $s$ satisfy

$$0 < n/s < f'(0) \,, \tag{3.1.5}$$

then there exists a unique positive value $k_0$ such that $sf(k_o)/n = k_0$.

The proof of the theorem is referred to, for example, in Koopmans (1965). If (3.1.5) holds, then one can describe the phase diagram of the Solow model as in Fig. 3.2.

The following theorem guarantees stability of the equilibrium.

**Theorem 3.1.3.** (*Stability of the Equilibrium*). The system is globally stable (Arrow and Hurwicz 1958). Furthermore, the equilibrium is asymptotically stable in the region $k > 0$.

The asymptotical stability can be proved by finding the Lyapunov function: $V = U^2$, where $U = k - k_0$ (Burmeister and Dobell 1970). Economic development can be described as follows. In the long run the economy will always smoothly converge to the unique equilibrium capital/labor ratio from any positive starting point. Moreover, along the balanced growth path, capital expands at the same rate as the population grows. This is the simple and beautiful picture portrayed by the Solow growth model. A historical review of the development of the model can be found in Solow's 1987 Nobel Prize Lecture (Solow 1988).

## 3.2 Classifications of Two-Dimensional Differential Equations

If $f(x) = Ax$ where $A$ is a constant matrix, the system $dx/dt = Ax$ is called a linear autonomous differential equation. We know that the unique equilibrium point $x = 0$ is stable if $\mathrm{Re}\,(z) \leqslant 0$ for each eigenvalue $z$ of $A$, and that $z$ is a simple eigenvalue whenever $\mathrm{Re}\,(z) = 0$ and asymptotically stable if and only if $\mathrm{Re}\,(z) < 0$ for each $z$.

To illustrate some of the concepts developed thus far, let us study a linear system involving two variables

$$\frac{dx}{dt} = f(x, y) = ax + by ,$$
$$\frac{dy}{dt} = g(x, y) = cx + dy . \tag{3.2.1}$$

Since the system is linear, we look for solutions of the form $(x, y)^{\mathrm{T}} = (u, v)^{\mathrm{T}} \exp(zt)$, where $u$ and $v$ are constant. We thus have

$$zu = au + bv , \quad zv = cu + dv . \tag{3.2.2}$$

(3.2.2) have solutions when $|A - zI| = 0$ where $A$ is the corresponding matrix of (3.2.1), i. e., when $z$ is the eigenvalue of the matrix and $(u, v)^{\mathrm{T}}$ is the corresponding eigenvector. The eigenvalue $z$ is determined by

$$z^2 - Tz + W = 0 , \tag{3.2.3}$$

where

$$T = a + d , \quad W = ad - bc ,$$

are the trace and the determinant of the matrix, respectively.

In general, (3.2.3) admits two distinct solutions $z_1$ and $z_2$. Thus, the solution of (3.2.1) is of the form

$$x = c_1 \exp(z_1) + c_2 \exp(z_2) ,$$
$$y = c_1\Theta_1 \exp(z_1) + c_2\Theta_2 \exp(z_2) , \tag{3.2.4}$$

where $c_1$, $c_2$ are determined by the initial conditions, and the coefficients $\Theta_1$, $\Theta_2$ are the roots of the equation

$$b\Theta^2 + (a - d)\Theta - c = 0 .$$

From these expressions one sees straightforwardly that qualitative characteristics of the solution are mainly determined by the eigenvalues. Since the eigenvalues $z$ satisfy the quadratic equation (3.2.3) with roots $z_1$ and $z_2$, we can distinguish the following cases (e. g., Britton 1986):

i) Let $z_1$, $z_2$ be real and distinct and of the same sign. Then the equilibrium is stable if they are negative, and unstable if they are positive. The equilibrium in this case is known as a stable or unstable node. The trajectories are of the form shown in Fig. 3.3.

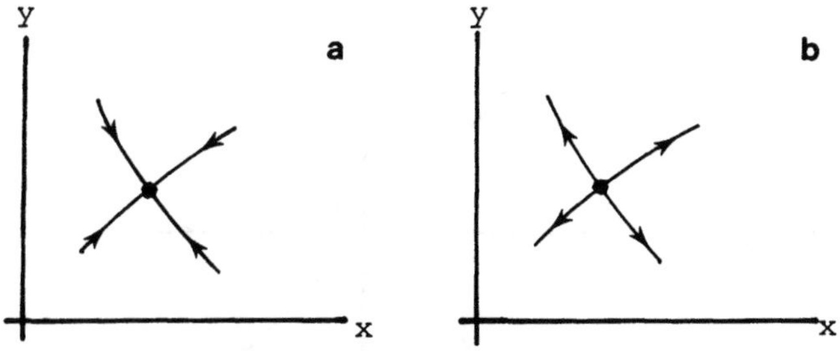


Fig. 3.3. Two eigenvalues are real and of the same sign, (a) stable node, (b) unstable node

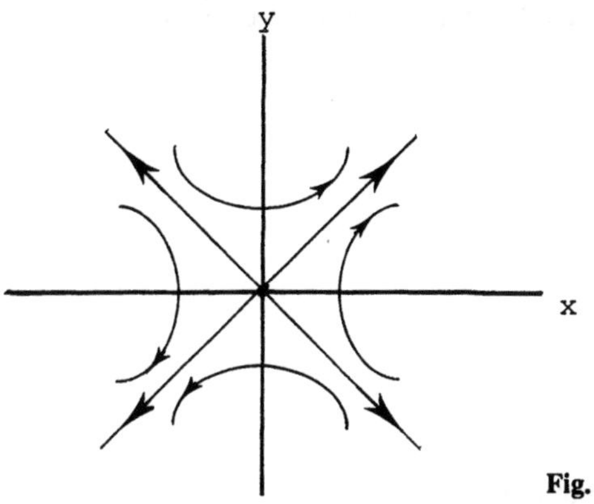


Fig. 3.4. Saddle point

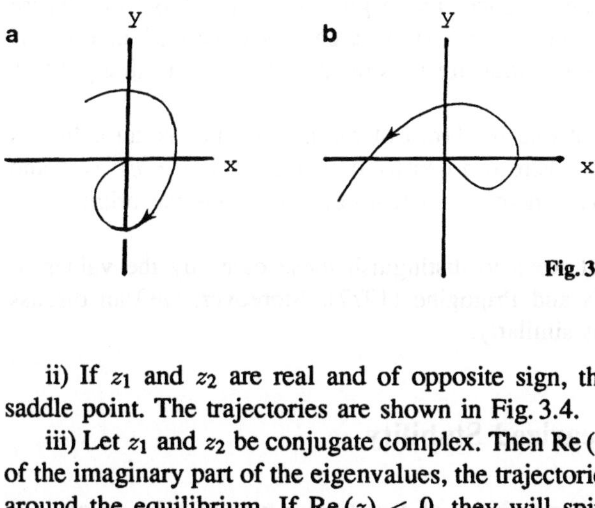


Fig. 3.5. (a) stable focus, (b) unstable focus

ii) If $z_1$ and $z_2$ are real and of opposite sign, then the equilibrium is called a saddle point. The trajectories are shown in Fig. 3.4.

iii) Let $z_1$ and $z_2$ be conjugate complex. Then $\mathrm{Re}(z_1) = \mathrm{Re}(z_2) = \mathrm{Re}(z)$. Because of the imaginary part of the eigenvalues, the trajectories in the phase plane will move around the equilibrium. If $\mathrm{Re}(z) < 0$, they will spiral into the point as shown in Fig. 3.5a, whereas if $\mathrm{Re}(z) > 0$, they will spiral out, as shown in Fig. 3.5b. The equilibrium is known as a stable or unstable focus. If $\mathrm{Re}(z) = 0$, we have the limiting case of a center, where the trajectories are closed, as illustrated in Fig. 3.6.

iv) Let us now consider the case of two equal eigenvalues. Without loss of generality, let $a = d$ and $b = c = 0$. If this does not hold, a linear transformation, or

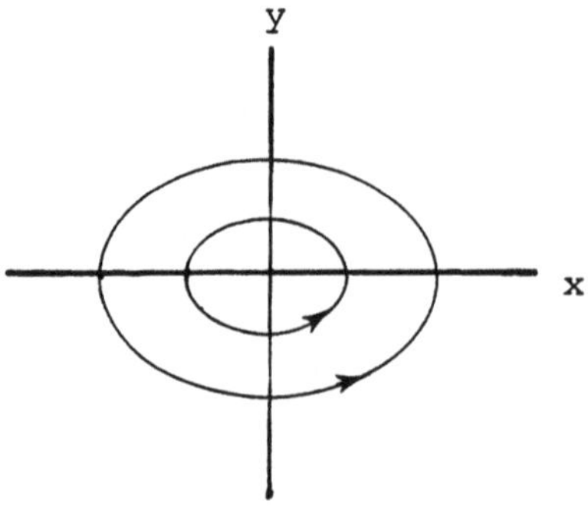

Fig. 3.6. Center

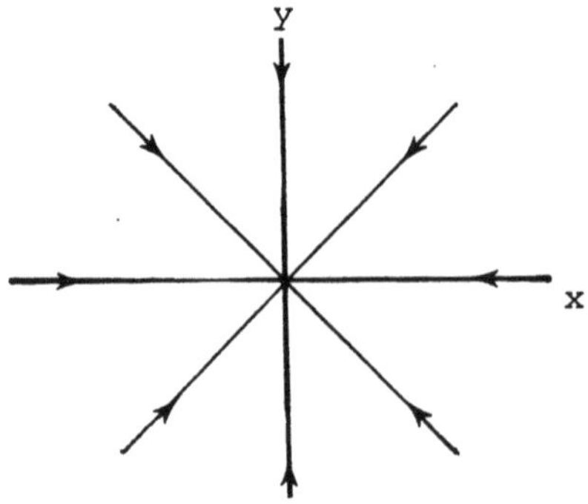

Fig. 3.7. The case of $a = d$ and $b = c = 0$

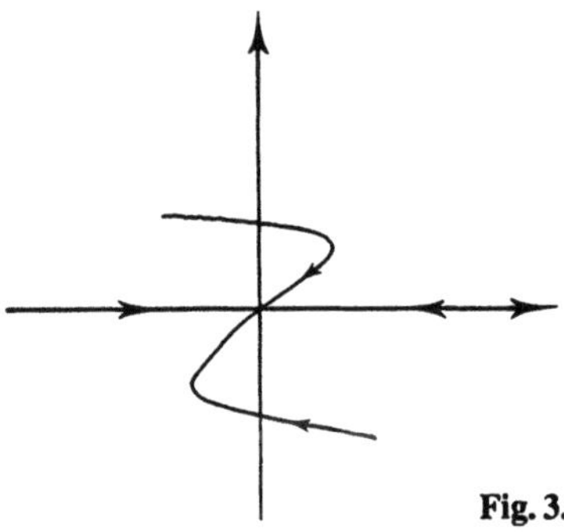

Fig. 3.8. The case of $b$ or $c = 0$

rotation of axes in the phase plane, reduces the original system to this form. There are then two possibilities. First, let $b = c = 0$. This case is illustrated in Fig. 3.7. Second, let $b$ (or $c$) $= 0$. The trajectories are illustrated in Fig. 3.8. Both equilibria are kinds of nodes.

v) The last possibility is that one (or both) of the eigenvalues is zero. In this case, as the matrix $A$ is singular, there is a non-trivial solution of $A(x, y)^{\mathrm{T}} = 0$, and all points $q(x, y)^{\mathrm{T}}$ are equilibria, where $q$ is a real constant. Thus the origin is no longer an isolated singularity.

It should be mentioned that ways to distinguish these cases by the values of $T$ and $W$ are given in Nicolis and Prigogine (1977). Moreover, we can discuss high-dimensional linear systems similarly.

## 3.3 The Principle of Linearized Stability

As we are usually interested in the stability of a particular type of solution of

$$\frac{dx}{dt} = f(x) , \tag{3.3.1}$$

it is often convenient to study the behavior of these equations around a particular "reference" state (e. g., equilibrim $x_0$). We may regard a solution $u(t)$ as an unperturbed solution that is continuously perturbed by the action of external disturbances or by internal fluctuations, $X(t)$. The latter leads from the solution $u(t)$ to a new solution

$$x(t) = u(t) + X(t) . \tag{3.3.2}$$

(3.3.2) defines a new coordinate system in phase space, centered on the reference state rather than at the point $(0, \dots, 0)$. Let us consider the case when $u$ is a time-independent equilibrium, $x_0$. The system can be rewritten as

$$\dot{X} = g(X) = AX + N(X) , \tag{3.3.3}$$

where $A$ is a matrix and $N(X) = \mathrm{O}(X)$ as $X \to 0$ (i.e., $|N(X)|/|X| \to 0$ as $|X| \to 0$). $N(X)$ represents the nonlinear terms in the equations. We assume here that $g(X)$ is sufficiently smooth to admit such a representation and $g(0) = 0$.

The stability of the reference state is now reflected by the stability of the "trivial solution", $X = 0$. The following well-known results illustrate relations of stability between nonlinear and their corresponding linearized systems.

**Theorem 3.3.1.** The linearized system for (3.3.3) is

$$\dot{X} = AX . \tag{3.3.4}$$

If the matrix $A$ has (a) all eigenvalues with a negative real part, or (b) at least one eigenvalue with a positive real part, then the trivial solution has the same character in the nonlinear system (3.3.3) as in the linearized system (3.3.4) in a sufficiently small neighborhood of the origin. However, if the linearized system has one or more eigenvalues with zero real part and no eigenvalues with positive real part, then the nonlinear system may be stable, unstable or asymptotically stable, depending on the nonlinear terms.

The proof is referred to, for instance, in Coddington and Levinson (1955), and Arrowsmith and Place (1982).

We will not repeat stability conditions for linear systems because they can be found in any standard textbook on differential equations.

Here, we would like to define an economic growth model with money to illustrate how the concepts just defined can be applied to economic problems. As this model is used later, we review it in detail (see Zhang 1989).

It is assumed that the production side is identical to the Solow model. Wealth may be held in alternative forms. The money, costlessly issued by the government, is assumed to serve as *numeraire*. Money is desired for transactions and investment purposes. The demand for money is affected by the distribution of income and wealth. However, for simplicity we assume that per capita demand for money is a function of per capita money income, per capita money wealth, and the expected yield on capital. The money market is assumed to be always at equilibrium. That is, the demand for money is always equal to its supply. The demand for money is assumed to take on the following form

$$m = G(y, w, r) , \tag{3.3.5}$$

where $m$ (= $M/L$ where $M$ is money stock and $L$ is labor force) is per capita money stock, $G$ is a continuous function with regard to the arguments, $y$ is per capita value of output, $w$ (= $pk + m$ where $p$ is price) per capita money wealth, and $r$ the expected money yield on the capital good ($r = f'(k) - d + E[dp/dt/p]$, where $k$ is per capital stock, $d$ is the depreciation rate, $E[dp/dt/p]$ is the expected inflation rate). Following Keynesian tradition, money is decided both for speculation and transaction purposes, and the function $G/p = g(k, r)$ is not homogeneous in $k$.

The real wealth, $W$, and real disposable income, $Y_d$, are defined as: $W = K + M/p$, and $Y_d = F(K, L) - dK + d(M/p)/dt$, respectively. As $F(K, L) = C + dK + dK/dt$ where $C$ is consumption, one has $Y_d = dW/dt + C$. That is, real net disposable income is equal to the change in real wealth plus real consumption. It is assumed that real consumption is always a constant fraction of real net disposable income: $C = cY_d$, $0 < c < 1$, where $c$ is the propensity to consume. From these assumptions one has: $dW/dt = sY_d$, where $s = (1 - c)$. The real per capita money $x$ is defined as $x = M/pL$. According to the definition, we have

$$\dot{x} = \left(\frac{\dot{M}}{M} - \frac{\dot{L}}{L} - \frac{\dot{p}}{p}\right) x = \left(z - n - \frac{\dot{p}}{p}\right) x , \tag{3.3.6}$$

where $n$ is the fixed population growth rate, and $z$ is the constant rate of increase in the nominal stock of money. The parameter $z$ is fixed by the government.

To formulate the dynamics of price, let us assume perfect myopic foresight, which states that the expected inflation rate is equal to the actual rate, $E[dp/dt/p] = dp/dt/p$. Consequently, we have: $dp/dt = p[r - f'(k) + d]$. On the other hand, from the portfolio equilibrium condition (3.3.5), we can solve $r$ as a function of $k$ and $x$ with $r = u(k, x)$ where $u_1 > 0$, $u_2 < 0$. The price dynamics is given by

$$\frac{\dot{p}}{p} = u(k, x) - f'(k) + d ,$$

from which (3.3.6) is rewritten as

$$\dot{x} = \left[f'(k) + v - u(k, x)\right] , \tag{3.3.7}$$

where $v = z - d - n$.

It is not difficult to identify the following equation for capital

$$\begin{aligned} \dot{k} = sf(k) - (n + sd)k - c(z - \dot{p}/p)x = sf(k) \\ - (n + sd)k - c\left[f'(k) - d + z - u(k, x)\right]x . \end{aligned} \tag{3.3.8}$$

The system which consists of (3.3.7) and (3.3.8) is called the Tobin model (see Zhang 1989). In what follows, our discussion is limited to the domain of values of the parameters such that the system is economically meaningful. Under appropriate conditions, the existence of a positive unique equilibrium can be guaranteed (Burmeister and Dobell 1970). In the equilibrium, the rate of price change may be either positive, negative, or zero, depending on the sign of $(z - n)$. It can be proved

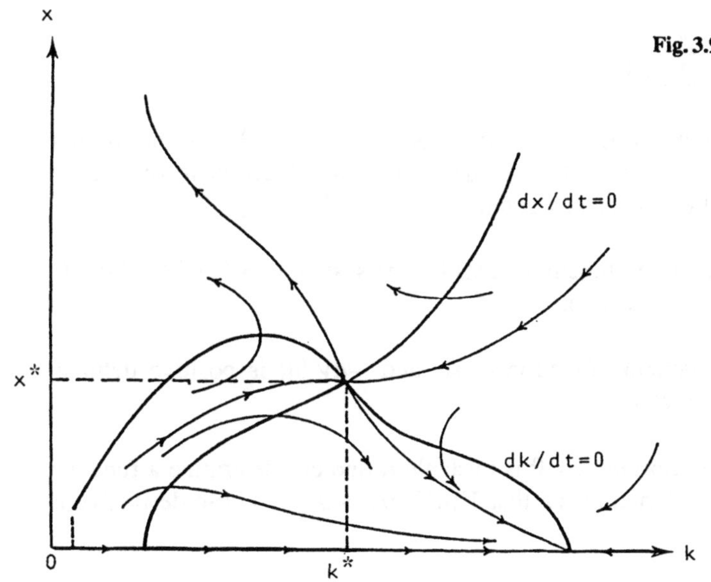


Fig. 3.9. The local dynamic behavior

that the equilibrium value of $k$ without money ($x = 0$) is higher than the equilibrium capital/labor ratio with money. The equilibrium is a saddle point. The behavior of the system near the equilibrium can be illustrated as in Fig. 3.9.

These results can easily be checked by applying the analytical results just described.

## 3.4 Lyapunov's Direct Method

The investigation of stability based on the system of linearized equations amounts to the direct integration of this system. For systems involving many economic variables this can often be an awkward way of proceeding, especially if the reference state is space and/or time dependent. In contrast, the method known as Lyapunov's second, or direct, method provides stability conditions that are: (a) independent of the integration of the linearized system, (b) applicable to reference solutions of all kinds, including space and/or time-dependent ones, and (c) applicable directly to the nonlinear systems, like (3.1.1). It should be noted that in the cases where the nonlinear terms are important in determining stability, there are few general theorems stating which possibility occurs, and it is usually necessary to consider each new system separately. These are the reasons why the Lyapunov theorem plays an especially important role in stability theory. The method has found wide application in economics (e. g., Arrow and Hahn 1971).

**Definition 3.4.1.** (*Lyapunov Function*). A function $V\colon R^m \to R$ is positive definite if (a) $V(0) = 0$ and (b) $V > 0$ everywhere else in some open region $G \in R^m$ containing the origin. For any solution $x = x(t)$ of $dx/dt = f(x)$, the function $V(x) = V(x(t))$ is a function of $t$ and its derivative is defined by

$$\frac{dV}{dt} = \sum_{i=1}^{m} \frac{dx_i}{dt}\frac{\partial V}{\partial x_i} = \sum_{i=1}^{m} f_i(x)\frac{\partial V}{\partial x_i} ,$$

if $V$ is sufficiently smooth. A Lyapunov function $V\colon R^m \to R$ for the system $dx/dt = f(x)$ is a function with continuous derivatives which are positive definite and such that $dV/dt \leqslant 0$ on $G$ for any solution $x$ of $dx/dt = f(x)$.

**Theorem 3.4.1.** If a Lyapunov function exists for the system (3.3.1) with $f(0) = 0$, then the equilibrium at zero is stable.

**Theorem 3.4.2.** If a Lyapunov function exists and $-dV/dt$ is positive definite, then zero is asymptotically stable.

**Theorem 3.4.3.** The equilibrium $x = 0$ is unstable if one can determine a function that assumes values in the domain such that $V(dV/dt) > 0$ and whose derivative is definite.

The proofs of these theorems can be found in Coddington and Levinson (1955) or Arrowsmith and Place (1982).

For example, consider the system

$$\begin{aligned} \frac{dx}{dt} &= y + xh(x,y) , \\ \frac{dy}{dt} &= -x + yh(x,y) , \end{aligned} \tag{3.4.1}$$

where $h$ is continuous near the origin and $h(0,0) = 0$. The linearized system has a center at the origin. Consider the function $V(x,y) = x^2 + y^2$. This is positive definite, and $dV/dt = 2Vh(x,y)$. Hence, the origin is asymptotically stable if $h$ is negative definite in its neighborhood and unstable if $h$ is positive definite. This is a non-local result.

We will use the Lyapunov direct method to prove stability of a Tâtonnement price adjustment process of the Arrow-Debreu system. The following example is predominantly based on Hahn (1982).

We assume that the economy has $N$ goods, $H$ households and $F$ firms. Define $x^h \in R^N$ as the net trade vector of household $h$; $x = \Sigma_h x^h$. Also, $y^f \in R^N$ is an activity of firm $f$ where positive components denote outputs and negative ones, inputs; $y = \Sigma_f y^f$. Let $z$ be the aggregate excess demand vector and $s$ the aggregate excess supply vector defined by

$$z = x - y = -s . \tag{3.4.2}$$

For convenience, let $Z$, $X$, $Y$ and $S$ denote the vectors $z$, $x$, $y$ and $s$ with their first component deleted. Let $p \in R_+^N$ be a price vector and $P$ the vector $(1/p_1)p$ with its first component deleted. It is assumed $p_1 > 0$. The endowment of $h$ is written as $w^h \in R_+^N$. Define

$$w = \Sigma w^h , \quad w^* = (w^1 , \dots , w^H) .$$

We are concerned here with economies which have continuously differentiable excess supply (demand) functions. It is well known that as a result of the rational behavior of the households and firms, we can determine the excess supply and demand as functions of $p$ and $w^*$, i.e., $s(p, w^*)$, $z(p, w^*)$. They are homogeneous at degree one in $p$ and obey Walras' law, respectively

$$s(p, w^*) = s(1, P, w^*) ,$$
$$ps(p, w^*) = 0, \quad \text{all} \quad p \in R_+^N . \tag{3.4.3}$$

Let $D = \{p \mid p > 0, \ \Sigma p_i = 1\}$ and let $G$ be the boundary of $D$.

**Definition 3.4.1.** (*Equilibrium*). $p^* \in D/G$ is an equilibrium if for each $i$, (i) $s_i(p^*, w^*)$ is non-negative, and (ii) $p_i^* s_i(p^*, w^*) = 0$.

Under appropriate conditions the existence of a unique equilibrium is guaranteed. We consider the following Tâtonnement

$$\frac{dp_i}{dt} = 0 \quad \text{if} \quad p_i = 0 \quad \text{and} \quad S_i(P) > 0 , \ i = 2, \ldots , N ,$$
$$\frac{dp_i}{dt} = -k_i s_i(p) , \quad k_i > 0 , \quad \text{otherwise} . \tag{3.4.4}$$

The solution to (3.4.4) can be proven to be bounded. Let us define

$$V(p) = \Sigma \frac{(p_i - p_i^*)^2}{k_i} . \tag{3.4.5}$$

Under the Walras law, one has: $dV/dt = 2p^* s(p)$. It can be proved that if all goods are gross substitutes, then $p^* s(p) < 0$ if $p$ is not equal to $kp^*$ for $k > 0$. One thus obtains the following result.

**Theorem 3.4.1.** If all goods are gross substitutes, then the unique equilibrium is globally asymptotically stable under (3.4.5).

The process just mentioned can be generalized (Hahn 1982). Moreover, different possible adjustment processes have been suggested in the literature (see Arrow and Hahn 1971).

## 3.5 Structural Stability

The stability concepts developed above are related to the way in which a dynamic system reacts to perturbations in initial conditions. However, sometimes we are interested in properties of functional forms of dynamic systems. For example, it is important to know about the stability of the money supply function itself. The concept of structural stability is related to qualitative properties of functions.

To explain the real world, we build economic models. However, the result may be very sensitive to the smallest change in the model. In such a case, an arbitrarily small change in the model leads to another model with essentially different properties. To illustrate the concept of structural stability, consider the well known predator-prey system which has been applied in economics by several authors. The system consists of two differential equations as follows

$$\frac{dx}{dt} = \alpha(y_1 - y)x \ ,$$
$$\frac{dy}{dt} = \beta(x - x_1)y \ . \tag{3.5.1}$$

In the dynamic urban literature, this model has been employed to describe the dynamics of a small urban area (Dendrinos and Mullally 1985, Zhang 1988c). In (3.5.1), $x$ stands for land use density, and $y$ for the land rent, and $\alpha$, $\beta$, $x_1$ and $y_1$ are appropriate parameters. This is a simple demand-supply model of speculative land rent under foresight, with particularly congruent expectations from demanders and suppliers.

Another application of this type of model to economics is given by Goodwin (1967) (see also, Gabisch and Lorenz 1986). As the predator-prey system will be mentioned several times later on, it may be useful to describe the Goodwin model in detail.

The Goodwin model is built to describe the class struggle as follows. Consider two households: workers and capitalists. Workers spend all their income $wL$ on consumption; capitalists save all their income, $Y - wL$, with $Y$ being production. The goods price is normalized to unity. Let $K$ denote capital, $a = a_0 \exp(gt) = Y/L$ denote labor productivity growing at the constant rate $g$, $k = K/Y$ the capital output ratio, and $N = N_0 \exp(nt)$ denote labor supply growing at the rate $n$. The wage income share of national income is $wL/Y = w/a$. Hence the profit share is $1 - w/a$. As the savings are determined by $S = Y - wL = (1 - w/a)Y$, investment is: $dK/dt = S = (1 - w/a)Y$, or $(dK/dt)/K = (1 - w/a)/k$, where we neglect any depreciation. Provided that the capital-output ratio $k$ is constant, it follows that $Y dK/dt = K dY/dt$. We thus have

$$\frac{dY/dt}{Y} - \frac{dK/dt}{K} = g \ ,$$

and combined with $(dK/dt)/K = (1 - w/a)/k$ we obtain: $(dL/dt)/L = (1 - w/a)/k - g$. Introducing the labor bill share, $y = w/a$, and the employment rate, $x = L/N$, we can show that

$$\frac{dx}{dt} = x\left[\frac{1-y}{k} - (g+n)\right] \ ,$$
$$\frac{dy}{dt} = y\left[\frac{dw/dt}{w} - \frac{da/dt}{a}\right] \ ,$$

where $(da/dt)/a = g$. The wage rate is assumed to be a fast variable, which is determined by a Phillips curve relation, i. e.,

$$\frac{dw}{dt} = wf(x), \ \lim_{x \to 1} f(x) = +\infty, \ \lim_{x \to 0} f(x) < 0, \ f' > 0.$$

Approximating this relation linearly by $(dw/dt)/w = -r + bx$ yields $(dy/dt)/y = -r + bx - g$. We thus derive the Goodwin model as follows:

$$\frac{dx}{dt} = x\left[\frac{1}{k} - (g+n) - \frac{y}{k}\right],$$
$$\frac{dy}{dt} = y\left[-(g+r) + bx\right].$$

Evidently, the Goodwin model conforms to (3.5.1). The general discussion about properties of (3.5.1) should apply to the Goodwin model. The formal identity of the Goodwin model with the Lotka-Volterra predator-prey system establishes an analogy between the class struggle and the struggle of competitive species. This model, with its interaction of the employment rate and the wage bill share, is strongly reminiscent of the models of classical political economics. The model, sometimes referred to as a neo-Marxian model, has stimulated recent attention to the classical economists such as Ricardo, Smith, and Marx. The modern extensions of this model are referred to, for example, in Desai (1973), Velupillai (1978), Shah and Desai (1981), van der Ploeg (1983, 1987), Flaschel (1984) and Zhang (1988a).

The model is very simple and can exhibit oscillations. However, the property of structural instability limits its applications. As is well known, even small perturbations in the functional forms will change the qualitative properties of the system. The proposed model thus cannot be transferred to the real process under consideration, because when we constructed the model the real situation was idealized and simplified, and the parameters were determined only approximately. The question then arises of how to choose those properties of the model of a process which are not very sensitive to small changes in the model, and thus may be viewed as properties of the real process. Structural stability implies such properties.

Although the fundamental ideas for the concept of structural stability were introduced by Poincaré, the modern development of the concept was initiated by the work of Andronov and Pontrjagin in 1937. Smale (1967) made significant progress with the theory of structural stability for phase spaces with small dimension. He showed that for phase spaces of large dimension, systems exist in the neighborhood of which there is no structural stable system. This result means that the problem of complete topological classification of differential equations with high-dimensional phase space is hopeless, even when restricted to generic equations and nondegenerate cases.

To explain the concept of structural stability, let us consider the differential equation $dx/dt = f(x)$ for a given vector field, $f$, on the manifold, $M$.

**Definition 3.5.1.** Two systems are said to be topologically orbitally equivalent if there exists a homeomorphism of the phase space of the first system onto the phase space of the second, conserving oriented phase curves of the first system onto oriented phase curves of the second. No coordination of the motion on corresponding phase curves is required.

**Definition 3.5.2.** Let $M$ be a compact manifold (of class $C^{k-1}$, $k > 0$). Let $f$ be a vector field of class $k$ (if $M$ has a boundary, then it is assumed that $f$ is not tangent to it). The system $(M, f)$ is said to be structurally stable if there exists a neighborhood of $f$ in the space $C^1$ such that every vector field in this neighborhood defines a system topologically orbitally equivalent to the initial one, and the homeomorphism of the equivalence is close to the identity homeomorphism.

A system which is not structurally stable is defined to be structurally unstable. The system (3.5.1) is an unstable system. We now provide another example of structurally unstable systems. The equation of a pendulum with friction is: $dx/dt = dy/dt = -x - ry$. If $r = 0$, then all phase curves are closed. If $r > 0$, they spiral towards the singular point 0, which is a local type. Consequently, a small change in the friction coefficient $r$ changes the behavior of the phase curve qualitatively if the coefficient had been zero before the change. It does not change the qualitative picture if the coefficient had been positive.

We can define the concept more simply, which is enough for our purpose. A system $dx/dt = f(x)$ is structurally stable if there exists a homeomorphism from the orbits of $dx/dt = f(x)$ to the orbits of $dx/dt = f(x) + p(x)$ for sufficiently small perturbations $p(x)$.

Let $M$ denote the interior of a closed curve without contact to any of the vector fields to be considered and let $G$ be the set of all such $C^k$ vector fields. We have the following theorem.

**Theorem 3.5.1.** A function $f(x)$ in $G$ is structurally stable if and only if every equilibrium point and every periodic orbit is hyperbolic and there are no connections between saddle points. Also, the set of structurally stable systems is open and dense in $G$.

These concepts and Theorem 3.5.1 are referred to in Chow and Hale (1982). The theorem provides a necessary and sufficient condition for identifying structural stability of a dynamic system. However, the results are not so easy to apply as may appear because it is difficult to check the conditions for real problems.

## 3.6 Conservative Systems

In this section we will define the concept of conservative systems and examine the properties of such systems. We will show the relation between conservative systems and structural stability.

Consider a dynamic system $dx/dt = f(x)$. The fundamental property of a conservative system is the existence of a function for the dependent variables which is a constant of the motion equation and plays the role of "energy". The system is conservative if there exists a function $G(x)$, known as a first integral, or simply an integral, of the system, such that

$$\frac{dG(x)}{dt} = \sum_{i=1}^{n} \frac{dx_i}{dt}\frac{\partial G}{\partial x_i} = 0 .$$

In physical terms, conservative systems are characterized by the fact that during evolutionary processes a "volume" element in phase space changes only its shape but retains its volume in the course of time, while in dissipative systems the volume elements shrink as time increases. This difference is illustrated in Fig. 3.10. In dissipative systems, trajectories are attracted to a fixed point, and volume shrinks, but in conservative systems the points rotate around an elliptic fixed point and volume is conserved.

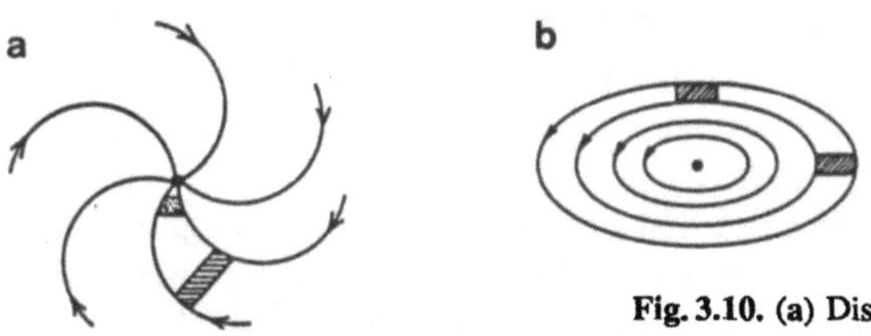


**Fig. 3.10.** (a) Dissipative systems, (b) conservative systems

Precisely, the system of ordinary differential equations $dx/dt = f(x,t)$, $x \in R^n$, is called dissipative if there are numbers $R > 0$ and $t_1 > 0$ such that for all solutions $x(.)$ of the system it is the case that $|x(0)| \leqslant R$ always implies that $|x(t)| < R$ for all times $t \geqslant t_1$. About dissipative systems, we have the following important theorem.

**Proposition.** The dissipative system $dx/dt = f(x,t)$, $x \in R^n$, has a solution of period $p > 0$ if (i) the function $f$ is p-periodic with respect to $t$; and (ii) for every initial value $x_0 \in R^n$ there is a unique solution $x(.)$ with $x(0) = x_0$ which exists for all times $t \in [0, \infty]$. Here $x(t)$ depends continuously on $x_0$.

*Proof.* To prove the proposition, we use the following Brouwer fixed point theorem.

Let $A$: $X \to X$ be a compact operator on $R^n$. Suppose that for some fixed natural number $m$ the set $A^m(X)$ is bounded. Then $A$ is a fixed point.

Construct the shift operator $A$: $R^n \to R^n$ by $Ax_0 = x(p)$. Here $x(.)$ is the solution of the system. Then $A^m x_0 = x(mp)$. Set $G = \{x \in R^n : |x| < R\}$. Hence $A^m x_0 \in G$ for all $x_0$ belongs to the closure of $G$ and sufficiently large $m$. Thus $A$ has a fixed point, to which the desired periodic solution corresponds. □

Conservative systems often have oscillatory solutions and have therefore been widely used to model phenomena such as oscillations in prey and predator populations, urban land rent and land use density interactions, unemployment and economic growth dynamics, and so on.

To show that system (3.5.1) is conservative, let us make the following transformation:

$$u = \frac{x}{x_1} , \quad v = \frac{y}{y_1} , \quad \sigma = \frac{\beta x_1}{\alpha y_1} , \quad t^* = \alpha t ,$$

under which (3.5.1) is rewritten as:

$$\begin{aligned} \frac{du}{dt^*} &= u(1-v) , \\ \frac{dv}{dt^*} &= \sigma v(u-1) , \end{aligned} \tag{3.6.1}$$

The following first integral can easily be identified as

$$G(u,v) = \sigma(u - \log u) + v - \log v = A .$$

where $A$ is a constant. Since $G(u,v)$ does not change as we move along a trajectory or solution curve of the equations, these trajectories are defined by the curves $G(u,v) = A$ for different values of the constant $A$. It follows from this that the singular point (1,1) cannot be a stable focus. For if it were, then all curves in a neighborhood of it would tend to it, and hence would have $G(u,v) = G(1,1)$, since $g$ is a continuous function. But this implies that $G$ is constant in the neighborhood of (1,1), which contradicts its definition. It also follows by similar arguments that there are no stable or unstable limit cycles surrounding the point (1,1). All trajectories starting in the positive quadrant are bounded, so the only possibility is that the phase plane consists of closed trajectories around the singular point, each with a different value of the energy $G(u,v)$. The model is thus orbitally stable but not stable.

In any attempt at modeling, certain effects have to be neglected. We have to be sure that such neglect does not seriously affect the final solution. That is, for the model to be a good one for reflecting the problem, it is necessary to require the model to be structurally stable.

In order to show how the behavior of (3.6.1) can be affected by small perturbations, let us add a term $-ru^2$ to the first of the equations (3.6.1) to obtain

$$\begin{aligned} \frac{du}{dt} &= u(1-v) - ru^2 , \\ \frac{dv}{dt} &= \sigma v(u-1) . \end{aligned} \tag{3.6.2}$$

If the parameter $r$ is extremely small, it is natural to require that the new term will not have a significant effect on the solution of the original system. However, an eigenvalue analysis shows that the equilibrium $(1, 1-r)$ is a stable focus in the linearized system and is therefore a stable focus in the nonlinear system, however small $r$ may be. In fact, $V(u,v) = \sigma(u - \log u) + v - (1-r)\log v$ is a global (in the positive quadrant) Lyapunov function for the system. The solutions now spiral in to the equilibrium, and the system can no longer be put forward as a model for urban oscillations.

In fact, the general perturbation problem to (3.6.1)

$$\frac{du}{dt} = u(1-v) + \varepsilon f_1(u,v) ,$$
$$\frac{dv}{dt} = \sigma v(u-1) - \varepsilon f_2(u,v) ,$$

where $f_i$ ($i = 1, 2$) are perturbation functions, has been studied by Freedman and Waltman (1975). By the use of the implicit function theorem they proved that even though $\varepsilon$ is arbitrarily small, it is possible to observe stable/unstable cycles for some specified functions $f$, which are qualitatively different from the original periodic solution.

Conservative systems tend to be amenable to analysis, but they have some major disadvantages as models for real systems. As all conservative systems are structurally unstable (Britton 1986), they should be used with great care. In the remainder of this section, we provide an example of conservative economic models.

The dynamic model is suggested by Andersson and Zhang (1988a). The model is defined within the framework of the Leontief input-output system. The economy consists of $n$ sectors without joint production. The growth rate of the economy is controlled by the government. It consists of the dynamics of prices and outputs

$$\frac{dp}{dt} = Ax + g(p,x)Bx - x ,$$
$$\frac{dx}{dt} = p - A^{\mathrm{T}}p - g(p,x)B^{\mathrm{T}}p , \tag{3.6.3}$$

in which $p = (p_1, \ldots, p_n)^{\mathrm{T}}$ is the "normalized" price vector, $x = (x_1, \ldots, x_n)$ the "normalized" output, and $A = (a_{ij})$ and $B = (b_{ij})$ are the technological coefficient and investment coefficient matrices, respectively. In (3.6.3), $g(p,x)$ is the growth rate of the economic system. The growth rate is determined by the government, which maximizes a "social" utility function subject to technological capacities.

The price equations of (3.6.3) mean that if the demand for the $i$th good exceeds the supply of this good, then the price of the $i$th good should be increased, and vice versa. This equation takes the behavior of the consumers into account. Similarly, the quantitative equations mean that if the total cost per unit of the $i$th product is greater than the price of this good, then the $i$th sector should decrease its output in order to decrease loss.

**Theorem 3.6.1.** The dynamic system (3.6.3) is conservative.

*Proof.* It can be ascertained that the following function plays the role of the first integral:

$$V(p,x) = p^{\mathrm{T}}p + x^{\mathrm{T}}x . \qquad \square$$

It should be noted that more general dynamic adjustment processes are suggested by Andersson and Zhang (1988a). We will not discuss these in detail.

Although the conservative systems display no attracting regions in phase space, i. e., no attracting fixed points, no attracting limit cycles, and no strange attractors, we can still find "chaos" with a positive $K$-entropy, i. e., there are "strange" or "chaotic" regions in phase space, but they are not attractive and can be densely interwoven with regular regions in such systems (see Schuster 1988). We are not concerned with such complicated problems in this book.

## 3.7 Bifurcation Theory

> Mathematical discoveries, small or great ... are never born of spontaneous generation. They always presuppose a soil seeded with preliminary knowledge and well prepared by labor, both conscious and subconscious.
>
> *Henri Poincaré*

The primary objective of this section is to introduce some aspects of bifurcation theory.

There are two distinct aspects of bifurcation theory: static and dynamic. Static bifurcation theory is concerned with the changes that occur in the structure of the set of zeros of a function as parameters in the function are varied. In differential equations, the equilibrium solutions are the zeros of the vector field. Therefore, methods in static bifurcation theory are directly applicable. Dynamic bifurcation theory is concerned with the changes that occur in the structure of solutions of differential equations as parameters in the vector field are varied.

A change in the qualitative properties could mean a change in stability of the original system, and thus the system must assume a state different from the original design. In vague terms, the values of the parameters where this change takes place are called bifurcation values. Knowledge of the bifurcation values is absolutely necessary for a complete understanding of the system. Consider the following evolution equations

$$\frac{dx}{dt} = f(x,r)\,, \tag{3.7.1}$$

where $x$ is defined in some space, $r$ represents a parameter vector, and $f$ is a function vector which satisfies certain requirements. There may exist different types of solutions such as: (i) steady solutions, (ii) periodic solutions, (iii) sub-harmonic solutions, and (iv) asymptotically quasi-periodic solutions, and so on.

Consider the equilibrium equations $f(x,r) = 0$. Sometimes, derivatives of $f$ are required, and it will always be assumed that $f$ has as many derivatives as necessary if this is not explicitly stated. We may consider equilibrium states as functions of the parameters. Usually multiple equilibria may exist for given values of the parameters. The basic question to discuss is how the equilibrium depends upon the parameters.

For convenience, let both $x$ and $r$ belong to $R^1$. The (static) bifurcation problem is equivalent to the study of the curves $f(x,r) = 0$ and their singular points. The main tool for the proof of the existence of solutions in bifurcation theory is the implicit function theorem which holds for vector-valued functions of multiple parameters (see, e. g., Chow and Hale 1982). For a one-dimensional problem, the theorem may be stated as follows.

**Lemma** (*Implicit Function Theorem in* $R^1$). Let $f(x_0, r_0) = 0$ and $f$ be $C^1$ in some open neighborhood of $(x_0, r_0)$ in the $(x, r)$-plane. Then if $f_x \neq 0$, there exists $\alpha$, $\beta > 0$ such that (i) the equation $f(x, r)$ has a unique solution $x = x(r)$ with $x_0 - \beta < x < x_0 + \beta$ whenever $r_0 - \alpha < r < \alpha + r_0$; and (ii) $x_r(r)$ exists and $x_r(r) = -f_r(x(r), r)/f_x(x(r), r)$.

We can classify points on solution curves as follows (e. g., Iooss and Joseph 1980, Britton 1986).

**Definition 3.7.1.** (*One Dimension Problem*)

i) A regular point $(x_0, r_0)$ of $f(x, r) = 0$ is a point where either $f_x \neq 0$ or $f_r \neq 0$. A regular turning point is a regular point at which $r_x(x)$ changes sign. Here, $f_x = 0$ but $f_r \neq 0$ at the point (see Fig. 3.11a).
ii) A singular point is a point which is not regular, $f_r = f_x = 0$ at the equilibrium.
iii) A bifurcation point is a singular point through which pass two or more branches of $f(x, r) = 0$.
iv) A double point is a singular point through which pass two and only two branches of $f(x, r) = 0$ possessing distinct tangents, and where all second derivatives of $f$ do not simultaneously vanish. A turning double point is a double point at which $r_x$ changes sign on one branch (Fig. 3.11b).
v) A cusp point is a point of second-order contact between two branches of the curve (Fig. 3.11c).
vi) A conjugate point is an isolated singular point solution of $f = 0$.
vii) A higher-order singular point is a singular point at which all three second derivatives of $f$ are zero.

Bifurcation theory studies both the existence and the stability of equilibrium solutions, since an unstable equilibrium solution of the equations will not be observed in reality. There is an intimate relationship between changes of stability and bifurcation. For a more rigorous treatment of bifurcation theory, see Sattinger (1973), Iooss

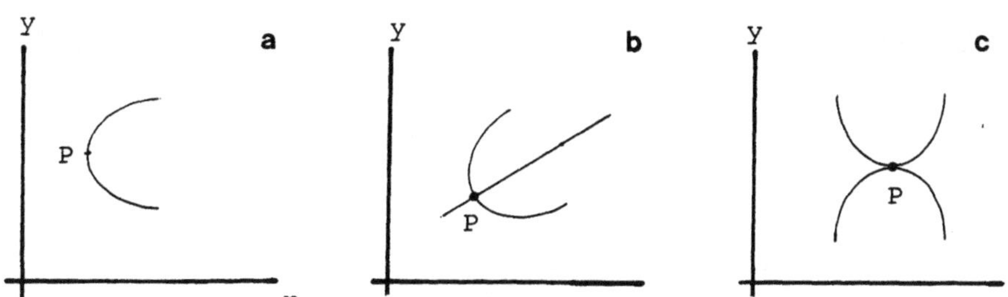


**Fig. 3.11.** Classification of bifurcations, (a) a regular turning point, (b) a turning double point, (c) a cusp point

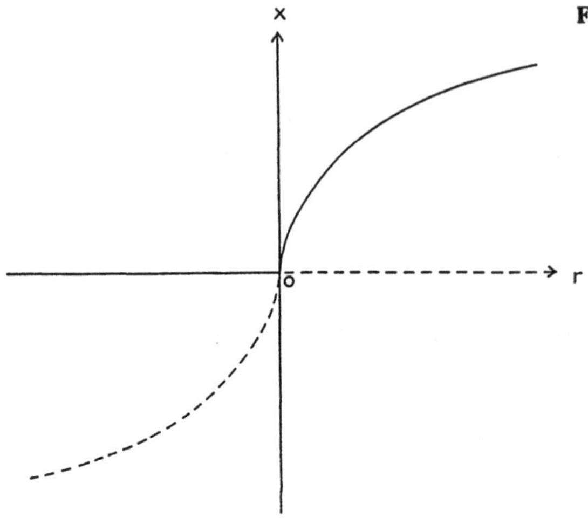


Fig. 3.12. The bifurcation diagram of (3.7.2)

and Joseph (1980) or Chow and Hale (1982). In what follows, we will provide a few examples to illustrate the concept of bifurcation.

Consider the equation

$$\frac{dx}{dt} = rx - x^4 = f(x,r) \, . \tag{3.7.2}$$

The bifurcation diagram is shown in Fig. 3.12, in which the line represents stability and the dashed line represents instability.

For the equation

$$\begin{aligned} \frac{dx}{dt} &= y - x(x^2 + y^2 - r) \, , \\ \frac{dy}{dt} &= -x - y(x^2 + y^2 - r) \, , \end{aligned} \tag{3.7.3}$$

the point $r = 0$ is a bifurcation point with the flow depicted in Fig. 3.13. The periodic orbit $x^2 + y^2 = r$ bifurcates from the equilibrium solution (0,0) and there are exchanges of stability as illustrated in the figure. This type of bifurcation (the Hopf bifurcation) is a consequence of the dynamics.

For the equation

$$\begin{aligned} \frac{dx}{dt} &= y \, , \\ \frac{dy}{dt} &= x - x^2 + ry \, , \end{aligned} \tag{3.7.4}$$

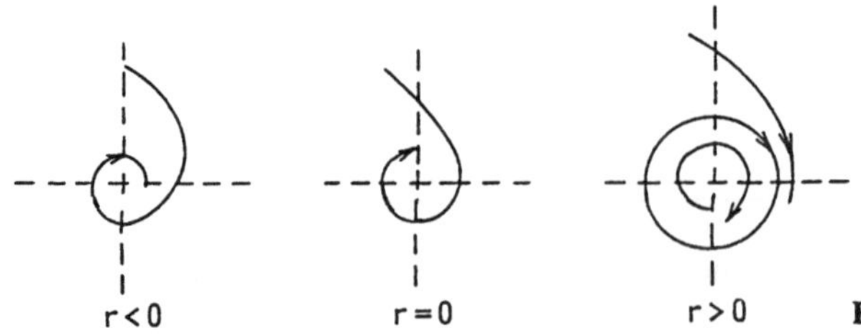


Fig. 3.13. The Hopf bifurcation

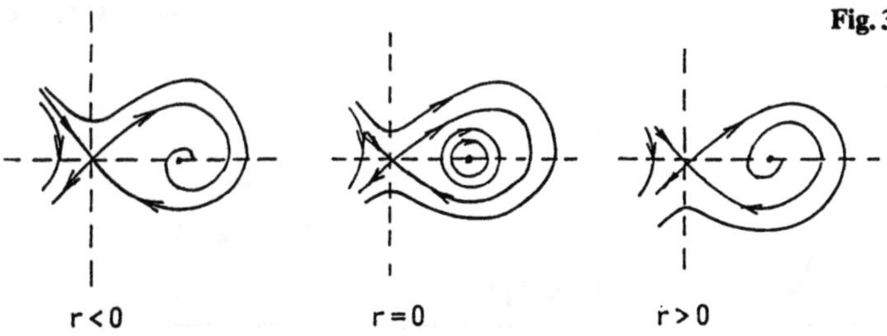


Fig. 3.14. Bifurcations of (3.7.4)

the point $r = 0$ is a bifurcation point with the orbits described in Fig. 3.14 (see Chow and Hale 1982). The equilibrium point (0,1) exchanges its stability properties as $r$ passes from negative to positive values.

Consider the equation

$$\frac{dx}{dt} = x(u - x^2 - by^2 + dx^4) ,$$
$$\frac{dy}{dt} = y(v + cx^2 + y^2) ,$$

where $b > 0$, $c > 0$, $bc > 1$, $d \neq 0$ are fixed parameters, and $u$ and $v$ are bifurcation parameters varying in a neighborhood of (0,0). A complete description of the bifurcations that occur in this equation is summarized in the following theorem (see (Chow and Hale 1982).

**Theorem 3.7.2.** There is a neighborhood $U$ of $(x, y) = (0,0)$ and a neighborhood $V$ of $(u, v) = (0,0)$ such that when the neighborhood $V$ is divided into regions as shown in Fig. 3.15, the flow for the equation for $x \geqslant 0$, $y \geqslant 0$ is the one depicted in Fig. 3.16 provided that there is at most one periodic orbit for any $(u, v)$ between $L_2'$ and $L_2''$. The curves $L_1$, $L_2$, $L_2''$, $L_3$ are given by

$$L_1 : \quad u + bv + O(|v|) = 0 ,$$
$$L_2' : \quad u(1 + c) + v(1 + b) + dh'v^2 + O(v^2) = 0 , .$$
$$L_2'' : \quad u(1 + c) + v(1 + b) + dh''v^2 + O(v^2) = 0 ,$$
$$L_3 : \quad uc + b + O(|v|) = 0 ,$$

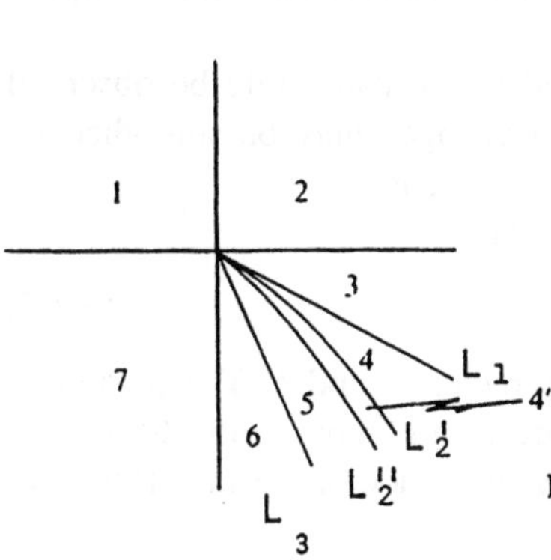


Fig. 3.15. Bifurcation set

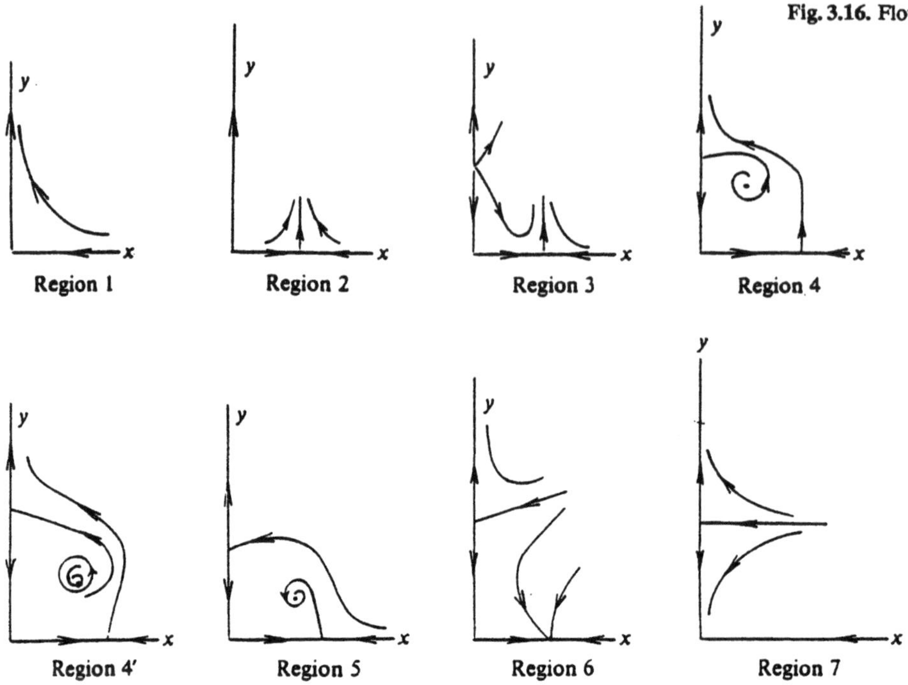


Fig. 3.16. Flows

where $v \leqslant 0$, $h'$ and $h''$ are computable constants. All of the bifurcations are of the saddle-node type except the ones on $L_2'$ and $L_2''$, where a Hopf bifurcation occurs at $L_2''$ and a heteroclinic orbit is exhibited on $L_2'$.

We now define some other useful concepts in bifurcation theory. The concept of cascading bifurcation is defined as a sequential bifurcation of solutions of nonlinear equations, as the parameter increases (see Fig. 3.17). Each bifurcation may result in more complicated behavior. The Landau-Hopf picture is considered as an appropriate example. The scenario is as follows: a time-independent (spatially homogeneous) state bifurcates into other time-independent (spatially heterogeneous) states. A new such state then bifurcates into an oscillating state (Hopf bifurcation). Then the limit cycle bifurcates into a torus. Landau conjectured that these kinds of transition are continued in such a way that systems exhibit subsequent bifurcations to tori of higher and higher dimensions.

It may be argued that the bifurcations described above can rarely be observed in practice, since the sharp transitions or structural changes may be smoothed by imperfections and other disturbances that are always present.

For instance, consider a general nonlinear problem

$$F(x, r, h) = 0 \,. \tag{3.7.5}$$

The solutions $x = x(r, h)$ depend on two scalar parameters $r$ and $h$. The parameter $r$, called the bifurcation parameter, is the magnitude of an "input" into the system modelled by (3.7.5) and the small parameter $h$ is an amplitude of the imperfections.

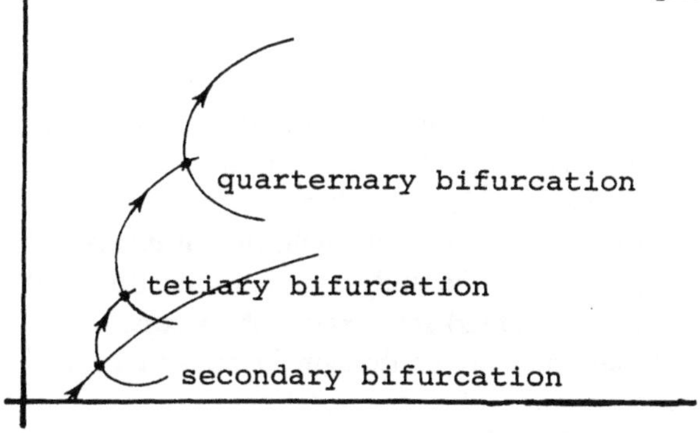


**Fig. 3.17.** A cascade of supercritical bifurcations

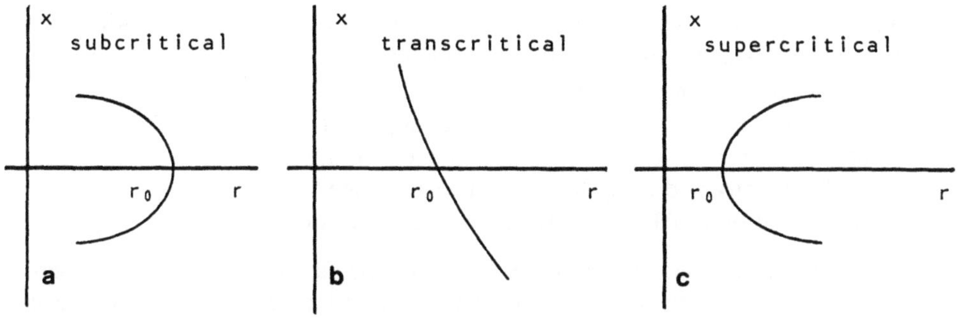


**Fig. 3.18.** Simple bifurcation points

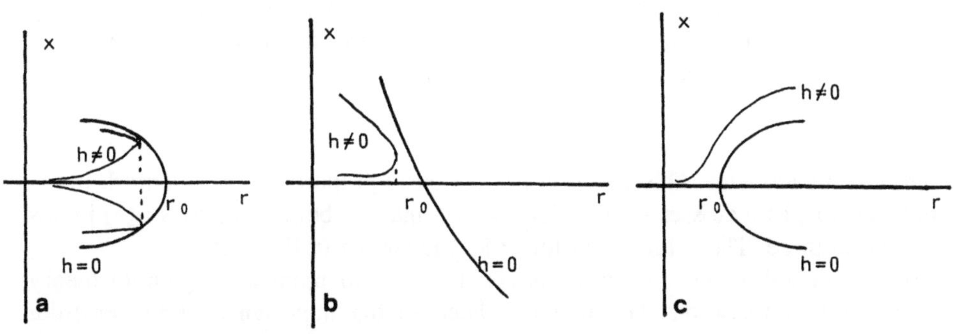


**Fig. 3.19.** Influence of imperfections

If $h = 0$, $F(x, r, 0)$ is a bifurcation problem. We refer to (3.7.5) when $h$ is not equal to 0 as the perturbed or imperfect bifurcation problem.

Consider the case when $h = 0$ and $r$ is a bifurcation point of $F(x, r, 0)$. Figure 3.18 illustrates three types of bifurcations that occur near the simple bifurcation point $r = r_1$. Typical response diagrams for the solutions of the imperfect bifurcation problem are sketched in Fig. 3.19. In comparison to $r_1$, a new critical parameter $r = r_c < r_1$ is defined. The point $r_c$ is called a limit point.

## 3.8 Singularity Theory

Many economic problems can be related to the study of properties of smooth functions. Rational behavior of a household or producer in perfect markets can be described by functions which depend upon prices. Comparative statics is applied to deal with how demand or supply shifts if market prices are changed. Production and utility functions play an elementary role in the study of the behavior of household and producer. Singularity theory is concerned with the study of smooth functions and their classification. This theory has enjoyed a remarkable degree of success. Catastrophe theory – one of the most important streams of thought in modern applied mathematics – is its special case.

Consider a smooth function $f\colon R^n \to R^m$ and assume that $f$ has a critical point at the origin, i. e., $Df(0) = 0$. Singularity theory addresses the following questions:

i) The determinacy problem: what is the local character of $f$ in a neighborhood of the origin? Basically, this question amounts to asking "at what point is it safe to truncate the Taylor series of $f$?"
ii) The unfolding problem: what are the essential perturbations of $f$? That is, what perturbations of $f$ can occur which change the qualitative nature of $f$ and which cannot be transformed away by a coordinate change?
iii) The classification problem: can we classify the types of singularities of $f$?

Elementary catastrophe theory solves these problems when $m = 1$; its generalization to singularity theory solves the first two, and gives relatively complete information on the third for small values of $n$, $m$.

We use an example to illustrate the way in which singularity theory can be applied. We discuss the pitchfork bifurcation from the point of view of singularity theory. The example is examined in detail in Golubitsky and Schaeffer (1984).

Let us consider the equation

$$x^3 - rx = 0 \,, \tag{3.8.1}$$

where $r$ is a parameter. This equation has the basic property of the pitchfork bifurcation. That is, as $r$ crosses some value $r_0(=0)$, the number of solutions $n(r)$ jumps from one to three. The solution set for (3.8.1) is shown in Fig. 3.20.

As mentioned above, two basic issues are focused upon in singularity theory approaches to bifurcation. The first is related to the importance of higher-order terms. The underlying question may be phrased: To what extent do the low-order terms in the Taylor series expansion of a bifurcation problem $f(x, r)$ determine its

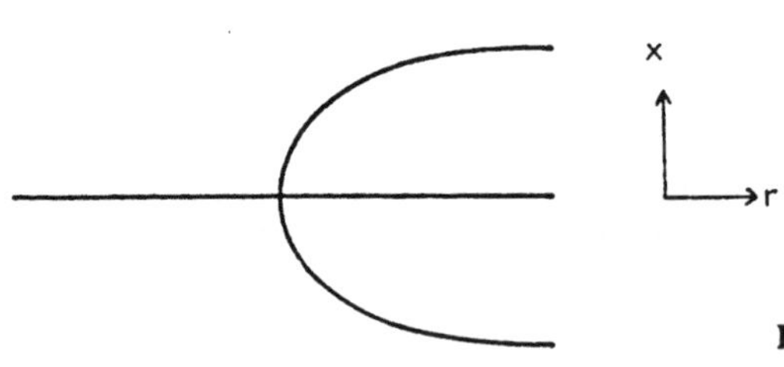


Fig. 3.20. The pitchfork bifurcation

qualitative behavior, regardless of the higher-order terms that may be present? For the case of the pitchfork bifurcation, let $f(x,r)$ be a bifurcation problem such that when $(x,r) = (x_0,\ r_0)$ we have

$$f = f_x = f_{xx} = f_r = 0\,, \quad f_{xxx} f_{xr} < 0\,. \tag{3.8.2}$$

Obviously, (3.8.1) satisfies the requirements. In this case, $n(r)$, the number of solutions of $f(x,r)$, jumps from one to three as $r$ crosses $r_0$. This can be proved using the implicit function theorem. However, in singularity theory approach, one proves considerably more. We try to prove that any $f$ satisfying (3.8.2) may be transformed by an appropriate change of coordinates into the standard model for the pitchfork, $x^3 - rx = 0$. More precisely, if $f$ satisfies (3.8.2), then there exists: (i) a local diffeomorphism of $R^2$ of the form $(x,r) \to (X(x,r), Y(r))$ mapping the origin to $(x_0, r_0)$; and (ii) a nonzero function $S(x,r)$ such that

$$S(x,r) f\big(X(x,r), Y(r)\big) = x^3 - rx\,, \tag{3.8.3}$$

near the origin, where $X_x(x,r) > 0$ and $Y' > 0$. Since $S$ is nonzero, the solutions of $f(x,r) = 0$ differ from those of $x^3 - rx = 0$ only by the diffeomorphism. That is, the higher-order terms in $f$ have no effect on the qualitative behavior of the model in the small – they may be transformed away entirely by a change of coordinates.

Equation (3.8.3) leads to the definition of a fundamental concept called equivalent in the singularity theory. Two bifurcation problems $f$ and $g$ are equivalent if they may be related through an equation

$$S(x,r) f\big(X(x,r), Y(r)\big) = g(x,r)\,,$$

where $S$ is nonzero and positive and $(X,Y)$ is a local diffeomorphism which preserves the orientations of $x$ and $r$.

If $f$ and $g$ are equivalent, then the two multiplicity functions are related as follows

$$n_f\big(Y(r)\big) = n_g(r)\,, \tag{3.8.4}$$

which is one of the most important consequences of equivalence.

The treatment of the pitchfork is representative of the general singularity theory approach to the determinacy problem. For further explanation, let us call $x^3 - rx = 0$ a normal form for the pitchfork bifurcation. Any bifurcation problem $f(x,r)$ which at a special point $(x_0, r_0)$ satisfies

$$f = f_x = f_{xx} = f_r = 0\,, \quad f_{xxx} > 0\,, \quad f_{xr} < 0\,, \tag{3.8.5}$$

is equivalent to this normal form. We say that (3.8.5) solves the recognition problem for this normal form. Equivalent bifurcation problems have the same qualitative properties; more precisely, qualitative properties are those which are unchanged by equivalence.

The second issue arises from, for example, the study of how bifurcation problems may depend on parameters. In a bifurcation problem $f(x,r)$, small variations of an

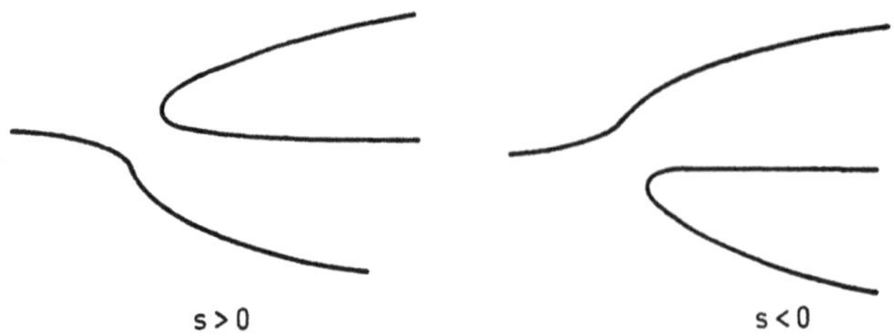


Fig. 3.21. Perturbations of the pitchfork

auxiliary parameter usually lead to dramatic changes in the bifurcation diagram at a singularity of $f$. Let us consider the perturbed pitchfork

$$F(x,r,s) = x^3 - rx + s = 0 \,, \tag{3.8.6}$$

The bifurcation diagrams are shown in Fig. 3.21 for the equation with $s$ not equal to zero.

In the classical literature there appear to be two distinct ways in which auxiliary parameters arise in bifurcation problems. For instance, the original formulation of an economic model may involve many auxiliary parameters. In other cases, the parameters arise from the more subtle issue of imperfect bifurcation. In the singularity theory approach, the occurrence of parameters is handled as follows. Given a bifurcation problem $f$, firstly construct a certain distinguished family of perturbations of $f$. Let $F(x,r,s_1 \ldots, s_k)$, or simply $F(x,r,s)$, be a $k$-parameter of bifurcation problems. We call $F$ a perturbation of $f$ if

$$F(x,r,0,\ldots,0) = f(x,r) \,. \tag{3.8.7}$$

To solve the classification problem, we seek a $k$-parameter family $F$ with perturbations of $f$, having the property that any perturbation of $f$ whatever is equivalent to $F$ for some $s$ near the origin. That is, given any perturbing term $hp(x,r,h)$, there are parameter values $s_1(h), \ldots, s_k(h)$ such that for small $s$, $f + hp$ is equivalent to $F$. We call such a $F$ a universal unfolding of $f$. It should be noted that the number $k$ of parameters required for a universal unfolding depends on the specific function $f$ under consideration. For example, it can be shown that

$$F(x,r,s) = x^3 - rx + s_1 + s_2x^2 \,, \tag{3.8.8}$$

is a universal unfolding of the pitchfork.

Finding a universal folding of $f$, we will explore the parameter space $R^k$ of the folding to enumerate the various bifurcation diagrams: $\{(x,r) : F(x,r,s) = 0\}$. For the universal unfolding (3.8.8) of the pitchfork, there are essentially four different bifurcation diagrams, as illustrated in Fig. 3.22, which can occur as $s$ varies (Golubitsky and Schaeffer 1984).

It should be noted that even if three or more parameters are introduced into the model, no new behavior can occur. This results from the fact that (3.8.8) is a universal unfolding of the pitchfork. Singularity theory methods tell the exact number of parameters required to describe the most general perturbation of a bifurcation problem. The singularity theory approach also touches on stability problems. The reader is further referred to Golubitsky and Schaeffer (1984, 1988).

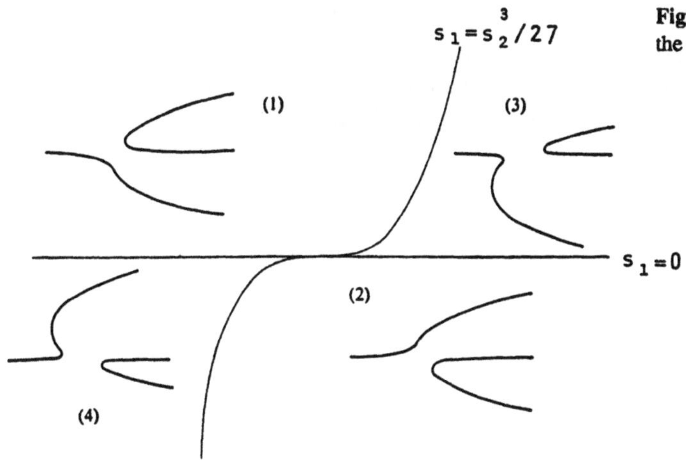


**Fig. 3.22.** Universal folding of the pitchfork

## 3.9 Catastrophe Theory

Let us consider the dynamic system

$$\frac{dx_1}{dt} = f_1(x, r) , \quad i = 1, \ldots , n , \tag{3.9.1}$$

where $x_i$ represents state variables, and $r$ represents parameters (usually called control variables in catastrophe theory).

An assumption that plays an important role in catastrophe theory – although a great many attempts aim to relax it – is that the system (3.9.1) derives from a potential $V$ in the following way:

$$\frac{dx_i}{dt} = -\frac{\Sigma \partial V(x, r)}{\partial x_i} = f_i(x, r) , \tag{3.9.2}$$

where $V$ is the "potential" function. This assumption implies that (3.9.1) consists of a gradient system.

A general classification of the solutions of (3.9.2) can be carried out upon the basis of the breakdown in structural stability. One may find the points where there is a change in the stability properties of the steady states. These points are hypersurfaces in the parameter space, along which either a branching of solutions of the equations take place, or where $V$ attains its absolute minimum at no less than two distinct points. In other words, by crossing these hypersurfaces one switches from a region where certain dynamics take place to a region where the dynamics are qualitatively different.

There are a large number of applications of catastrophe theory in the sciences. Examples of applications to social systems can be found, for example, in Wilson (1981). In Chap. 4, we will also provide some examples. In what follows, we study structural changes of the dynamic systems which are specified by functional forms of $V(x, r)$ where $x \in R_n$ and $r \in R_k$ (see Gilmore 1981).

First, we are concerned with the local properties of $V(x,r) : R^n \times R^k \rightarrow R_1$. The local properties are determined by a sequence of theorems – the implicit function theorem, the Morse lemma and the Thom theorem of functional analysis.

The implicit function theorem guarantees that if the gradient $\Delta V$ is nonzero at a particular point, then it is possible to choose a smooth (derivatives of arbitrarily high order exist) change of variables:

$$y_j = y_j(x) , \quad j = 1, \dots , n , \tag{3.9.3}$$

such that $V$ can be written as

$$V = y_1 + c , \tag{3.9.4}$$

where $c$ is a constant.

**Definition 3.9.1.** (*Morse Critical Points*). The equilibrium points, or critical paths, of a smooth function $V(x)$ are the points at which $\Delta V = 0$. The critical points at which $\det V_{ij} \neq 0$, where $V_{ij} = \partial^2 V/\partial x_i \partial x_j$, are called isolated, nondegenerate, or Morse critical points.

If a point is a Morse critical point, then the Morse lemma guarantees the existence of a smooth change of variables such that the potential can be written locally as a quadratic form:

$$V = \sum_{i=1}^{n} \Theta_i y_i^2 , \tag{3.9.5}$$

where $\Theta_i$ are the eigenvalues of the stability matrix $V_{ij}$ evaluated at the equilibrium. By absorbing a "length scale" into the new coordinates according to $z_i = |\Theta_i|^{1/2} y_i$, the quadratic form (3.9.5) is transformed into the Morse canonical form

$$V = -z_1^2 - \dots - z_i^2 + z_{i+1}^2 + \dots + z_n^2 = M_i^n(z) . \tag{3.9.6}$$

The function $M_i^n(z)$ is called a Morse $i$-saddle. Only Morse 0-saddles have a local minimum at the equilibrium, so that only the 0-saddles are locally stable.

**Definition 3.9.2.** (*Non-Morse Critical Points*). The critical points of $V(x)$ at which $\det V_{ij} = 0$ are called nonisolated, degenerate, or non-Morse critical points.

If the potential depends on one or more control parameters $r$, the stability matrix $V_{ij}$ and its eigenvalues $\Theta$ also depend on these control parameters. Then one or more of the eigenvalues may assume the value of zero for certain values of the control parameters. In this case, $\det V_{ij} = 0$. Hence, the conditions required for the validity of the Morse lemma no longer hold. If $m$ eigenvalues $\Theta_1(r), \dots , \Theta_m(r)$ vanish at $r = r_0$, the Thom splitting lemma may be used to split the potential into a non-Morse part and a Morse part:

$$V(x,r) = f_N\big(y_1(x,r), \ldots, y_m(x;r);r\big) + \sum_{j=m+1}^{n} \Theta_j(r)\big(y_j(x)\big)^2 . \tag{3.9.7}$$

The $m$ "bad" coordinates $y_i(x;r)(i = 1, \ldots, m)$, associated with the $m$ vanishing eigenvalues $\Theta_i(r)$, are functions of $x$ and $r$. The "good" coordinates $y_{m+j}(x)(j = 1, \ldots, n-m)$, associated with the nonvanishing eigenvalues $\Theta_{m+j}(r)$, are smooth functions only of the original state variables $x$. At $(x_0, r_0)$ the stability matrix $\partial^2 f_N/\partial y_i \partial y_j (1 \leqslant i,\ j \leqslant m)$ vanishes (all matrix elements are zero while the $(n-m) \times (n-m)$ stability matrix of the Morse function is nonsingular. Under suitable conditions ($k \leqslant 5$ and no special or symmetry conditions are present in the family of potentials $V(x;r)$) the Thom theorem guarantees the existence of a smooth change of variables such that the potential can be written in the canonical form

$$V = \mathrm{CG}(m) + \sum_{j=m+1}^{n} \Theta_j y_j^2 , \tag{3.9.8}$$

where the function $\mathrm{CG}(m)$ is called a catastrophe germ. In Table 3.1 we list all the canonical catastrophe germs for $k \leqslant 5$. These correspond to only one ($m = 1$) or two ($m = 2$) vanishing eigenvalues.

It should be noted that the decomposition (3.9.7) is valid in a neighborhood of $(x_0, r_0)$ in $R^n \times R^k$, but does not give specific forms for $f_N$; the decomposition (3.9.8) is only valid in a neighborhood of $x_0$ in $R^n$, but gives specific forms, called catastrophe germs, for $f_N$. In fact, Thom has proved a more desirable decomposition. If $x_0$ is a non-Morse critical point of $V(x;r)$ for $r = r_0$, then in an open neighborhood of $(x_0, r_0)$ in $R^n \times R^k$

$$V = \mathrm{Cat}\,(m,k) + \sum_{j=m+1}^{n} \Theta_j(r) y_j^2 . \tag{3.9.9}$$

The function $\mathrm{Cat}(m,k)$ is called a catastrophe function, or simply a catastrophe. A catastrophe consists of two parts: the catastrophe germ $\mathrm{CG}(m)$ and the perturbation $\mathrm{Pert}(m,k)$, i. e., $\mathrm{Cat}(m,k) = \mathrm{CG}(m) + \mathrm{Pert}(m,k)$. Table 3.1 lists canonical forms for

**Table 3.1.** Thom's elementary catastrophes

| Name | k | Germ | Perturbation |
|---|---|---|---|
| $A_2$ | 1 | $x^3$ | $a_1x$ |
| $A_{\pm 3}$ | 2 | $\pm x^4$ | $a_1x + a_2x^2$ |
| $A_4$ | 3 | $x^5$ | $a_1x + a_2x^2 + a_3x^3$ |
| $A_{\pm 5}$ | 4 | $\pm x^6$ | $a_1x + a_2x^2 + a_3x^3 + a_4x^4$ |
| $A_6$ | 5 | $x^7$ | $a_1x + a_2x^2 + a_3x^3 + a_4x^4 + a_5x^5$ |
| $D_{-4}$ | 3 | $x^2y - y^3$ | $a_1x + a_2y + a_3y^2$ |
| $D_{+4}$ | 3 | $x^2y + y^3$ | $a_1x + a_2y + a_3y^2$ |
| $D_5$ | 4 | $x^2y + y^4$ | $a_1x + a_2y + a_3x^2 + a_4y^2$ |
| $D_{-6}$ | 5 | $x^2y - y^5$ | $a_1x + a_2y + a_3x^2 + a_4y^2 + a_5y^3$ |
| $D_{+6}$ | 5 | $x^2y + y^5$ | $a_1x + a_2y + a_3x^2 + a_4y^2 + a_5y^3$ |
| $E_{\pm 6}$ | 5 | $x^3 \pm y^4$ | $a_1x + a_2y + a_3xy + a_4y^2 + a_5xy^2$ |

1- and 2-variable catastrophes. The catastrophe function reduces to the catastrophe germ when the physical control variable $r_i$ equals $r_{i0}$ or the mathematical control parameters $a_j (j = 1, \ldots, v)$ equal zero.

We have described the local characteristics of a potential by a sequence of theorems. In Chaps. 4 and 8 we will use some results from (elementary) catastrophe theory.

## Appendix: Remarks on Bifurcation Theory

As bifurcation theory plays a very important role in synergetic economics, we should like to explain some results of this theory.

First, we discuss some general theorems which state the conditions under which bifurcation occurs. To this end, we write (3.7.1) in the form

$$\frac{dx}{dt} - L(r)x + N(x,r) = 0 , \tag{3.A.1}$$

where $L$ is the linearized operator and $N$ comprises all contributions that are nonlinear in $x$. An equivalent way to formulate the problem is to define $r$ in such a way that there is an $r$-independent part $J_0$ and a continuous part proportional to $r$:

$$J_0 x - rx + N(x,r) = 0 \tag{3.A.2}$$

where

$$J_0 = \frac{d}{dt} - L_0 .$$

**Theorem 3.A.1.** The number $r_c$ can be a bifurcation point of (3.A.2) only if it is an eigenvalue for the operator $J_0$.

The converse to this theorem need not be true. For a square matrix $L(r)$, we say that $r_c$ is an eigenvalue of algebraic multiplicity $k$ if

$$\det |L - zI| = (z - z_c)^k h(z) ,$$

where $h(z_c)$ is not equal to zero.

**Theorem 3.A.2.** If $z_c$ (not equal to zero) is an eigenvalue of odd multiplicity of $J_0$ in (3.7.2), then $z_c$ is a bifurcation point for this equation.

To generalize Theorem 3.A.2, let us consider the case where the variable vector $x \in X$, the parameter vector $r \in M \in R^m$, $r = (r_1, r_2, \ldots, r_m)$, $F : X \times M \to Z$ and

$$\begin{aligned} F(x,r) &= Bx - \sum_{j=1}^{N} r_j A_j x + N(x,r) \\ &= L(r)x + N(x,r) , \end{aligned} \tag{3.A.3}$$

where $B$ and $A_j$ ($j = 1, \dots, n$) are bounded linear operators, $N(0,r) = 0$, $D_x N(0,r) = 0$. A point $r$ is defined as an eigenvalue of $(B, A_1, \dots, A_n)$ if zero is an eigenvalue of $L(r)$.

**Theorem 3.A.3.** Let $X$ and $Z$ be Banach spaces, $M$ an open interval, and $F \in C^m(M \times X, Z)$ for $m \geqslant 2$. If $r_0$ is a simple eigenvalue of $(B, A_1, \dots, A_n)$ with eigenvector $y_0$ not equal to zero, then $(r, x) = (r_0, 0)$ is a bifurcation point of $F(x, r) = 0$. Moreover, there exist $C^{m-1}$ functions

$$r^*(v) = r_0 + O(|v|) ,$$
$$x^*(v) = v y_0 + O(|v|) \quad \text{as} \quad v \to 0 , \tag{3.A.4}$$

for real $v$ near zero such that

$$F(r^*(v), x^*(v)) = 0 .$$

All zeros of $F$ near $(r, 0)$ are either the trivial solution $x = 0$ or given by (3.A.4). If $F$ is analytic in the neighborhood or $(r_0, 0)$, then so are $r^*(v)$, $x^*(v)$ near $v = 0$.

The theorem is referred to in Chow and Hale (1982).

# 4. Multiple Equilibria and Structural Changes in Economic Systems

> The further development of analytical economics along the lines of comparative dynamics must rest with the future. It is to be hoped that it will aid in the attack upon diverse problems ... even ... the majestic problems of economic development.
>
> *P.A. Samuelson* (1947)

One of the most important subjects in economic analysis is the examination of the effects of changes in external parameters upon the behavior of economic variables. The analysis of such effects is called comparative analysis. Comparative statics analysis and comparative dynamics analysis differ according to whether the analysis is completed for a static or a dynamic model. When the system is stable, the comparative dynamics analysis is called the correspondence principle by Samuelson. We call comparative analysis as explored in Samuelson's *Foundations*, traditional comparative analysis. In a sense, this book is intended to study problems of comparative analysis which are neglected in traditional comparative analysis.

## 4.1 Catastrophe Theory and Comparative Statics Analysis

As mentioned in Chap. 2, the whole story of Samuelson's *Foundations of Economic Analysis* is fundamentally based upon two types of very general hypotheses. The first is that the conditions of equilibrium are equivalent to the maximization (minimization) of some magnitude. This hypothesis often implies the validity of (traditional) comparative statics analysis from which many meaningful economic theorems can be derived. The second is that the system is in "stable" equilibrium or motion. As discussed in Chap. 2 from the second hypothesis, the correspondence principle between comparative statics and dynamics is valid. Although most parts of the book examine the behavior of dynamic systems when the second hypothesis fails, it is important to examine the behavior of economic systems when the first hypothesis is relaxed. This section is concerned with the first hypothesis.

This hypothesis is applied to static comparative statics. This general method has been employed in (e. g., micro and welfare) economics to derive meaningful results. When the equilibrium values of variables can be regarded as the solutions of an optimization problem, it is possible to determine unambiguously the changed directions of the solutions in respect to shifts of parameters. Let us provide two simple examples to illustrate the method (Samuelson 1947).

Consider a firm with a given demand curve depending on price and output and a given production cost. The firm is assumed to be subject to a tax rate $r$ per unit output. The profit of the firm is defined by

$$D = xp(x) - C(x) - rx\,, \tag{4.1.1}$$

where $x$, $p$ and $C$ represent output, price and the lowest total production cost, respectively. For each given tax rate, the firm will decide upon output level. There is an equilibrium output for a given tax rate. Our present concern is how the output decided by the firm changes according to shifts in tax rate.

It is assumed that the firm selects the output which maximizes its profit. The equilibrium value will emerge as a solution of maximizing $D$ in (4.1.1). For a regular maximum, necessary and sufficient conditions are: $D_x = 0$, $D_{xx} < 0$. From $D_x = 0$ we have

$$\left[xp(x) - C(x)\right]_x - r = 0\,, \tag{4.1.2}$$

from which we can determine equilibrium $x = g(r)$. Differentiating (4.1.2) with respect to $r$ yields

$$\left[xp(x) - C(x)\right]_{xx} \frac{dx}{dr} = 1\,. \tag{4.1.3}$$

However, since $D_{xx} = \left[xp(x) - C(x)\right]_{xx}$, which is negative as required, from (4.1.3) we have

$$\frac{dx}{dr} < 0\,. \tag{4.1.4}$$

We thus conclude that if the firm is assumed to be always in equilibrium before and after the tax is imposed, an increase of tax will always yield a fall in output. Here, we do not specify the functional forms of $p(x)$ and $C(x)$ in the above discussion. Our single requirement is that the problem has a regular maximal solution. This is sufficient to determine the change direction of $x$ after the tax is imposed. Hence, for the given information that the firm maximizes profit, we can forecast the behavior of the firm when tax policy is changed. This example serves as a typical illustration of what we mean by comparative statics analysis.

Now, we examine what will happen if we relax the hypothesis made in the comparative statics analysis.

Consider an optimizatian problem

$$\min f(x, r)$$

where $x$ represents the variables and $r$ the parameters. The minimum of $f$ occurs when

$$\operatorname{grad} f = 0\,. \tag{4.1.5}$$

The solution to (4.1.5) gives the equilibrium point which minimizes the potential function $f(x, r)$. As $r$ varies, the optimal solution determines a surface in the space

$(x, r)$, which presents possible equilibrium states of the system. According to traditional comparative statics analysis, for a smooth, slow and small change in $r$, a corresponding smooth, small change in $x$ can be anticipated. The resulting equilibrium trajectory in $(x, r)$ space is smooth and cannot be folded in any way.

Let the second derivative of $f(x, r)$ in equilibrium be zero or the Hessian matrix be singular. In these cases, the states determined by grad $f = 0$ may be not optimal. Such equilibrium points are known as singularities and it is at and near such points that unusual system behavior can be observed. As shown in Chap. 3, the elementary purpose of catastrophe theory is to classify the kinds of singularity which can occur. Catastrophe theory is concerned with sudden and discrete changes in system state variables which result from a slow, smooth and small change in one or more parameters. For a number of control variables in the vector $r$ up to or equal to 4, the types of singularity, in a topological sense, are relatively few (see Chap. 3).

For simplicity, consider the cusp catastrophe – one of Thom's elementary catastrophes. It is the mostly widely applied in science because of its simplicity and typical behavior.

Consider a potential function

$$f(x, r) = \frac{x^4}{4} + r_1 \frac{x^2}{2} + r_2 x \,. \tag{4.1.6}$$

The stationary values are found by setting $df/dx$ to zero:

$$x^3 + r_1 x + r_2 = 0 \,. \tag{4.1.7}$$

Such an equation can have either one or three real roots. If

$$\left(\frac{-r_1}{3}\right)^3 > \left(\frac{r_2}{2}\right)^2 ,$$

then the equation has three real roots. Otherwise, it has only a single real root. The boundary of the region for single and multiple solutions is defined by

$$4r_1^3 + 27r_2^2 = 0 \,. \tag{4.1.8}$$

This produces the cusp-shaped curves on the control manifold – the $(r_1, r_2)$ plane (see Fig. 4.1).

As shown in Fig. 4.1, outside the cusp-shaped region there is only one root and this is always a minimum. Inside the region, there are three real roots, and always one maximum (an unstable state) and two minima, as can be checked by examining the second derivative of $f$. The shaded region is the catastrophe set and the boundary is the bifurcation set where a local minimum disappears. This can be seen at points 3 and 7 in Fig. 4.1, where the disappearing minimum merges into the local maximum to create a point of inflexion on the boundary. The $r_1$ axis, for $r_1 < 0$, represents the conflict set: where there are two minima of equal value (point 5 on Fig. 4.1). In the case of cusp catastrophe, $r_1$ is termed the "splitting factor" and $r_2$ the "normal factor" (Zeeman 1977). The reason is that it is the value of $r_1$ which determines whether a trajectory is in a region where the surface is folded. If $r_1 > 0$, then the surface is single valued, while if $r_1 < 0$ it is double valued. In the case of the

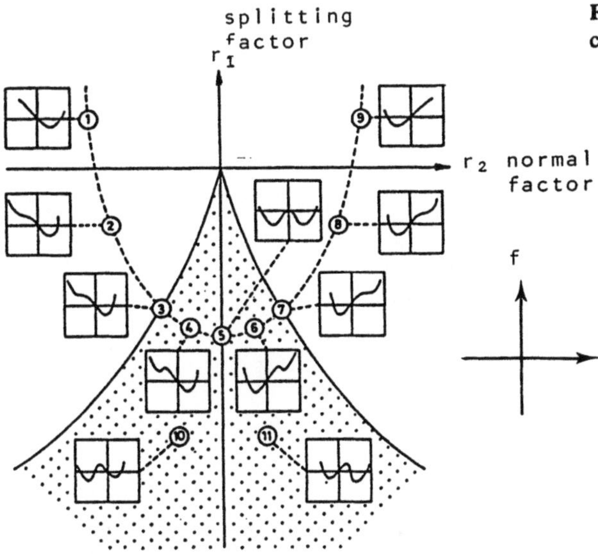


**Fig. 4.1.** The control manifold for the cusp catastrophe

normal factor $r_2$, $x$ changes monotonically as $r_2$ changes, and continuously except for jumps at the bifurcation points.

The surface of equilibrium $(x, r)$-values is shown in Fig. 4.2.

Three types of behavior which we are not accustomed to from traditional economic analysis are shown above in Fig. 4.2. They are: (1) a sudden jump (or catastrophe); (2) hysteresis – a reverse path to some point different from the original; and (3) divergence – a small difference in approach towards a cusp point leads the

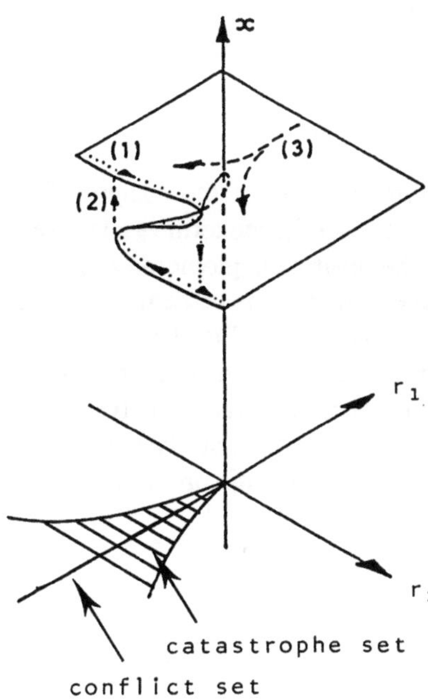


**Fig. 4.2.** The cusp catastrophe

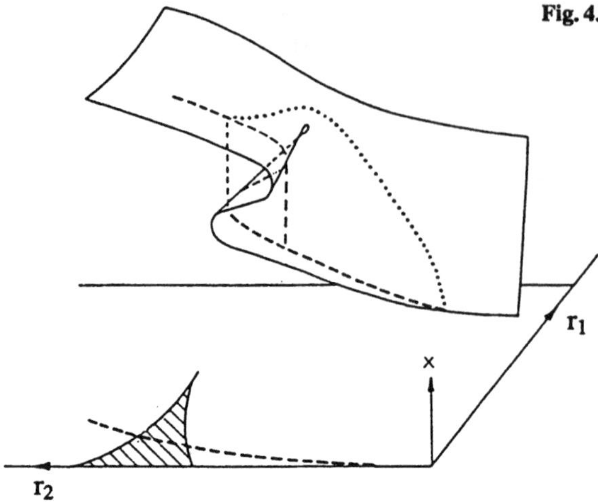


Fig. 4.3. The cusp catastrophe in Isard's model

system to the upper or lower surface and hence to a very different state. This kind of behavior cannot be explained by traditional comparative static analysis. Thus, if we relax the assumption in traditional comparative statics analysis, the behavior of systems is no longer characterized by unique and smooth reactions to small shifts of parameters. Multiple equilibria and sudden changes can occur.

Now, we provide an example of applying catastrophe theory to theories of urban and regional structural change by Isard (1977). In Isard's model, the state variable $x$ represents the population of a city or a region, and the control variables $r_1$ and $r_2$ represent the increase in productivity of each unit of population and the direct contribution of a marginal unit of population to total welfare, respectively. The potential function is assumed to be taken in canonical form as

$$f(x,r) = -\frac{x^4}{4} + r^1 \frac{x_2}{2} + r_2 x + C\ , \qquad (4.1.9)$$

where $C$ is a constant.

The potential function is interpreted as a welfare function of the society. The problem is to find solutions which maximize welfare. In (4.1.9), the term $r_2 x$ is taken as the direct contribution to welfare and $r_1 x^2/2$ as the positive agglomerative gains. The term $-x^4$ is interpreted as representing negative externalities or deglomeration forces. The possibilities of different trajectories are illustrated in Fig. 4.3.

From the discussion above, we see that catastrophe theory can be used to deal with comparative statics problems which traditional comparative statics analysis fails to solve. Similarly as we will show, bifurcation theory and singularity theory are useful to economic analysis because they can help us to analyze the problems which cannot be solved by the methods of traditional economic analysis.

## 4.2 Modeling Regional Dynamics

To show how the concept of bifurcation can be used to explain the dynamics of economic evolution, we will consider a dynamics of regional development.

In recent years, sudden and unexpected discontinuities have become key words in the literature on regional development (e. g., Wilson 1981, Andersson and Batten 1988). Evidence of such behavior in the evolution of cities as a global family has been investigated. An illuminating example is the analysis by Mees (1975). He took the hypothesis of Piremnne (1925) as the starting point for an analysis of sudden changes in the specification pattern of a set of trading regions. The hypothesis stated that the central cause of the revival of European cities and towns in the late middle ages was the emergence of free trade and consequent improvements to the transportation systems. Based upon these studies, Andersson (1986) claimed that the sequence of fundamental changes in the world economy over the last millennium can be explained by the changing structure of logistical systems. In other words, the great structural changes of production, location, trade, culture and institutions are triggered by slow steady changes in the associated logistical networks. Logistical networks are those systems in space which can be used for the movement of commodities, information, people and money in association with the production or consumption of commodities. The following example shows how the concept of logistical networks can explain qualitative aspects of regional evolution.

The model for logistical revolutions in the world is presented according to Andersson (1986) and Andersson and Batten (1988). It is assumed that the observed fluctuations in city development can be captured, or at least approximated in qualitative terms, by a third-order system of differential equations

$$\frac{dy}{dt} = -T\left(\frac{y^3}{3} - ry - x\right), \quad \text{“fast equation”},$$

$$\frac{dx}{dt} = -T^{-1}y, \quad \text{“slow equation”}, \tag{4.2.1}$$

where $r$ is a control parameter and $T$ an adjustment speed coefficient. In (4.2.1), $y$ can be interpreted, for example, as a city's capacity for producing commodities, and $x$ as the city's accessibility to networks of transport and communication. The system is a reformation of the celebrated Van der Pol equation. It turns out that discontinuous changes in the value of $y$ may be produced as the value of $x$ moves slowly into critical parametric domains. Fig. 4.4 illustrates a typical cycle in which shifts may take place repetitively.

Abrupt rises and falls in capacity are well documented, and may be precipitated by gradual changes in local accessiblity. A key point to grasp with this type of analysis is that changes in the value of the “fast” variable may take place relatively quickly. Thus if one were to observe the state of the system just before and just after the change, one may be tempted to conclude that the “slow” variable was not influential. In all cases the “slow” phase will prevail for the majority of the time, but the “fast” phase translates the whole system into a fundamentally different regime.

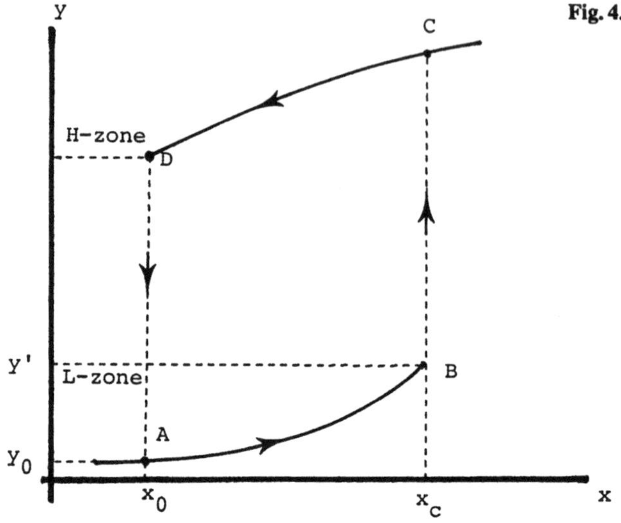


**Fig. 4.4.** A cycle of fast and slow variables

The slow expansion of network infrastructure ($x$) through the instrument of investment in physical capital will follow the trajectory located in the $L$-zone of Fig. 4.4. Initially the system is located at $A$. As $x$ is changed, eventually a point $b$ is reached beyond which the very nature of the city's productive capacity changes quite markedly. At this point, the equilibrium loses its stability and a "phase transition" is underway. In this nonequilibrium phase, the speed of change is determined by constraints on capital, labor, and other resource needs which are applicable to the emerging production regime.

A key feature of this type of nonlinear analysis is its cyclic nature. If network investment stops once the $H$-zone is reached, and depreciation is then the dominant interactive effect, the system may follow the solution trajectory depicted on the $H$-zone until it finally returns to the initial state at $D$ and then drops back down to the $L$-zone. It should be noted that it is not so easy to understand the unusual nature of the critical points at $B$ and $D$. The underlying process may be referred to as divergence, because a smooth but small change in the network infrastructure capacity can cause unexpectedly large fluctuations in the equilibrium value of commodity production. This comes about through a discontinuous change of state or phase transition. The transition takes place no matter how slowly the network capacity increases. This implies that urban development may be triggered simply by the addition of one small but important link in the network. Slight differences in transportation conditions may result in large differences in the final commodity capacity if the urban structure is located at a critical state.

A question naturally arises whether such an esoteric mathematical analysis can explain the rise and fall of various cities over the past years. We think that the model can be used to illustrate the problem qualitatively. It can help us deepen our insight into real dynamic features of urban evolution.

Consider an example of possible applications of the model. According to Andersson (1986), the development of cities and their interregional economic relations in the world from the years 1000 A.D. until 2000 A.D. may be perceived in terms of four revolutions: (i) emerging in Italy during the 11th century and ending in Northern Europe during the 16th century, (ii) emerging in Spain, Portugal and Italy during the 16th century and ending in Northern Europe during the 19th century, (iii) emerging in England during the 18th century and ending in the developing countries, probably during the 21st century, and (iv) emerging in Japan, Switzerland, West Germany and Sweden at the end of the 20th century.

We may view the reopening of trade routes and the renewed possiblity of passage through Europe and Asia as phases of slowly improving network infrastructure – as reflected by the easing of trade barriers, dangers en route, transportation cost and other restrictions on movement. This corresponds to the First Logistical Revolution.

We are now witnessing the emergence of the Fourth Logistical Revolution, associated with the growth of information processing and communication capacity as well as the growth of the knowledge base. The improvement to the transportation system, especially to the air transportation network, is steadily reducing the importance of physical contiguity of places and regions.

Sweden may be a typical case for the analysis of the Fourth Logistical Revolution (see Andersson and Batten 1988). For explanation, we subdivide the labor force into four types of occupations – knowledge handling jobs, administration and information jobs, personal services, and goods handling jobs. The location of knowledge-intensive production units is very sensitive to the availability of labor with high levels of education and appropriate qualifications. The importance of universities and other advanced educational and scientific institutions in regional development is increasing. We might thus perceive that the key characteristic of the Fourth Logistical Revolution is the slow expansion of the knowledge base $x$. Here, the relation between $x$ and the production capacity $y$ is described by (4.2.1). In this case, the trajectory being followed is very similar to the one described in Fig. 4.4.

It should be mentioned that divergence of the industrial structural change can already be observed. For instance, in 1977, very few firms allocated more than 17% of their expenditure to R&D. This group of firms accounted for no more than 7% of total value of the Swedish manufacturing industry. According to a recent estimate, firms allocating more than 17% to R&D accounted for more than one quarter of the total capitalization of Swedish manufacturers in 1985. Such a change is nothing short of revolutionary. The products involved in this expansion are chemicals, aircraft, aerospace products, instruments, electronics, "high-tech" tools, machines and robots (Andersson and Batten 1988).

## 4.3 Some Examples of Structural Changes

This section provides some examples of applying catastrophe theory and bifurcation theory to social systems. These applications focus on qualitative characteristics of social dynamic systems. All of the examples below can also be found in Wilson (1981).

### 4.3.1 Business Cycles in the Kaldor Model

An early application of catastrophe theory to economics stems from Varian (1979), who presented an augmented version of the Kaldor model

$$\frac{dy}{dt} = a\left[C(y,w) + I(y,k) - y\right] ,$$
$$\frac{dk}{dt} = I(y,k) - I_0 , \qquad (4.3.1)$$

where $y$ is national income, $C$, consumption, $I$, gross investment, $w$, wealth, and $I_0$, "replacement" investment. The parameter $a$ is adjustment speed. The consumption function $C(y, w)$ is specified as follows

$$C(y,w) = c(w)y + D(w) . \qquad (4.3.2)$$

The savings $S(y, w)$ are thus equal to

$$S(y,w) = y - C(y,w) . \qquad (4.3.3)$$

The investment function $I(y, k)$ is assumed to grow logistically with $y$ and to decrease with increasing $k$. The equilibrium of the system is given by

$$S(y,w) + I(y,k) , \quad I(y,k) = I_0 . \qquad (4.3.4)$$

Examples of various curves and the equilibrium points are plotted in Fig. 4.5.

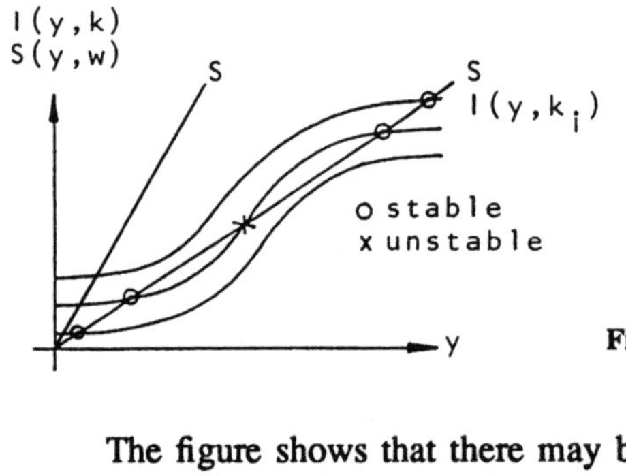


**Fig. 4.5.** Determination of equilibria

The figure shows that there may be one low (or high) stable equilibrium value of $y$, or three equilibrium values. In the case of multiple equilibria, the outer two are stable, while the inner one is unstable. This case generates folds in $dy/dt = 0$. As the saving function is dependent on $w$, different wealth values may generate various kinds of equilibrium. It can be shown that the resulting possible equilibrium states take the cusp form shown in Fig. 4.6.

To illustrate how a (hysteric) cycle may be generated, we assume that the variable $y$ is located on the top of the manifold in Fig. 4.6. At this moment, the dynamics of the system are completely determined by the dynamics of slow variables. For simplicity, let $w$ be fixed. Suppose that there is a small exogenous disturbance of the equilibrium $E$. If the disturbance is very small, the system returns to $E$ rapidly according to the dynamics of $k$. However, once $k$ is increased such that the point

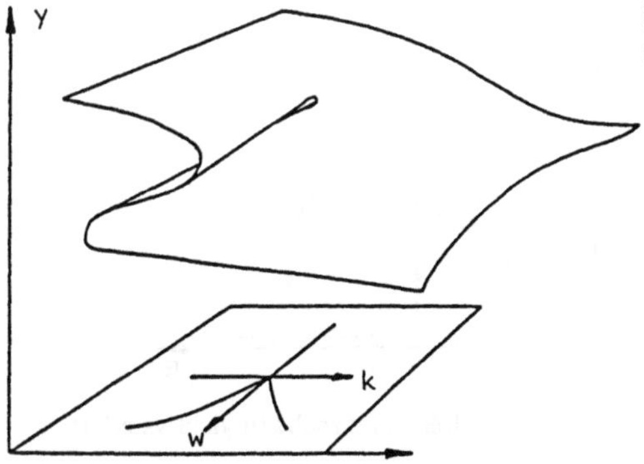


**Fig. 4.6.** Structural changes in the Kaldor model – taking $k$ and $w$ as slow variables

$B$ is crossed, a catastrophe occurs and income jumps down to the lower branch. A slow movement along $dy/dt = 0$ is initiated until the bifurcation point $C$ is reached where another catastrophe occurs and where $y$ jumps back to the upper branch.

### 4.3.2 Resource Management

We consider a fisheries management problem treated in Clark (1976). Let the resource population be denoted by $x(t)$ and the rate of removal of the population by $h(t)$. If we assume the natural growth of the population to be dependent on $x$ and to be generally described by $F(x)$, then one has

$$\frac{dx}{dt} = F(x) - h(t) . \tag{4.3.5}$$

Let $E$ be the amount of effort devoted to fishing in units so that the removal rate can be taken as

$$h = Ex . \tag{4.3.6}$$

That is, the removal rate is linearly proportional both to the effort and the stock. Let the equilibrium of (4.3.5) be $x^*$. Then the sustainable yield, $Y$, is given by $Y = Ex^*$.

Consider the case when $F(x)$ is a depensation curve, where $F(x)/x$ is an increasing function of $x$ for some range $0 < x < K^*$. It exhibits critical depensation when $F(x)$ has some small value of $x$, say $0 < x < K_0 < K^*$. $K_0$ is called the minimum viable population level. As the behavior of systems appears to be very similar in the critical and non-critical depensation cases, we only investigate the system with a noncritical depensation function.

The basic diagrams of this case are presented in Fig. 4.7.

There are three equilibria. It can be shown that the origin is stable if $E > E^+$ ($= F'(0)$). We assume $E > F'(0)$ in the remainder of the discussion. The variable $x^+$ is always unstable and generates the (unstable) yield, which is represented by the dashed part of the yield-effort curve (see Fig. 4.8). As $E$ increases from a low level, there is an equilibrium point and a corresponding yield $Ex^*$ ($= E^*$). This reaches a maximum, say at $E^M$. We now examine what happens if fishing effort is increased continuously. When $E^M$ is passed, small shifts only cause small changes. However,

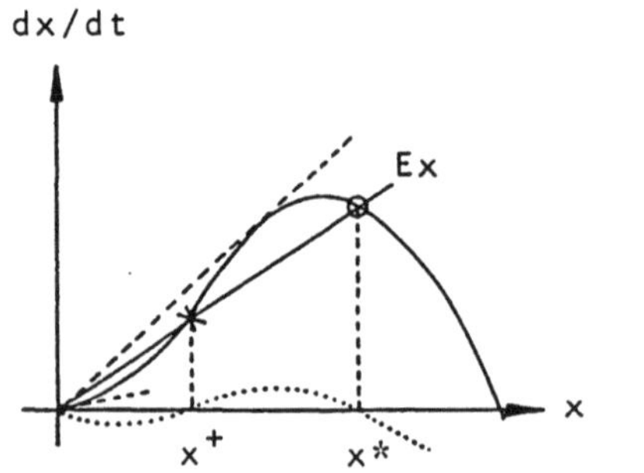

Fig. 4.7. The non-critical depensation case

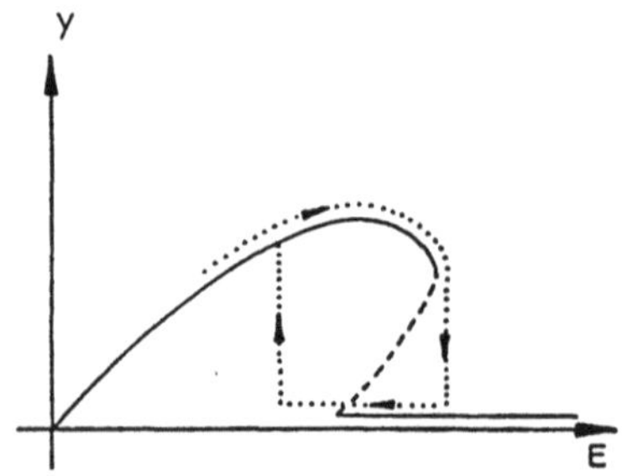

Fig. 4.8. A fold catastrophe in yield vs effort

if $E$ passes $E^*$, the yield will suddenly drop to zero. Let $E$ decrease again. As the origin is a stable equilibrium point when $E^M > E^+$, the situation cannot be recovered by this reduction. If $E$ is reduced to a level less than $E^+$, then the origin becomes unstable, $x$ increases, and $E$ can be slowly increased to $E^M$ again. Thus, there is a hysteresis effect, which is explicitly shown in Fig. 4.8.

### 4.3.3 Dynamic Transportation Modal Choice and Bifurcation

Consider an application of bifurcation theory to transportation problems. The model is suggested by Deneubourg, de Palma and Kahn (1979). They consider an unlabeled origin-destination pair between which a number of trips, $D$, has to be distributed between two modes (for example, bus and car). Let $x_i$ denote the number of trips by mode $i (i = 1,2)$. The dynamics of modal choice is given by

$$\frac{dx_i}{dt} = D_i - x_i \, , \tag{4.3.7}$$

subject to $x_1 + x_2 = D$. If we represent the attractiveness of mode $i$ by $A_i(x)$, then it is proposed that $D_i$ take the form

$$D_i(x) = \frac{DA_i(x)}{A_1(x) + A_2(x)} \, . \tag{4.3.8}$$

The dynamics is dependent on the way in which $A_i$ is specified. Here, we examine a simple case in which the attractiveness is proportional to speed. Assume

$$A_i = v_i \, , \quad i = 1,2 \, ,$$

where $v_i$ is the average speed of mode $i$. It is assumed that there is no direct interaction between the modes. We consider mode 1 as car and mode 2 as bus. It is assumed that car speed decreases with usage due to congestion, while bus speed improves initially as supply responds to demand, and then decreases. These assumptions are expressed by

$$v_1 = \frac{1}{a + bx_1} \, ,$$

$$v_2 = \frac{dx_2^n}{c + sx_2^r} \, .$$

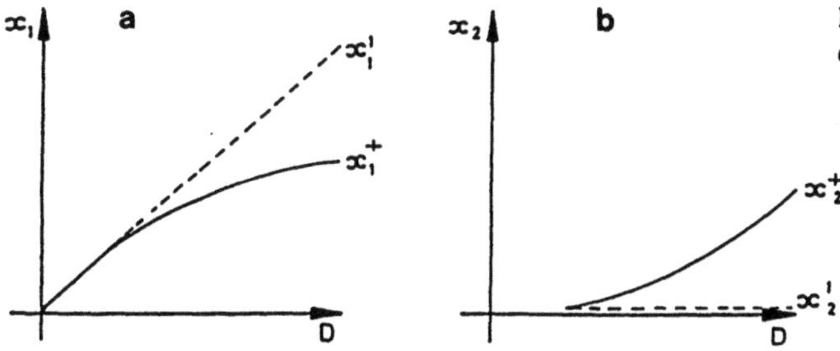


**Fig. 4.9.** Modal choice bifurcation diagrams

For simplicity, choose $n = r = s = 1$. The dynamics are given by

$$\frac{dx_1}{dt} = \frac{D/(a+x_1)}{1/(a+x_1)+dx_2/(c+x_2)} - x_1 \,,$$
$$\frac{dx_2}{dt} = \frac{D\,dx_2/(c+x_1)}{1/(a+x_1)+dx_2/(c+x_2)} - x_2 \,. \tag{4.3.9}$$

There are three equilibrium values of (4.3.9). Appropriately fixing other parameters, we show the behavior of the system, taking $D$ as a bifurcation parameter, in Fig. 4.9.

More complicated cases were analyzed in Deneubourg, de Palma and Kahn (1979). For instance, they studied the case in which attractiveness is dependent on the effects of publicity and imitation: the more people use a mode, the more popular it becomes.

### 4.3.4 Multiple Equilibria in Wilson's Retail Model

Wilson's retailing model (Wilson 1981) provides an important example of earlier applications of catastrophe theory to regional science and geography. The model studies the behavior of consumers when they travel from residences (or workplaces) to shopping centers. The purpose of the model is to investigate how consumers' behavior is affected by a given spatial distribution of shopping centers. The supply side of the system decides the location and size of centers. It is assumed that there is a trade-off between the benefits of center size and the increasing cost of travel to larger centers, and the supply side balances the costs of supply against the revenue attracted at a particular location when they determine center location and size.

Let the spatial system consist of $n$ zones ($i = 1, \ldots, n$). $S_{ij}$ is the flow of cash from residents of zone $i$ to shops in zone $j$. The proposal is that it be determined by

$$S_{ij} = \frac{e_i P_i W_j^a \exp(-bc_{ij})}{\Sigma_k W_k^a \exp(-bc_{ik})} \,,$$

where $e_i$ is the average per capita expenditure on shopping goods by the residents of zone $i$, $P_i$ is the population of zone $i$, $W_j$ is defined as the attractiveness of shops in zone $j$ – commonly taken as being measured by the center's size, $c_{ij}$ is the cost of travel from $i$ to $j$ in suitable units, and $a$ and $b$ are constants.

Let $k$ be a suitable constant to describe the cost per unit of supply. The dynamics are specified as

$$\frac{dW_j}{dt} = D_j - kW_j \,, \tag{4.3.10}$$

where the adjustment speed is chosen as unity, and $D_j$ is the total revenue attracted to center $j$

$$D_j(W) = \Sigma_i S_{ij}(W) \,. \tag{4.3.11}$$

The dynamics may be generally written as

$$\frac{dW_j}{dt} = W_j M_j(W) \,, \tag{4.3.12}$$

where $M_j$ is a nonlinear function of $W = (W_1, \ldots, W_n)^{\mathrm{T}}$. It is interesting to investigate the way in which equilibrium values may be affected by parameters $k$, $a$, $b$, $e_i$ and $P_i$. Figures 4.10–12 (adapted from Wilson 1981) represent the effects of $a$, $b$, and $k$, respectively, on equilibrium patterns.

As the analysis is complicated, it may be helpful to mention some conclusions by Wilson on the behavior of the model. High $a$ and low $b$ cause the system to have a relatively small number of large centers, and vice versa. If the cost parameter is large, the number of centers tends to decrease. For a particular zone at a particular time, there is a surface in the parameter space on one "side" of which development is possible in that zone, and on the other side of which development is not possible.

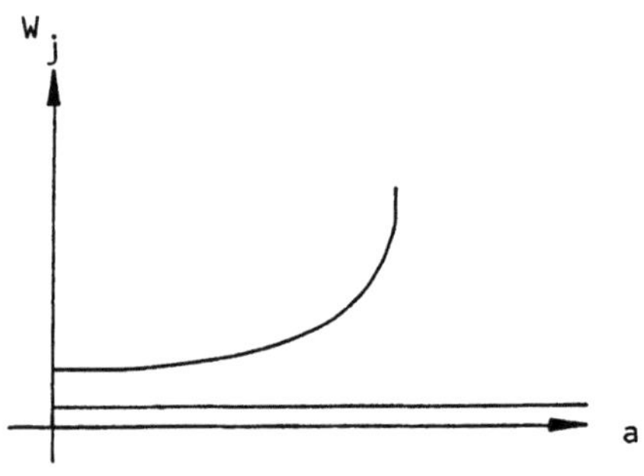


**Fig. 4.10.** Effect of $a$ on $W_j$

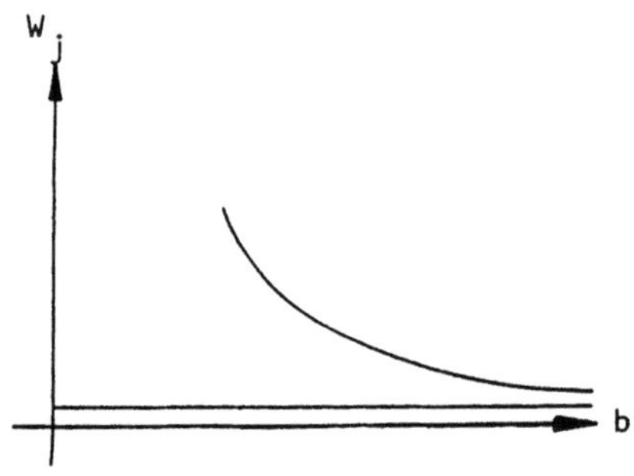


**Fig. 4.11.** Effect of $b$ on $W_j$

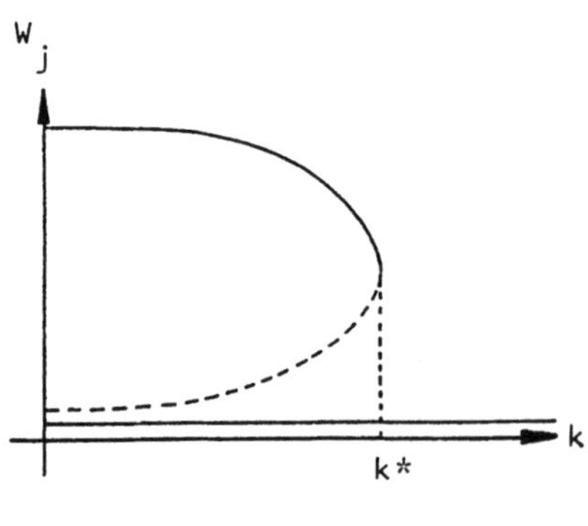


**Fig. 4.12.** Sudden change in shopping center size

These are called DP and NDP states for that zone, respectively. The values of every parameter can be classified in this sense. For instance, $a = 1$ is a critical state: for $a < 1$, zones are always in the DP state, while this is not the case for $a > 1$. When a zone switches from the NDP to the DP state, if development actually takes place, this will be recorded as a "jump" in that particular $W_j$ value. Such a jump may cause a secondary jump in the other $W_k$ variables.

## 4.4 A Bifurcation Analysis for an Economic Growth Model

This section applies the bifurcation analysis method of Iooss and Joseph (1980) to an economic growth model recently proposed by the author (Zhang 1989 and 1989a). This model focuses on the role of intellectuals in economic growth.

It is assumed that there is only one commodity which can be used as consumption and capital. There is one production sector whose output can either be used for investment in production or for consumption in households. It is assumed that in production processes, three inputs – (physical) capital, knowledge (human capital) and physical labor – are required.

The total labor force, denoted by $L$, is assumed to grow at a fixed rate $n$. The labor force is divided into knowledge and physical workers, denoted by $L_1$ and $L_2$, respectively, where $L_1 = n_1 L$, $L_2 = n_2 L$, $n_1 + n_2 = 1$. We assume that $n_1$ and $n_2$ are constant.

The production structure is described by the following production function

$$Y = F(G, L_2, K) = A(G) L_2^{\alpha} K^{\beta} , \tag{4.4.1}$$

where $Y$ is the output, $G$ is the knowledge (human capital), $K$ the physical capital, and $\alpha$ and $\beta$ ($\alpha + \beta < 1$) are positive coefficients. For simplicity, we specify $A(G)$ as: $A(G) = G^{\gamma}$, where $\gamma$ is a positive constant. Moreover, we require $\alpha + \beta + \gamma = 1$. This means that scale effects of the production are neutral with respect to the inputs.

The output per unit of the labor force is equal to $Y/L$. We assume that the consumption levels of the physical workers and the intellectuals are positively proportional to $Y/L$. The consumption level of a physical worker is assumed to be equal to $c_1 Y/L$, and that of an intellectual to be equal to $c_2 Y/L$, where $c_1$ and $c_2$ are positive constants.

The capital accumulation is given by

$$\dot{K} = Y - \frac{c_1 L_1 Y}{L} - \frac{c_2 L_2 Y}{L} - rK , \tag{4.4.2}$$

where $r$ is a fixed depreciation rate of capital. For the whole economy the consumption rate is given by $(c_1 n_1 + c_2 n_2)$; the savings rate is equal to $1 - (c_1 n_1 + c_2 n_2)$.

Now, we will discuss knowledge accumulation. The knowledge accumulation is affected by the behavior of the intellectuals and the physical workers. The intellectuals affect knowledge accumulation through direct education and R&D activities, while the physical workers do so through the effects of the "learning by doing" process (Arrow 1962). This is a strict assumption. For instance, the physical workers can participate in R&D activities; the intellectuals may take part in production activities. Potential dynamics of the growth of knowledge is suggested as follows

$$\dot{G} = pY + H(c_1 Y/L, L_1, G) - \mu G \,, \tag{4.4.3}$$

where $pY$ describes the effects of "learning by doing" of the physical workers, $H$ is a contribution function of the intellectuals to knowledge growth, and $\mu$ is a fixed depreciation rate of the growth of knowledge. The behavioral interpretation of (4.4.4) is given in Zhang (1989a). We specify $H$ as

$$H = \frac{c_1 Y/L}{a_1 + c_1 Y/L} L_1^d G^\Theta \,, \quad d + \Theta = 1 \,, \tag{4.4.4}$$

where $a_1$, $d$ and $\Theta$ are non-negative parameters. We interpret $a_1$ as a measure of efficiency of the intellectuals. If $a_1$ is equal to zero, then the level of consumption does not affect the growth of knowledge. If it is infinitely large, then $H$ becomes zero, and the intellectuals can contribute nothing to the growth of knowledge. In general, we should have: $0 < a_1 < \infty$. We can see that if the level of consumption $c_1 Y/L$ is sufficiently high, then the level of consumption will not affect the growth of knowledge. The function $H$ increases with respect to $L_1$ and $G$, though it is "neutral" to $L_1$ and $G$.

The dynamics of the system consists of the evolutionary equations (4.4.2) and (4.4.3). The system can be rewritten as

$$\begin{aligned} \dot{k} &= y - c_1 n_1 y - c_2 n_2 y - rk - nk \,, \\ \dot{g} &= py + H(k, g) - \mu g - ng \,, \end{aligned} \tag{4.4.5}$$

where $k = K/L$, $g = G/L$, $y = n_2{}^\alpha g^\gamma k^\beta$, and

$$H(y, g) = \frac{n_1^d c_1 g^\Theta y}{a_1 + c_1 y} \,. \tag{4.4.6}$$

The properties of the system are examined in detail by Zhang (1989a). Now, we will complete a bifurcation analysis with regard to $n$.

First, we guarantee the existence of a unique equilibrium and find the stability conditions of the equilibrium. An equilibrium is defined as a solution of

$$\begin{aligned} y - c_1 n_1 y - c_2 n_2 y - rk - nk &= 0 \,, \\ py + H(y, g) - \mu g - ng &= 0 \,. \end{aligned} \tag{4.4.7}$$

From the first equation of (4.4.7), we can obtain

$$g = \left( \frac{r' k^{1-\beta}}{n_2^\alpha} \right)^{1/\gamma} \,. \tag{4.4.8}$$

where $r' = (r + n)/(1 - c_1 n_1 - c_2 n_2)$. Substituting (4.4.8) into the second equation of (4.4.7), we have

$$pr'k + n_1^d c_1 r' k \left( \frac{r' k^{1-\beta}}{n_2^\alpha} \right)^{\Theta/\gamma} /(a_1 + r' c_1 k) - (\mu + n) \left( \frac{r' k^{1-\beta}}{n_2^\alpha} \right)^{1/\gamma} = 0 \,. \tag{4.4.9}$$

To guarantee the existence of solutions of (4.4.9), function $C$ is defined as

$$C(k) = pr'k - (\mu + n)\left(\frac{r'k^{1-\beta}}{n_2^\alpha}\right)^{1/\gamma} + n_1^d c_1 r'k\left(\frac{r'k^{1-\beta}}{n_2^\alpha}\right)^{\Theta/\gamma} /(a_1 + r'c_1 k) .$$

It can be proved that $C(0) = 0$, $C(\infty) = -\infty$, $C'(0) > 0$. The existence of solutions can be guaranteed by these properties of $C(k)$. To see this, we assume that there are multiple equilibria. From the properties of $C(k)$ we see that if $C(k)$ has no unique solution, then $C(k)$ has at least three solutions. We define $c(k) = (a_1 + r'c_1 k)C(k)/k$. $c(k)$ also has at least three solutions. This implies that there is as positive value of $k$ such that $c'' = 0$. On the other hand, it can be shown that if $\alpha/\gamma - 1 > 0$, $(1-\beta)\Theta/\gamma - 1 < 0$, $c''$ is always negative for any positive $k$. Therefore, a necessary condition for uniqueness is that $\alpha/\gamma - 1 > 0$, $(1-\beta)\Theta/\gamma - 1 < 0$.

It should be noted that uniqueness may be identified under more general conditions. The necessary condition just obtained holds if $\alpha > \gamma > \Theta > 0$. That is, the value of the parameter of capital in the production function is larger than that of knowledge in the production function, while the value of the parameter of knowledge in the production function is larger than that in the contribution function of the intellectuals. If this condition does not hold, we may have multiple equilibria. In what follows, we assume the existence of a unique equilibrium which is denoted by $(k_0, g_0)$.

We now find the stability conditions of the equilibrium. It can easily be shown that the two eigenvalues ($q_1$ and $q_2$) of the Jacobian at the equilibrium are given by

$$q^2 + m_1 q + m_2 = 0 , \tag{4.4.10}$$

where

$$\begin{aligned}
m_1 &= r + n - \frac{n'\beta y}{k} - N(k, g) , \\
m_2 &= \left[\frac{n'\beta y}{k} - (r + n)\right] N(k, g) - \frac{n'\gamma y M(k, g)}{g} , \\
n' &= 1 - (n_1 c_1 + n_2 c_2) , \\
M(k, g) &= \frac{p\beta y}{k} + \frac{a_1 \beta H}{k(a_1 + c_1 y)} , \\
N(k, g) &= \frac{p\gamma y}{g} - (\mu + n) + \left(\frac{\Theta + a_1\gamma}{a_1 + c_1 y}\right)\frac{H}{g} .
\end{aligned}$$

$n'$ and $M$ are positive. As at the equilibrium $py/g + H(k, g)/g = \mu + n$, we have $N = py[\gamma - 1]/g + [\Theta + a_1\gamma/(a_1 + c_1 y) - 1]H/g$, where $a_1\gamma/(a_1 + c_1 y) < 1$. If $\Theta$ is sufficiently small, $N$ is negative. If knowledge does not affect the knowledge growth function $H$ (i.e., $\Theta = 0$), $N$ is negative. In the case of $\Theta = 1$, $N = py[\gamma - 1]/g + a_1 c_1 \gamma y/(a_1 + c_1 y)^2$. If there is no effect of the learning by doing process upon the growth of knowledge (i.e., $p = 0$), $N$ is positive, while if the learning by doing process strongly affects the growth of knowledge, $N$ may be negative.

It can be seen that if $m_1 < 0$, the equilibrium is unstable. In the case where $m_1 > 0$, if $m_1 < 4m_2$, the system is stable. In the case where $m_1^2 > 4m_2$, if $m_2$

is positive, the system is stable; if $m_2$ is negative, the system is unstable. When $m_2 = 0$, the system is neutral. It is not easy to give a definite conclusion about the stability because the expressions for $m_1$ and $m_2$ are too complicated. Zhang analysed the effects of shifts in different parameters on the equilibrium. As we are mainly concerned with applications of bifurcation theory, we will discuss the case of $m_2 = 0$.

As $m_2$ may be either positive or negative, it is not too strict to assume the existence of appropriate parameters such that the equality $m_2 = 0$ holds, which is true if $(n'\beta y/k - r - n)N = n'\gamma y M/g$. As $M$ is positive, if $(n'\beta y/k - r - n)$ is negative, $N$ is negative too. We denote by $n_0$ the value of $n$ at which $m_2 = 0$ holds. We choose $n$ as a bifurcation parameter.

For convenience, we rewrite the system in the local form near the equilibrium. Let

$$U_1 = k - k_0\,, \quad U_2 = g - g_0\,.$$

Then the system can be written in the form

$$\dot{U}_1 = f^1\,, \quad \dot{U}_2 = f^2\,, \tag{4.4.11}$$

where

$$\begin{aligned} f^1(U,n) &= r_1(n)U_1 + \frac{n'\gamma y U_2}{g} + \frac{n'\beta(\beta-1)yU_1^2}{k^2} \\ &\quad + \frac{n'\beta\gamma y U_1 U_2}{kg} + \frac{n'\gamma(\gamma-1)yU_2^2}{g^2} + \mathrm{O}(|U|^3)\,, \\ f^2(U,n) &= MU_1 + N(n)U_2 + M_k U_1^2 + M_g U_1 U_2 + N_g U_2^2 + \mathrm{O}(|U|^3)\,, \end{aligned} \tag{4.4.12}$$

in which $U = (U_1, U_2)^{\mathrm{T}}$, $r_1 = (n'\beta y/k - r - n)$, $M_k$, $M_g$, and $N_g$ are the partial derivatives of $M$ and $N$ with regard to $k$ and $g$, respectively. In (4.4.12), only $r_1$ and $N$ are dependent on $n$. It can be identified that $M_k$ and $N_g$ are negative, and $M_g$ is positive.

For appropriate values of the parameters we have: (i) one eigenvalue is equal to zero; (ii) the two eigenvalues are zero with Reisz index two; and (iii) the two eigenvalues are zero with index one. It is possible to observe bifurcations in these three cases. For simplicity, we are only concerned with case (i).

As $m_2 = 0$ and $N$ is negative, $q_1(n_0) = 0$ and $q_2(n_0) = -m_1 < 0$. As the equilibrium is at the critical state, it is impossible to analyse the effects of a change in $n$ upon the behavior of the system by traditional comparative statics analysis.

We parameterize the bifurcating branches by

$$U_1 = \varepsilon\,, \quad U_2 = \varepsilon w(\varepsilon)\,, \quad n = n_0 + \varepsilon z(\varepsilon)\,, \tag{4.4.13}$$

where $\varepsilon$ is the expansion amplitude parameter, and $w(\varepsilon)$ and $z(\varepsilon)$ are continuous functions of $\varepsilon$ to be determined. We are interested in the behavior of $U$ when $\varepsilon z(\varepsilon)$ is not equal to zero.

At $\varepsilon = 0$, we have $f(= (f^1, f^2)^{\mathrm{T}}) = 0$, and $m_2 = 0$. We will construct solutions of $f = 0$ for $\Theta$ not equal to zero. Substituting (4.4.13) into (4.4.12), we obtain

$$f^1(U,n) = \varepsilon m^1(w,z,\varepsilon)\ , \quad f^2(U,\varepsilon) = \varepsilon m^2(w,z,\varepsilon)\ , \tag{4.4.14}$$

where

$$m^1 = r_1 + \frac{n'\gamma y w}{g} + \varepsilon\left[-z + \frac{n'\beta(\beta-1)y}{k^2} + \frac{n'\beta\gamma y w}{kg} + \frac{n'\gamma(\gamma-1)yw^2}{g^2}\right] + \mathrm{O}(|\varepsilon|^2)\ ,$$

$$m^2 = M + Nw + \varepsilon\left[-z + M_k + wM_g + N_g w^2\right] + \mathrm{O}(|\varepsilon|^2)\ . \tag{4.4.15}$$

We seek the steady bifurcating solutions of $m^i(w,z,\varepsilon) = 0$ $(i = 1,2)$. We first require $m^1(w_0, z_0, 0) = 0$, $m^2(w_0, z_0, 0) = 0$, i.e., $r_1 + n'\gamma y w_0/g = 0$, $M + Nw_0 = 0$. Since $m_2 = 0$, we find a non-zero solution of the equation with respect to $w_0(= -M/N = -gr_1/n'\gamma y)$. As $N$ is negative, $w_0$ is positive.

Let $w(h) = w_0 + \varepsilon w_1(\varepsilon)$ and substitute this $w(\varepsilon)$ into (4.4.15) to solve

$$m^i(w,z,\varepsilon) = \varepsilon s_i(w_1,z,\varepsilon) = 0\ , \quad i = 1,2\ , \tag{4.4.16}$$

where

$$s_1 = \frac{n'\gamma y w_1}{g} - z + \frac{n'\beta(\beta-1)y}{k^2} + \frac{w_0 n'\beta\gamma y}{kg} + \frac{w_0^2 n'\gamma(\gamma-1)y}{g^2} + \mathrm{O}(\varepsilon)\ ,$$

$$s_2 = Nw_1 - z + M_k + w_0 M_g + w_0^2 N_g + \mathrm{O}(\varepsilon)\ . \tag{4.4.17}$$

From $s_i(w_{i0}, z_0, 0) = 0$, we solve $w_{10}$ and $z_0$ as

$$w_{10} = n''\left[\frac{n'\beta(1-\beta)y}{k^2} - \frac{w_0 n'\beta\gamma y}{kg} + \frac{w_0^2 n'\gamma(1-\gamma)y}{g^2} + M_k + w_0 M_g + w_0^2 N_g\right]\ ,$$

$$z_0 = Nw_1 0 + M_k + w_0 M_g + w_0^2 N_g\ , \tag{4.4.18}$$

in which $n''(= 1/(n'\gamma y/g - N)) > 0$. Hence, we obtain

$$n = n_0 + \varepsilon z_0 + \mathrm{O}(\varepsilon^2)\ ,$$

$$k = k_0 + \varepsilon\ ,$$

$$g = g_0 + \varepsilon w_0 + \varepsilon^2 w_{10} + \mathrm{O}(\varepsilon^3)\ , \tag{4.4.19}$$

for which all of the parameters are explicitly given.

We will not discuss the stability of the new equilibrium because the expression is too complicated. We can determine the stability by the standard linearization method with regard to the equilibrium.

It should be noted that $dn/d\varepsilon$, $dk/d\varepsilon$, and $dg/d\varepsilon$ are dependent upon the non-linear terms of the system. That is, the linearized terms cannot determine the effects of the parameter upon the behavior of the variables. The correspondence principle fails because the assumption of stability is not satisfied. However, instability does not mean the destruction of the system. A new equilibrium, which may be either stable or unstable, is established.

As $w_0$ is positive, one can conclude that the effects of a change in the population growth rate depend upon the sign of $z_0$. If $z_0$ is positive, an increase in $n$ will cause the capital per capita and the level of knowledge to become high; if $z_0$ is negative, an increase in the growth rate will reduce the capital and the level of knowledge. From the equation $z_0 = Nw_{10} + M_k + w_0 M_g + w_0^2 N_g$, where $Nw_{10}$ is uncertain, $M_k$ and $w_0^2 N_g$ are negative, and $w_0 M_g$ is positive, it is difficult to give a general conclusion about the effects of population growth.

We have analyzed the potentially unstable situation to illustrate the complexity of nonlinear systems. Further bifurcation analysis can be completed according to the method developed in Iooss and Joseph (1980).

## 4.5 Singularity Theory in Economic Analysis

We have provided some examples of applications of catastrophe theory and bifurcation theory to different economic problems. Now, we would like to mention some possible directions in which singularity theory may be applied to economic analysis.

It may be argued that many serious debates in economics are concerned with the existence of (stable) relationships among variables. For instance, in neoclassical growth theory it is generally assumed that there exist production functions which describe the technological structure of production. The existence of such functions is a question belonging to the field, as Hicks (1965) calls it, of the "static method in dynamic theory" (see Zhang 1989). In the economic literature, we can find many similar fundamental issues in this field. We have additionally the problems related to the existence, for instance, of a demand function of money whose existence is the most basic assumption in "monetarism", a consumption function for goods, a saving function, and a demand function for goods. All of these functions have been used in dynamic analysis.

Once we accept the assumption that there exist such stable relationships among variables, a further problem is to find out functional forms which are potentially useful. Theoretically, there may be many functions which can be used in the analysis. Even in practice, it is very significant to classify the possible functional forms because it is impossible to test all of these forms. From a theoretical point of view, we would like to find the simplest one among the possible functional forms which satisfy certain constraints. In what follows, we show that singularity theory may be applied to shed light on these problems. For simplicity, we are especially concerned with money demand functions.

In formulating a theory of the demand for money, Friedman (1953) stressed that money was one asset, one way of holding wealth, and that demand for it could be analyzed using the standard theory of consumer choice. He wrote the demand for money as

$$M = f\left(r_b, r_e, \frac{1}{P}\left(\frac{dP}{dt}\right), w, \frac{Y}{P}, u\right) ,$$

where demand for money depends on the expected yields for holding bonds and equities, the expected inflation rate, the ratio of human to non-human wealth, the real income $(Y/P)$ and variables affecting tastes and preferences (summarized by $u$). According to Friedman, the quantity theory comprised two assertions: (i) the empirical hypothesis that the demand for money is stable; and (ii) that there are important factors affecting the supply of money that do not affect the demand for money.

As this functional form is empirically determined, we have to choose some functions to test the theory. Initially, some information, for instance, first and second partial derivatives of the money demand with regard to the variables, may be known. With this knowledge, we determine what form of functions are to be used in an empirical study. In this case, singularity theory may tell us that only a few functions are useful in a qualitative sense. In other words, if the information about partial derivatives is correct, we have only a few functions which can be used in the analysis. Thus we may focus on these possible functions.

Another problem is that when using the money demand function, we always neglect some unimportant factors. It is necessary to require that the omission of these factors does not have serious effects upon qualitative properties of the function employed. Otherwise, it is impossible to empirically obtain a useful function. Singularity theory tells us what functions can be used for this purpose.

Finally, it should be emphasized that application of singularity theory is technically very difficult. As most results in the theory are mostly local, we have to be aware of its limitations in practical economic analysis.

## 4.6 Remarks

This chapter has mainly been concerned with structural changes in equilibrium solutions of various economic systems. We have shown that if the assumptions in the traditional comparative analysis are relaxed, multiple equilibria and sudden change may occur as parameters are shifted near their critical values. The behavior examined in this chapter is time-independent in the long term. In the next chapter, we will examine how time-dependent structures are created from time-independent equilibrium solutions as parameters change.

# 5. Economic Cycles

Nature is the realization of the simplest conceivable mathematical ideas.

*Albert Einstein*

In the previous chapter, we showed that small changes in external parameters may cause dramatic evolution of nonlinear dynamic economic systems when the systems are near critical points. Multiple equilibria are not the exception in such unstable systems. If we consider that small changes in parameters may be due to chance, these results imply that economic evolutionary paths cannot be decided by perfect historical determinism. Chance can change the development trajectory dramatically. However, the economic phenomena examined in Chap. 4 are time-independent. This chapter is concerned with those structural changes (due to small shifts of parameters) which result in regular time-dependent behavior – limit cycles.

## 5.1 Theories of Economic Cycles

Economic life changes ... partly because of changes in the data, ... there is another ... source of ... change ... which arises from within the system, and this kind of change is the cause of so important phenomena that it seems worthwhile to build a theory for it.

*J.A. Schumpeter* (1934)

The observed fluctuations in economic data vary greatly in amplitude and scope as well as duration. These phenomena are both national and international in scope, and are sometimes persistent – long enough to permit the development of cumulative movements in the downward as well as upward direction. Business cycles belong to the history of modern economies with interdependent markets, free enterprise, and private ownership of financial assets and capital goods. They developed in the era of great growth of industry, banking and credit. They are varied and changing, even while retaining their general characteristics of persistence and pervasiveness as well as specific regularities of amplitude and timing. For instance, Fig. 5.1. displays the data on six time series for the postwar U.S.: real GNP, the unemployment rate, the Baa bond rate, the percentage rate of change in the real money supply, the inflation rate in the GNP deflator, real output (GNP) per man-hour, and an estimate of the straight time real wage.

The variables – for real GNP, unemployment, and output per man-hour – in the figure show up as irregular, cumulative movements. There are the recessions of 1954, 1958, 1960, 1970, and 1974–1975, as characterized by reductions in real GNP and substantial increases in the unemployment rate. It should be noted that other (advanced) countries followed similar courses.

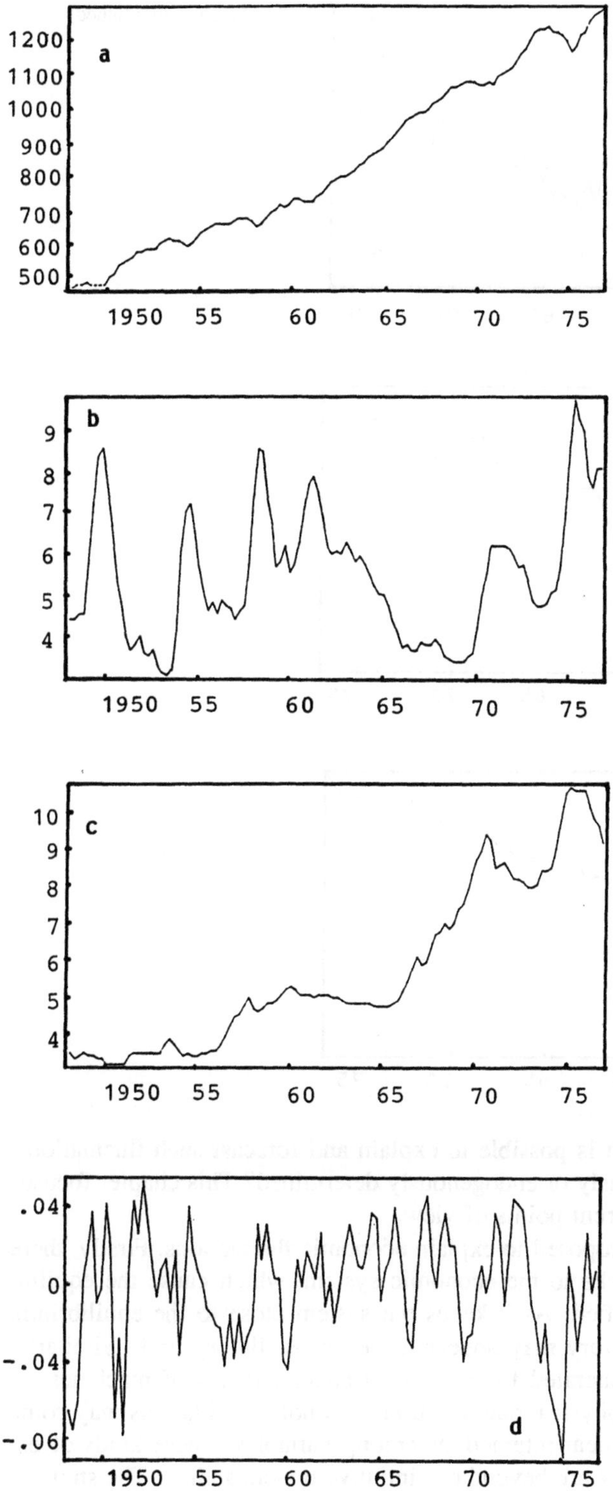


**Fig. 5.1a–g.** An example of economic fluctuations; (a) real GNP, (b) unemployment rate, (c) interest rate (Baa bond rate), (d) change in real money supply, (e) inflation rate, (f) output per worker hour, (f) real wage (NSA). (Source: Sargent 1979)

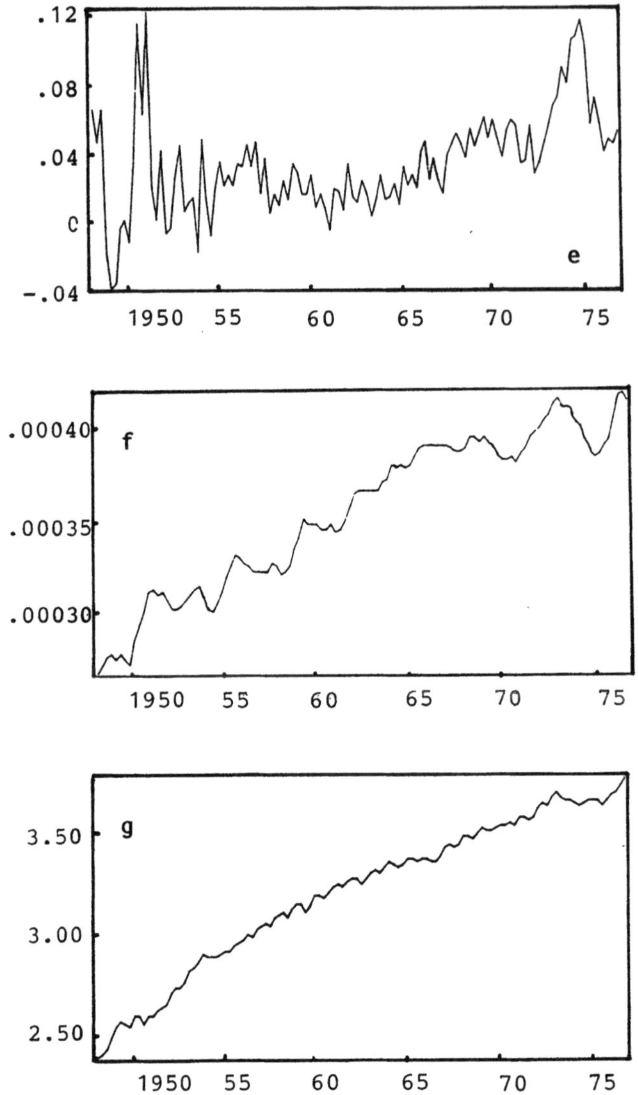


**Fig. 5.1** (continued)

Our question is whether it is possible to explain and forecast such fluctuations. Are business cycles exogenously or endogenously determined? This chapter focuses on these problems from different points of view.

Two reasons have been proposed to explain economic fluctuations. Firstly, there exist random exogenous shocks to the economic system, which cause the equilibrium of the system to be shifted. As it keeps the system close to the equilibrium, the resulting economic trajectory may appear to be an oscillatory frisk. Similarly, periodic behavior may be generated from physical characteristics of machines, or houses, which are subject to physical deterioration. Secondly, oscillations may come from the complexity of nonlinear interactions among variables. These kinds of oscillation are endogenous. They are beyond our intuitive understanding. The study of endogenous economic cycles is one of the most significant subjects in economics. This chapter is mainly concerned with business cycles in nonlinear systems.

Theoretical economists tend to agree that business cycles have mainly endogenous explanations involving recurrent fluctuations in interrelated monetary and real variables, prices and quantities, and expectations and realizations, though they tend to disagree principally on which factors should play star and support roles.

Looking back across monthly or quarterly data representing many different variables, we find that business cycles can be clearly distinguished from other fluctuations in that they are as a rule larger, longer, and more widely diffused. They dominate changes in the economy over spans of several years, in contrast to the seasonal and other variations which spend themselves in periods of a year or less. They reflect and interact with long growth trends which dominate developments across decades.

As shown in Zarnowitz (1985), interest in business cycles is itself subject to a wave-like movement, waxing during and after periods of turbulence and depression, waning in periods of substantial stability and continuing growth.

The classics of business cycle literature have made lasting contributions to the description and analysis of the motion of industrialized market economies. We will simply describe some of these theories. The role of discrepancies between the market and the "natural" interest rates on fluctuations was much explored by Knut Wicksell (1898). Hawtrey (1913) addressed the cumulative processes of inflationary expansions and deflationary contractions induced by bank credit fluctuations that were constrained by the availability of reserves under the gold standard. At below-equilibrium market rates, excessive bank credit produces over-investment in capital goods industries and imposes "forced saving" on those whose incomes lag behind inflation (Hayek 1933). Monetary changes are linked to real vertical maladjustments, which mean imbalances between production of capital and consumer goods or between aggregates of investment plans and savings decisions (Tugan-Baranovskii 1894, Spiethoff 1953). Long gestation and life periods of capital goods and some cyclical aspects of the acceleration principle were investigated (Aftalion 1913, Clark 1917). Under uncertainty, interdependent expectations of businessmen generate widespread errors of optimism in expansions and pessimism in contractions (Pigou 1927). Unpredictable shifts in demand or supply lead to horizontal maladjustments – overinvestment in some sectors (Robertson 1915). Fluctuations of business profits which result from oscillations of unit costs of labor and production helped explain the cyclical movements in investment and output (Mitchell 1913). Schumpeter (1939) saw economic growth itself as a cyclical process, reflecting technological progress and spurts of innovations. Keynes (1936) attributed to the trade cycle a sudden, sharp downturn, a protracted decline, and a gradual sluggish upturn. Here, the sharp downturn or "crisis" is explained by a sudden collapse in the marginal efficiency of capital. However, Keynes' analysis is only implicitly and partially dynamic. These illustrate a broad range of views held in traditional economics, though there are overlaps among these theories.

The essential agreement of these theories was that economic cycles are endogenous. That is, they deliberately concentrated on the internal dynamics of the system. They generally held that contemporary industrial economies are, as a result of such dynamics, subject to recurrent fluctuations with major regularities that can be explained economically. They viewed the role of the exogenous forces as secondary, even though acknowledging that the latter continually act as originators or disturbers of endogenous processes, with the power to accelerate, retard, interrupt, or reverse the endogenous movement of the economic system. Moreover, they generally appreciated the seriousness of the problem of economic instability, though they did not treat instability as a source of fluctuations as we do. For them, an economy is always in, or at least tending closely to, general equilibrium. This may be one of the reasons that for a long time business cycles were simply viewed by most economic theorists as merely temporary "frictional" interference with, and departure from, equilibrium.

The 1930s and 1940s saw a proliferation of formal models of essentially endogenous cycles in aggregate output, which used various versions of the investment accelerator and the consumption multiplier (Harrod 1936, Kalecki 1937, Samuelson 1939, Metzler 1941, Hicks 1950). A closely related but more general class of models is based on the capital adjustment (or "flexible accelerator") principle: current investment equals some fraction of the gap between the desired and actual capital. The desired stocks vary directly with output. Net investment depends positively on output and inversely on the initially available stock of capital (Kalecki 1935, Kaldor 1940, Goodwin 1951). The dynamics of these models come from lags, nonlinearities, or both, though few theoretical business cycle models make important use of nonlinearities. A substantive use of nonlinearities in the theory of business cycles has been systematically explored in the recent literature.

Recently, some analytical methods such as bifurcation theory, catastrophe theory, and singularity theory have been applied to analyze large economic fluctuations involving crises, depressions, rapid recoveries, and limit cycles. This chapter is developed along this direction.

The study of business cycles is almost coextensive with short term macroeconomics and it has a large interface with the economics of growth, money, inflation, and expectations. There are monetarist interpretations of business cycles, equilibrium models with price misperceptions and intertemporal substitutions.

The direction of the study has recently been led from "adaptive" to "rational" expectations.

The rational approach is generally monetarist in the sense of relying on monetary shocks, but the emphasis shifts from nominal demand changes and lagged price adjustments to informational lags and supply reactions. Various problems and complications arise, which lead to new attempts to explain the persistence of cyclical movements, the role of uncertainty and financial instability, real shocks, gradual price adjustments, etc. (see, e. g., Barro 1988).

## 5.2 Some Mathematical Results Related to Limit Cycles

### 5.2.1 The Poincaré-Bendixson Theorem and Its Applications to Economics

First, we consider a two-dimensional ordinary differential equation system

$$\frac{dx_1}{dt} = f(x_1, x_2) ,$$

$$\frac{dx_2}{dt} = g(x_1, x_2) , \qquad (5.2.1)$$

where $x = (x_1, x_2)^T$, $x$ is limited to $U \in R^2$, and functions $f$ and $g$ are sufficiently smooth with respect to $x$. A comprehensive investigation of limit cycles of two-dimensional systems is carried out by Ye et al. (1986).

A point $x^*$ is defined as a limit point of $x$ if there is a sequence such that $\lim W_t(x) = x^*$ as $t \to +\infty$, where $W_t(x)$ is the flow of the system.

**Theorem 5.2.1.** (*Poincaré-Bendixson*). A non-empty compact limit set of a continuously differentiable dynamic system in $R^2$, which contains no equilibrium point, is a closed orbit.

The following theorem is a consequence of Theorem 5.2.1.

**Theorem 5.2.2.** A closed trajectory of a continuously differentiable dynamic system in $R^2$ must enclose an equilibrium point with $dx/dt = 0$. Furthermore, if a trajectory is contained in a closed subset $D$ of $U$, then $L(x)(x \in D)$ is a non-empty, closed and connected set.

These theorems mean that if a subset $D$ of $U$ can be detected such that the limit set $L(x)$ is non-empty and compact and does not contain the equilibrium, then the limit set is a closed orbit which encloses the equilibrium. It should be noted that the above theorems do not exclude the possibility of multiple limit cycles.

Let $D$ be a simply connected domain in $U$. We have

**Theorem 5.2.3.** (*Bendixson*). Assume that $f$ and $g$ in (5.2.1) have continuous first order derivatives in $D$. If the sum $(\partial f/\partial x_1 + \partial g/\partial x_2)$ has the same sign for all $D$, then there is no periodic solution for (5.2.1) lying entirely in $D$.

This theorem provides the conditions for excluding limit cycles in the domain $D$.

**Theorem 5.2.4.** (*De Baggis*). Let the system be structurally stable. It then has only a finite number of limit cycles in $D$, which are alternately stable and unstable in the asymptotical sense.

There are many applications of the Poincaré-Bendixson theorem to economics (see Schinasi 1982, Semmler 1985, 1986). Chang and Smyth (1971) applied the Poincaré-Bendixson theorem to Kaldor's 1940-business cycle model. The Chang-Smyth model is defined as

$$\frac{dY}{dt} = \alpha\left[I(Y, K) - S(Y, K)\right] ,$$

$$\frac{dK}{dt} = I(Y, I) , \qquad (5.2.2)$$

where $Y$, $K$, $S$, and $I(Y, I)$ represent real income, capital, the consumption function, and the net investment function, respectively. It is assumed that $S_K < 0$, and $|I_K| > |S_K|$.

It can be shown that the product of the eigenvalues equals $\alpha(S_K I_Y - S_Y I_K)$ at an equilibrium point. This number must be positive in order to exclude the possibility of a saddle point. The sum of the eigenvalues is equal to $\alpha(I_Y - S_Y) + I_K$, which must be strictly positive if we require the equilibrium to be unstable. Chang and Smyth have established the following theorem.

**Theorem 5.2.5.** (*Chang and Smyth*). If (5.2.2) is defined in $\mathbb{R}_+^2$ and has the following properties: (i) $I_K < S_K < 0$, $I_Y > 0$, $S_Y > 0$; (ii) at equilibrium $(K_0, Y_0)$, $\alpha(I_Y - S_Y) + I_K > 0$ and $S_K I_Y < S_Y I_K$; (iii) $dK/dt = 0$ intersects the $K$-axis for a finite $K(0) > 0$; (iv) $dY/dt = 0$ intersects the $Y$-axis for a finite $Y_1 > Y_0$, and $\lim K = +\infty$ as $Y \to 0$; and (v) the system is structurally stable, then every trajectory either is a limit cycle or approaches a limit cycle.

This theorem is identified by applying the Poincaré-Bendixson theorem. It is worth mentioning that Kaldor's 1940 model is the first endogenous business cycle model, although the model is in a slightly different form from the Chang and Smyth model.

Theorem 5.2.5 provides the conditions for the existence of limit cycles. Nothing has been said about the uniqueness of oscillations, as the Poincaré-Bendixson theorem only establishes that at least one limit cycle exists, i. e., it may be possible that several alternately stable and unstable cycles simultaneously exist. This means that the selection of the limit cycle in which the system finally moves depends on the initial values of the variables involved.

Recently, Lorenz (1986) has tried to solve the uniqueness problem for the Chang-Smyth model by applying the Levinson-Smith theorem. Although his result is rather limited, it is important since the question of the uniqueness of limit cycles in nonlinear business cycle models has rarely been addressed in the literature.

Before reporting the result by Lorenz, we explain the Levinson-Smith theorem. Consider the so-called generalized Liénard equation

$$\frac{dx}{dt} = y - f(x) ,$$
$$\frac{dy}{dt} = -g(x) ,$$

or identically

$$\frac{d^2x}{d^2t} + f'(x)\frac{dx}{dt} + g(x) = 0 . \tag{5.2.3}$$

**Theorem 5.2.6.** (*Levinson and Smith* 1942). Equation (5.2.3) has a unique solution if the following conditions are satisfied

i) $f'$ and $g$ belong to $C_1$;
ii) For arbitrarily given $x_1 > 0$ and $x_2 > 0$ such that for $-x_1 < x < x_2$ we have $f'(x) < 0$, and $f'(x) > 0$;
iii) $xg(x) > 0$, for any $x$ not equal to 0;
iv) $\lim_{x \to \infty} F(x) = \lim_{x \to \infty} G(x) = \infty$ ,
where

$$F(x) = \int_0^x f'(v)dv \ , \quad G(x) = \int_0^x g(v)dv \ ;$$

v) $G(-x_1) = G(x_2)$.

We apply this theorem to (5.2.2). First, rewrite (5.2.2) in the form

$$\frac{d^2Y}{d^2t} = \alpha \left[ (I_Y - S_Y)\frac{dY}{dt} + (I_K - S_K)\frac{dK}{dt} \right] \ .$$

Differentiating the goods market adjustment equation with respect to time yields

$$\frac{d^2Y}{d^2t} - \alpha(I_Y - S_Y)\frac{dY}{dt} - \alpha(I_K - S_K)I = 0 \ . \tag{5.2.4}$$

This equation does not belong to the Liénard-type equation. To apply Theorem 5.2.6, we assume that the actual change in the capital stock is determined by savings decisions, i. e., $dK/dt = S$, where $S = S(Y)$ and assume that the expression $(I_K - S_K)$, denoted by $W(Y)$, is independent of the capital stock. Thus (5.2.4) is rewritten as

$$\frac{d^2Y}{d^2t} - \alpha W(Y)\frac{dY}{dt} - \alpha I_K S = 0 \ . \tag{5.2.5}$$

Under some other very acceptable assumptions, Lorenz (1986) proved that if the investment and savings functions are symmetric with respect to $Y$, then the solution of (5.2.5) is uniquely determined.

### 5.2.2 The Hopf Bifurcation Theorem

The Poincaré-Bendixson theorem is quite limited for economic systems, since it only holds for two-dimensional systems. As most economic systems are higher-dimensional, it is very desirable to find some analytical methods to identify oscillations of higher-dimensional systems. The Hopf bifurcation theorem is very useful for this purpose.

It may be said that one of the most important results in bifurcation theory is the Poincaré-Andronov-Hopf theorem, which is usually referred to as the Hopf bifurcation theorem. The term Hopf bifurcation refers to a phenomenon in which a steady state of an evolution equation evolves into a periodic orbit as a bifurcation parameter is varied. The Hopf bifurcation theorem provides sufficient conditions for determining when this behavior occurs. Except for Pontryagin's maximum principle, there may be no mathematical theorem which is applied so often. We find the following reasons for this popularity: (i) The conditions of the occurrence of Hopf bifurcations are usually easy to identify; (ii) The theorem can be applied to any number of dimensions and to general spaces; (iii) Hopf bifurcations are the only time-dependent behavior that is well understood in bifurcation theory; (iv) It provides an example of differences between linear and nonlinear phenomena; and (v) The theorem also furnishes a basis for systematically locating the periodic orbits "in large".

Now we present the standard Hopf bifurcation theorem. Consider an autonomous system

$$\frac{dx}{dt} = f(x,r)\,, \quad f(0,r) = 0\,, \tag{5.2.6}$$

where $f : R^n \times R \to R^n$ is $C^\infty$, $r$ is a bifurcation parameter, and $x = 0$ is a steady state of the system for all $r$.

Hopf (1942, or Marsden and McCracken 1976) showed that a one-parameter family of periodic solutions to (5.2.6) emanating from $(x,r) = (0,0)$ can be found if two hypotheses on $f$ are satisfied. Let $A(r)$ be the $n \times n$ Jacobian matrix of $f$ along the steady state solution. The first Hopf assumption is:

i) $A(0)$ has a pair of nonzero, purely imaginary simple eigenvalues $\pm iz_0$; and

ii) $A(0)$ has no other eigenvalues lying on the imaginary axis. (5.2.7)

It should be mentioned that (ii) can be generalized. It can be proved that periodic orbits exist if $A(0)$ has other eigenvalues on the imaginary axis, provided that none of these are an integer multiple of $\pm iz_0$. Furthermore, we claim that $A(r)$ has simple eigenvalues of the form $z_1(r) \pm iz_2(r)$, where $z_1(0) = 0$, $z_2(0) = z_0$, and $z_i$ is smooth with respect to $r$. This follows from the fact that $A(r)$ has real entries which depend smoothly on $r$ and that the purely imaginary eigenvalues are simple.

The second Hopf assumption is

$$z_1'(0) \quad \text{is not equal to zero}\,, \tag{5.2.8}$$

that is, the imaginary eigenvalues of $A(r)$ cross the imaginary axis with nonzero speed as $r$ crosses zero (see Fig. 5.2).

The Hopf bifurcation theorem states that there is a one-parameter family of periodic solutions to $dx/dt = f(x,r)$ if (5.2.7) and (5.2.8) hold.

It is also important to know whether the bifurcation is subcritical or supercritical and to know stability conditions for the bifurcating cycles. All of these questions are referred to, for example, in Marsden and McCracken (1976), Iooss and Joseph (1980), Chow and Hale (1982), or Golubitsky and Schaeffer (1984).

We provide an example to show the way in which Hopf bifurcations can be identified. Consider

$$\begin{aligned}\frac{dx_1}{dt} &= rx_1 - x_2 + \left(x_1^2 + x_2^2\right)x_1\,,\\ \frac{dx_2}{dt} &= x_1 + rx_2 + \left(x_1^2 + x_2^2\right)x_2\,.\end{aligned} \tag{5.2.9}$$

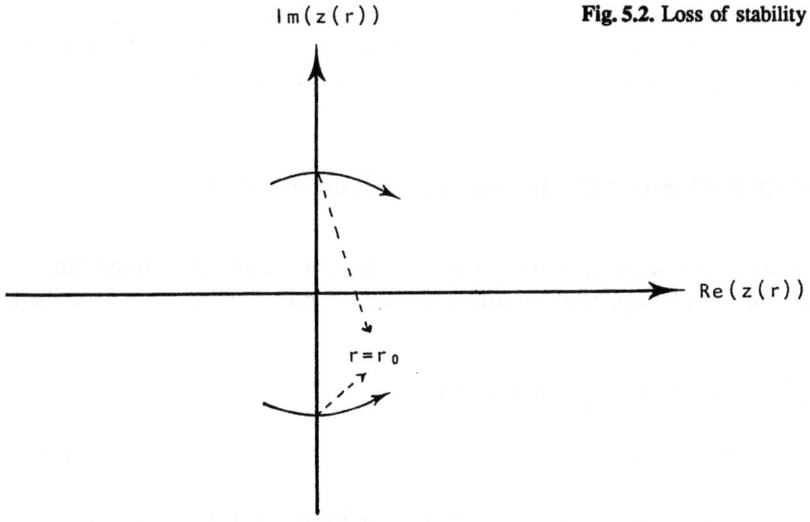


**Fig. 5.2.** Loss of stability as $r$ passes $r_0$

The origin (0,0) is an equilibrium point. From linear analysis we can obtain the following conclusions. For $r < 0$ the steady state is stable, while for $r > 0$ the steady state is unstable. However, for $r = 0$, the steady state is neutral. Thus, Hopf bifurcations may be observed as $r$ increases from zero to positive. Golubitsky and Schaeffer (1984) proved that the phase portraits of this system can be illustrated as in Fig. 5.3. For $r > 0$, there is exactly one periodic solution of (5.2.6). Moreover, this periodic solution is stable in the sense that all nearby orbits approach it. Thus, there has been an exchange of stability from the steady state $x = 0$ when $r < 0$ to the newly created periodic solution when $r > 0$.

The cubic terms (5.2.9) push $x$ towards the interior of the circles $|x|$ = const. For $|x|$ large, these dominate, thus forcing orbits towards the origin. On the other hand, when $|x|$ is small the linear terms dominate, and if $r > 0$ the linear terms force orbits away from the origin. The existence of a periodic solution results from the competition of these forces.

There are many papers to generalize the Hopf bifurcation theorem, for instance, to the cases of infinite dimensional systems, and there are several books applying it to various problems in science (e. g., Marsden and McCracken 1976, Guel and Rsler 1979, Hassard, Kazarinoff and Wan 1981). Moreover, this theorem has recently

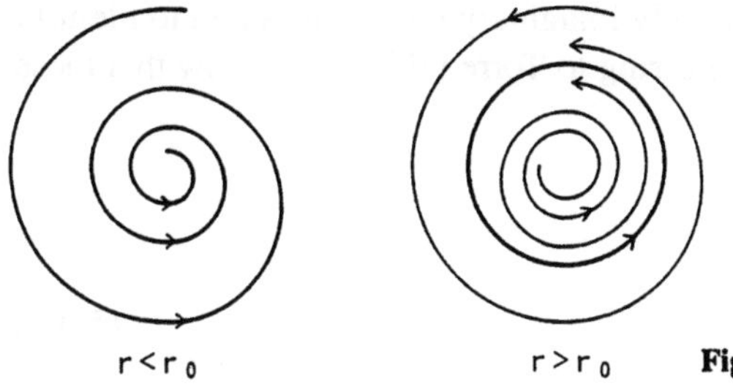


**Fig. 5.3.** Phase portraits for (5.2.9) with $r_0 = 0$

found wide applications in economics (see, for instance, Benhabib and Miyao 1981, Benhabib and Nishimura 1986, Zhang 1988d, 1989b, 1990a). The rest of this chapter provides some examples of applications of the theorem to economics.

## 5.3 The Simplified Keynesian Business Cycle Model

We now consider a dynamic economic system which is essentially suggested by Keynes in his *General Theory*. The simplified Keynesian business model is described by

$$\frac{dY}{dt} = \alpha\{I(Y,R) - S(Y,R)\} = \alpha F(Y,R) \, ,$$
$$\frac{dR}{dt} = \beta\{L(Y,R) - L_s\} \, , \tag{5.3.1}$$

in which all parameters and variables are positive and represent the following:

$Y$: the output level;
$R$: the rate of interest;
$I(Y,R)$: investment function ($I_Y > 0$, $I_R < 0$);
$S(Y,R)$: savings function ($S_Y > 0$, $S_R > 0$);
$L(Y,R)$: total demand for money ($L_Y > 0$, $L_R < 0$);
$L_s$ = the fixed supply of money;
$\alpha, \beta$ = positive adjustment parameters.

The system simply states that if investment is larger than savings, then output level tends to increase, and vice versa; if the money demanded is larger than that supplied, the interest rate tends to increase.

The requirements ($I_Y > 0$, $I_R < 0$, $S_Y > 0$, $S_R > 0$, $L_Y > 0$, $L_R < 0$) imply that investment is positively related to output level, and negatively dependent on interest rate; an increase in output or interest rate will make people save more; more money is demanded if output increases or interest rate falls.

The existence of a positive equilibrium ($Y_0$, $R_0$) determined by the intersection of $L(Y,R) = L_s$ and $F(Y,R) = 0$, is assumed. It is sufficient to limit the discussion to a local domain.

The existence of business cycles in this system was initially proposed by Torre (1977). Here, we re-examine the behavior of the model.

To apply the Hopf bifurcation theorem to the problem, we have to find the conditions for the existence of a pair of purely imaginary eigenvalues and to identify the loss of stability of the equilibrium. Referring to Torre (1977), we know that these conditions are established if

$$\alpha_0 = \frac{\beta L_R}{F_Y} \, ,$$

$$F_Y L_R - F_R L_Y > 0 \, , \quad F_Y > 0 \, , \tag{5.3.2}$$

hold at the equilibrium. As $\alpha$ is meaningful at any point in $\mathbf{R}_+$, there is a value of the parameter $\alpha$ such that the first equality in (5.3.2) is valid. As $F_Y = I_Y - S_Y$, $F_Y > 0$ means that the marginal investment in the product is larger than the marginal savings with regard to output. Further interpretation of (5.3.2) is given in Torre (1977).

**Theorem 5.3.1.** Let (5.3.2) hold. There then exist limit cycles – Hopf bifurcations – around $(Y_0, R_0)$. The critical value of the bifurcation parameter $\alpha$ is $\alpha_0$. The bifurcated cycle of period $2\pi/w(\varepsilon)$ is approximately given by

$$\begin{aligned} Y(\varepsilon,t) &= Y_0 + 2\varepsilon\alpha_0 F_R \cos\left[w(\varepsilon)t\right] + \mathrm{O}(\varepsilon^2) \,, \\ R(\varepsilon,t) &= R_0 - 2\varepsilon\left\{z_0 \sin\left[w(\varepsilon)t\right] + \alpha_0 F_Y \cos\left[w(\varepsilon)\right]\right\} + \mathrm{O}(\varepsilon^2) \,, \end{aligned} \tag{5.3.3}$$

where $z_0 = \{\alpha_0\beta(F_Y L_R - F_R L_Y)\}^{1/2}$, $\varepsilon$ is the expansion amplitude parameter, and

$$\begin{aligned} \alpha &= \alpha_0 + \varepsilon^2 x_2 + \mathrm{O}(\varepsilon^4) \,, \\ w(\varepsilon) &= z_0 + \varepsilon^2 w_2 + \mathrm{O}(\varepsilon^4) \,, \end{aligned} \tag{5.3.4}$$

where $x_2$ and $w_2$ are constant parameters. Moreover, if $x_2$ is positive, the periodic solutions are stable, while if $x_2$ is negative, they are unstable.

*Proof.* We use the bifurcation method of Iooss and Joseph (1980) to complete the proof. As the existence of periodic solutions has been identified by Torre (1977), we only need show the method for approximately calculating the periodic solution.

To write the system in a local form, we introduce $U_1 = Y - Y_0$, and $U_2 = R - R_0$, where $Y$ and $R$ satisfy (5.3.1). Denoted by $x$ are small perturbations of $\alpha$ from $\alpha_0$, i.e., $x = \alpha - \alpha_0$. Then (5.3.1) can be rewritten as

$$\frac{dU}{dt} = J(x)U + N(x,U,U) + \mathrm{O}(U^3) \,, \tag{5.3.5}$$

where $U = (U_1, U_2)^{\mathrm{T}}$, $J$ is the Jacobian at the equilibrium,

$$J = \begin{pmatrix} \alpha F_Y & \alpha F_R \\ \beta L_Y & \beta L_R \,, \end{pmatrix}$$

and $N$ is the quadratic terms of $U$. The two eigenvalues are determined by

$$z_{1,2}(x) = \frac{\beta L_R + \alpha F_Y}{2} \pm \left\{\frac{(\beta L_R + \alpha F_Y)^2}{4} - \alpha\beta(L_R F_Y - F_R L_Y)\right\}^{1/2} \,. \tag{5.3.6}$$

where $\alpha = \alpha_0 + x$. At $x = 0$, as (5.3.2) holds, there are a pair of purely imaginary eigenvalues, $\pm \mathrm{i}z_0$. If we denote $z(x)$ the eigenvalue which is equal to $\mathrm{i}z_0$ at $x = 0$, then $\mathrm{Re}\,[z_x(0)]$ is not equal to zero. Thus the loss of stability of the equilibrium is guaranteed. We have identified the conditions for Hopf bifurcations.

To obtain the explicit expression of the periodic solutions, we calculate the eigenvector $X$ and adjoint eigenvector $X^*$ with regard to $z(x)$ from

$$JX = z(x)X \,, \quad J^{\mathrm{T}}X^* = \bar{z}(x)X^* \,, \tag{5.3.7}$$

which satisfy $\langle X, X^* \rangle = 1$, $\langle \bar{X}, X^* \rangle = 0$, where $\langle , \rangle$ is the product operation in $C^2$. We have

$$X = [\alpha_0 F_R\,,\ \mathrm{i}z_0 - \alpha_0 F_Y]^{\mathrm{T}}\ ,$$
$$X^* = \left[ -\frac{\mathrm{i}\beta L_Y}{2\alpha_0 z_0 F_Y}\,,\ \frac{-z_0 + \mathrm{i}\alpha_0 F_Y}{2\alpha_0 z_0 F_Y} \right]^{\mathrm{T}} . \tag{5.3.8}$$

As $X$ and $\bar{X}$ are independent, $U$ can be expressed as a combination of them in the form of

$$U = \sigma(t)X + \bar{\sigma}(t)\bar{X}\ , \tag{5.3.9}$$

in which function $\sigma(t)$ is to be determined. Substituting (5.3.9) into (5.3.5), multiplying the resulted functions by $\bar{X}^*$, and then adding the equations, we have

$$\frac{d\sigma}{dt} = z(x)\sigma + r_0\sigma^2 + 2r_1|\sigma|^2 + r_2\sigma^2\ , \tag{5.3.10}$$

where $r_i$ are imaginary numbers. Here, we should not explicitly give $r_i$. As shown by Iooss and Joseph, the solution of (5.3.10) may be constructed by the following series

$$\begin{pmatrix} g(s,\varepsilon) \\ w(\varepsilon) - z_0 \\ x(\varepsilon) \end{pmatrix} = \sum_{i=1}^{\infty} \varepsilon^i \begin{pmatrix} g_n(s) \\ w_n \\ x_n \end{pmatrix}, \tag{5.3.11}$$

where

$$\sigma(t) = \sigma(s,\varepsilon)\,,\quad s = w(\varepsilon)t\,,\quad w(0) = z_0\ ,$$
$$x = x(\varepsilon)\,,\quad \varepsilon = \frac{1}{2\pi}\int_0^{2\pi} \exp(-\mathrm{i}s)\sigma(s,\varepsilon)ds\ .$$

The coefficients of low orders with respect to $\varepsilon$ are determined by

$$x_i = w_i = 0\,,\quad i = 1, 3, \dots, 2n-1, \dots\ ,$$
$$\sigma_1(s) = \exp(\mathrm{i}s)\ ,$$
$$\sigma_2(s) = \left[ r_0 \exp(\mathrm{i}2s) - 2r_1 - \tfrac{1}{3} r_2 \exp(-\mathrm{i}2s) \right] / \mathrm{i}z_0\ ,$$
$$\mathrm{i}w_2 - z_x(0)x_2 = 2\mathrm{i}\left( r_0 r_1 - 2|r_1|^2 - \tfrac{1}{3}|r_2|^2 \right) / z_0\ . \tag{5.3.12}$$

From (5.3.12), we can explicitly obtain $x_i$, $w_i (i = 1, 2, 3)$, $\sigma_1$ and $\sigma_2$.

Define a real number $D$ as

$$D = -\left[ \varepsilon\,\mathrm{Re}\left\{ z_x(0) \right\} + \mathrm{O}(\varepsilon^3) \right] \frac{dx}{d\varepsilon}\ , \tag{5.3.13}$$

where $x = x_2\varepsilon^2/2 + \mathrm{O}(\varepsilon^4)$. According to the factorization theorem in Iooss and Joseph (1980, Chap. VII), if $D$ is positive the cycle is unstable, while if $D$ is negative, it is stable. Thus we have identified the stability conditions in Theorem 5.3.1. □

We will not explicitly give the values of $r_i$, $w_i$, $x_i$ and other parameters because their expressions are too complicated. The behavior of the system is illustrated in

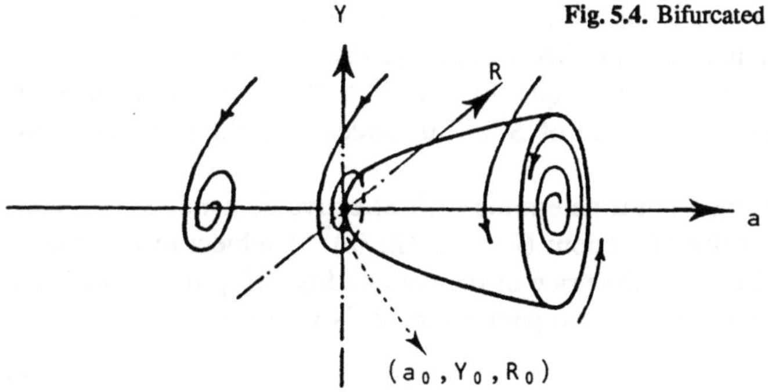


Fig. 5.4. Bifurcated cycles in the Keynesian model

Fig. 5.4. The cycle size is dependent on the bifurcation parameter. As the parameter is further away from its critical value, the radius of the cycle becomes larger.

The interest rate is sometimes higher than its point equilibrium value, and sometimes lower. Although it may arrive at $R_0$, the rate of interest cannot stay there permanently. As soon as it arrives at this equilibrium value, it tends to move away from it. It is driven by the nonlinear interactional forces of interest rate and output. Similarly, we can explain the behavior of output.

From (5.3.3) one has

$$\delta R(t) = -\alpha \sin\left[w(\varepsilon)t\right] - \frac{F_Y}{F_R}\delta Y(t) + O(\varepsilon) , \tag{5.3.14}$$

where $\delta R(t) = \{R(t) - R_0\}/2\varepsilon$, $\delta Y(t) = \{Y(t) - Y_0\}/2\varepsilon$. As $\delta Y(t)$ is a periodic function which is "independent" of $\delta R(t)$, we see that interactions between the two variables may appear to be very complicated.

## 5.4 Non-equilibrium in a Disequilibrium Model

In this section we apply the Hopf bifurcation theorem to an inventory model developed within the framework of disequilibrium macroeconomics. The model to be examined was initially suggested by Eckalbar (1985). Zhang (1989f) reexamined the behavior by introducing a nonlinear adjustment production function. The following discussion is based on the results of Zhang.

The economy consists of two sectors of household and firm, and three goods: money, labor and output. Output is storable by the firm but not by the household. The firm first has expectations about demand for the product, and then produces to cover the expected sales and to maintain a given stock-expected sales ratio. Production is subject to constraints of production capacity.

It is assumed that the households and firms meet in the labor market. The actual exchange quantity in the labor market is given by: $L = \min(L^*, L^d)$, where $L^*$ is the fixed quantity of labor offered by the households for selling, and $L^d$ is the quantity of labor the firms try to buy. It is assumed that $L^d = L^d(V, S^E)$, where $V$ is the firms' start of period finished-good stock and $S^E$ the firms' point sales expectation.

The current supply of output, $Q$, is assumed to equal $dL(d > 0)$ by appropriate choice of units. It is assumed that the firms adjust production in such a way that $V = f(S^{\mathrm{E}})$, where $f' > 0$ and $f''$ is not equal to zero. $f' > 0$ means that the level of desired stocks is an increasing function of $S^{\mathrm{E}}$. This agrees with the spirit of micro level stocks literature.

It is proposed that the households determine their effective demand, $S$, by maximizing the Cobb-Douglas utility function: $U = AS^b(M/p)^{1-b}$ subject to labor sales and budget constraints, where $p$ is the price of the commodity, $M/p$ the desired real balances, $A(A > 0)$ and $b(0 < b < 1)$ are parameters. $S$ is given by

$$S = a + cQ \,, \tag{5.4.1}$$

where $a = bM_0/p$, $c = wb/pd$, $w$ is the nominal wage and $M_0$ the households' initial money holding.

The firms produce what they think they can sell, $S^{\mathrm{E}}$, plus adjustment to correct stock levels, $f(S^{\mathrm{E}}){-}V$. Hence, the labor demand is given by $L^{\mathrm{d}} = [S^{\mathrm{E}}{+}f(S^{\mathrm{E}}){-}V]/d$. On the other hand, if the short side dominates the labor market, output cannot exceed $Q^* = dL^*$. The current output should be

$$Q = \min\left[S^{\mathrm{E}} + f(S^{\mathrm{E}}) - V, Q^*\right] . \tag{5.4.2}$$

In this disequilibrium model, the labor market may be in disequilibrium, but the firms can always supply output to satisfy the households' effective demand. Although the goods market is always in equilibrium, the firms may experience disequilibrium in stocks when the desired and actual stocks are different.

We assume that change in $V$ is equal to $Q - S$ and $S^{\mathrm{E}}$ is adaptively adjusted according to $S - S^{\mathrm{E}}$. The dynamics are given by

$$\begin{aligned} \frac{dV}{dt} &= Q - S = (1-c)\min\left[S^{\mathrm{E}} + f(S^{\mathrm{E}}) - V, Q^*\right] - a \,, \\ \frac{dS^{\mathrm{E}}}{dt} &= S - S^{\mathrm{E}} = a + c\min\left[S^{\mathrm{E}} + f(S^{\mathrm{E}}) - V, Q^*\right] - S^{\mathrm{E}} . \end{aligned} \tag{5.4.3}$$

A switching curve is defined by $S^{\mathrm{E}} + f(S^{\mathrm{E}}) - V = Q^*$. The $V - S^{\mathrm{E}}$ plane is divided by this curve into two regions:

$$\begin{aligned} W_1 &= \left[(V, S^{\mathrm{E}}) \mid S^{\mathrm{E}} + f(S^{\mathrm{E}}) - V < Q^*\right] , \\ W_2 &= \left[(V, S^{\mathrm{E}}) \mid S^{\mathrm{E}} + f(S^{\mathrm{E}}) - V > Q^*\right] . \end{aligned}$$

It can easily be seen that if the current trajectory belongs to $W_2$, the system is linear. This case is studied in Eckalbar (1985) in detail.

We are only concerned with the behavior of $(V, S^{\mathrm{E}}) \in W_1$. In this case, unemployment exists in the labor market. The dynamics are given by

$$\begin{aligned} \frac{dV}{dt} &= (1-c)\left[S^{\mathrm{E}} + f(S^{\mathrm{E}}) - V\right] - a \,, \\ \frac{dS^{\mathrm{E}}}{dt} &= a + (c-1)S^{\mathrm{E}} + cf(S^{\mathrm{E}}) - cV . \end{aligned} \tag{5.4.4}$$

The unique equilibrium is given by

$$S_0^{\mathrm{E}} = \frac{a}{1-c} < Q^* , \quad V_0 = f(S_0^{\mathrm{E}}) . \tag{5.4.5}$$

The eigenvalue $z_i$ is equal to

$$z_{1,2} = \frac{2c-2+cf'}{2} \pm \left[\frac{(2c-2+cf')^2}{4} - 1 + c\right]^{1/2} . \tag{5.4.6}$$

Let $c_0$ satisfy: $2c - 2 + cf' = 0$. As $c_0 = 2/(2 + f')$ and $f' > 0$, $0 < c_0 < 1$. Since $c = wb/pd$, where $0 < b < 1$, the condition $0 < c_0 < 1$ means that $pd < w$. However, from $Q = dL$ we see that $w < pd$ means that the firms' profit is positive. Accordingly, it is possible to find a meaningful value $c_0$ of $c$. At $c = c_0$, the eigenvalues are equal to $iv$ and $-iv$, respectively, where $v = (1 - c_0)^{1/2}$. We choose $c$ as bifurcation parameter with critical value $c_0$. As $c = wb/pd$, any changes in $w$, $b$, $p$ or $d$ will cause $c$ to shift. Let $x = c - c_0$ and denote the eigenvalue which equals $iv$ at $x = 0$ by $z(x)$. Taking the derivative of $z(x)$ with respect to $x$ yields

$$z_x(0) = \frac{(2+f')}{2} - \frac{\mathrm{i}}{2v} . \tag{5.4.7}$$

From (5.4.7), we see that the real part of $z_x(0)$ is positive. Hence, the loss of stability is identified. The Hopf bifurcation theorem holds at $x = 0$.

**Theorem 5.4.1.** For a small value of $x$ there exists a limit cycle near the equilibrium. The bifurcating cycle, with period $2\pi/s(\varepsilon)$, is given by

$$\begin{aligned} V(\varepsilon, t) &= V_0 + 2\varepsilon v^2(1 + f')\cos[s(\varepsilon)t] + O(\varepsilon^2) , \\ S^{\mathrm{E}}(\varepsilon, t) &= S_0^{\mathrm{E}} + 2\varepsilon v \left\{v \cos[s(\varepsilon)t] - \sin[s(\varepsilon)t]\right\} + O(\varepsilon^2) . \end{aligned} \tag{5.4.8}$$

where $\varepsilon$ is the expansion amplitude parameter and

$$s(\varepsilon) = v + \frac{\varepsilon^2 v f''^2}{12}(8 - 23c_0 + 24c_0^2 - 7c_0^3)x(1 + f')^2 + O(\varepsilon^4) . \tag{5.4.9}$$

Moreover, the bifurcation is supercritical.

The theorem is proved in Zhang (1988f). The periodic behavior is described in Fig. 5.5.

As we may keep the amplitude parameter $\varepsilon$ sufficiently small, we can always have the cycle within $W_1$.

To explain the behavior on the cycle, we separate the cycle into four parts as in Fig. 5.5b. Let the system be initially located at point $D$, at which the change rate of stocks is equal to zero. At point $D$, the expected sales begin to fall. As the output of the firms, $Q$, equals the demand of the households, $S$, while the expected sales are larger than the demand, the firms should expect that future sales will become lower than those presently expected. Thus, a decrease in $S^{\mathrm{E}}$ occurs. The system then leaves off at point $D$ and $S^{\mathrm{E}}$ continues decreasing. However, as $Q = S^{\mathrm{E}} + f(S^{\mathrm{E}}) - V$, decreasing $S^{\mathrm{E}}$ will reduce the value of $Q$. As the demand of the households is a function of the output, the decrease in the expected sales will indirectly result in

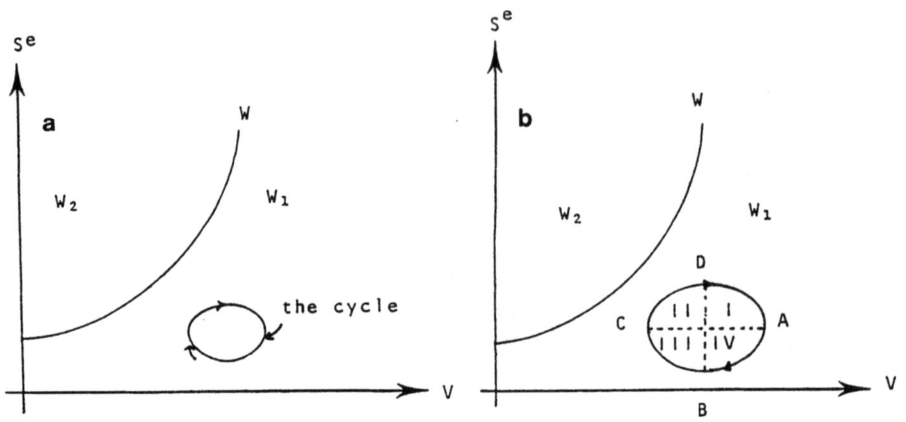


**Fig. 5.5a,b.** Economic fluctuations; (a) $c_0$ to $c_1$, (b) $c_1$ to $c_2$

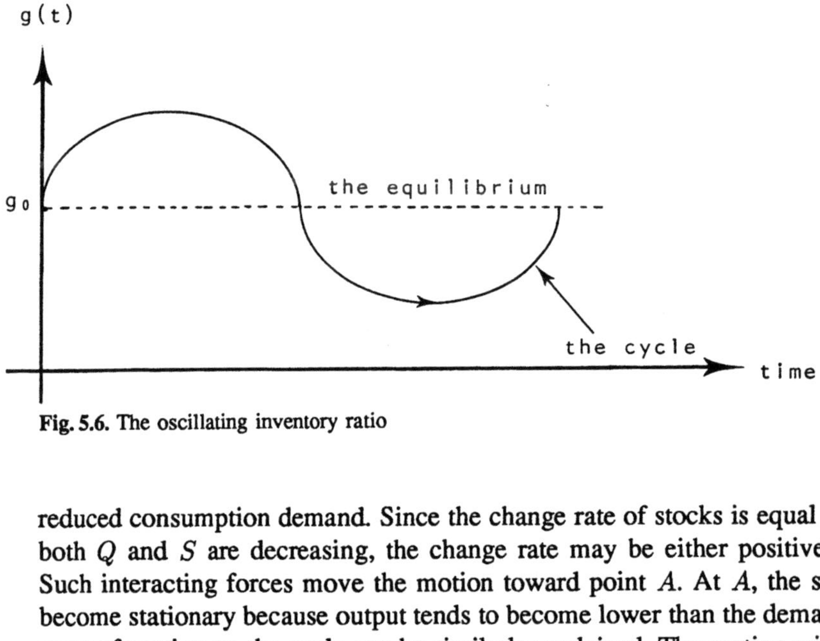


**Fig. 5.6.** The oscillating inventory ratio

reduced consumption demand. Since the change rate of stocks is equal to $Q - S$ and both $Q$ and $S$ are decreasing, the change rate may be either positive or negative. Such interacting forces move the motion toward point $A$. At $A$, the system cannot become stationary because output tends to become lower than the demand. The other parts of motion on the cycle can be similarly explained. The motion will be repeated until further bifurcations take place.

Since one of the virtues of the model is to explain the observed cyclical behavior of the stock-sales ratio, it is necessary to study this behavior. In $W_1$, the actual inventory scales ratio is given by

$$g(t) = \frac{V}{S} = \frac{V(t)}{a + cS^{\mathrm{E}}(t) + cf\{S^{\mathrm{E}}(t)\} - V(t)} .$$

Consequently, the ratio is also oscillatory as shown in Fig. 5.6.

## 5.5 Monetary Cycles in the Generalized Tobin Model

In Sect. 3.3, we presented the Tobin model (see also Tobin 1965, 1969). The equilibrium of this system is unstable. We will re-examine Tobin's analysis. The generalized model accepted in this section is referred to in Zhang (1990b). Although the following model, called the generalized Tobin model, is similar to the Tobin model in Sect. 3.3, they are different in the price dynamics, stability property, and some other respects. Here, we neglect depreciation, i. e., $d = 0$ in (3.3.8). (3.3.6) and (3.3.8) still hold for the generalized Tobin model. The variables to be used are the same as those defined in Sect. 3.3.

In the generalized Tobin model, it is assumed that the change in the price reflects both excess demand (or excess supply) and adaptive expectations. We adopt the Walrasian view that when there is excess demand the price rises, and when there is excess supply the price falls. By Walras's law, the excess demand for goods and services is equal to the (flow) excess supply of real balances. Without taking expected inflation into account, we can thus propose the following dynamics

$$\frac{dp}{dt} = \alpha p\left[x - g(k, q)\right] , \tag{5.5.1}$$

where $\alpha$ is a positive constant parameter, $q$ represents the expected inflation rate, and the function $g$ satisfies: $g_k = +\infty$, and $g_q = -\infty$ in the case of perfect substitutability between two assets – capital and money, and $g_k > 0$ and $g_q < 0$ in the case of imperfect substitutability.

It is assumed that the expected rate of price change may be different from the actual inflation rate. The dynamics are specified as

$$\frac{dq}{dt} = \beta\left[\frac{1}{p}\left(\frac{dp}{dt}\right) - q\right] , \tag{5.5.2}$$

where $\beta$ is the so-called "expectation coefficient".

We complete the building of the model. The generalized Tobin model consists of the following equations

$$\begin{aligned}
\frac{dk}{dt} &= sf(k) - (1 - s)(z - q)x - nk , \\
\frac{dx}{dt} &= x\left(z - \beta\left[x - g(k, q)\right] - n\right) , \\
\frac{dq}{dt} &= \beta\left(\alpha\left[x - g(k, q)\right] - q\right) ,
\end{aligned} \tag{5.5.3}$$

where the first two equations correspond to (3.3.6) and (3.3.8) (with $d = 0$), respectively.

A positive long-run equilibrium $(k_0, x_0, q_0)$ is determined as a solution of

$$\begin{aligned}
&sf(k_0) - (1 - s)n(z - q)x_0 - nk_0 = 0 , \\
&\alpha\left[x_0 - g(k_0, q_0)\right] = z + n = q_0 .
\end{aligned} \tag{5.5.4}$$

From (5.5.4) we have

$$x_0 = \frac{sf(k_0) - nk_0}{(1-s)n} ,$$

which exhibits the non-neutrality of money in the sense that the capital-labor ratio of the monetary model is lower than that of the non-monetary one. If $x_0 = 0$, then one has: $sf(k_0)/n = k_0$, which is identical to the solution of the Solow model. If $x_0$ is positive, then $sf(k_0)/n > k_0$, or $f(k_0)/k_0 > n/s$. It follows that non-neutrality is maintained.

As we are only concerned with stability of the equilibrium and local behavior of the system, we write the system in a local form near the equilibrium. Introduce

$$U_1 = k - k_0 , \quad U_2 = x - x_0 , \quad U_3 = q - q_0 , \tag{5.5.5}$$

where $(k, x, q)$ satisfies (5.5.3), and $U\left(= (U_1, U_2, U_3)^{\mathrm{T}}\right)$ are sufficiently small. Substituting (5.5.5) into (5.5.3) yields

$$\frac{dU}{dt} = AU + N(U,U) + \mathrm{O}(|U|^3) , \tag{5.5.6}$$

where $A$ is the Jacobian evaluated at the equilibrium and $N(U,U)$ is the quadratic term.

$$A = \begin{pmatrix} sf' - n & -(1-s)n & cx_0 \\ \alpha x_0 g_k & -\alpha x_0 & \alpha x_0 g_q \\ -\alpha\beta g_k & \alpha\beta & -\beta(1+\alpha g_q) \end{pmatrix} . \tag{5.5.7}$$

$N(U,U)$ is not explicitly given since we do not use it in the remainder of the study.

Introduce

$$\begin{aligned} -a_1 &= \text{trace}\, A = sf' - n - \alpha x_0 m - \beta(1+\alpha g_q) , \\ a_2 &= \begin{pmatrix} sf' - n & -(1-s)n \\ \alpha x_0 g_k & -\alpha x_0 \end{pmatrix} \\ &\quad + \begin{pmatrix} -\alpha x_0 & \alpha x_0 g_q \\ \alpha\beta & -\beta(1+\alpha g_q) \end{pmatrix} + \begin{pmatrix} sf' - n & cx_0 \\ -\alpha\beta g_k & -\beta(1+\alpha g_q) \end{pmatrix} , \\ a_3 &= -|A| . \end{aligned} \tag{5.5.8}$$

The eigenvalues $\Theta_i$ of the Jacobian are determined by

$$\Theta^3 + a_1\Theta^2 + a_2\Theta + a_3 = 0 . \tag{5.5.9}$$

The necessary and sufficient conditions for the stability are known as the Routh-Hurwitz criterion: (i) $a_i > 0$; (ii) $a_1 a_2 - a_3 > 0$. As discussed in Benhabib and Miyao (1981) and Zhang (1990b), the equilibrium may be either stable or unstable depending on parameter values. For instance, if we move from adaptive expectations towards perfect foresight, saddle-point instability may appear. For the sake of illustration, consider an increase in the stock of money at the equilibrium. The immediate impact of this is to increase the price level, and the real money stock tends to fall back to its original level, but the initial increase in money also tends to increase price expectations and reduce the capital stock. The latter two effects reinforce the fall of the money supply and may cause the money stock to overshoot its long-term equilibrium. As the money supply keeps falling beyond its equilibrium level, the effects on the two variables are reversed: the capital stock rises and expectations fall. Combined with the direct effect of the money stock on the money accumulation, the fall of the money stock will now be reversed. This discussion hints at the possibility of oscillations in the long run. We now prove precisely the existence of business cycles in the model.

The existence of Hopf bifurcations in the generalized Tobin model has been identified by Benhabib and Miyao (1981). Their results can be summarized as follows.

**Theorem 5.5.1.** If there exist a set of parameter values which guarantee the stability of the equilibrium, we can find a value of $\beta$, denoted by $\beta_0$, such that the Jacobian of the system has a pair of purely imaginary eigenvalues. Moreover, there exists a continuous function $v(\varepsilon)[v(0) = 0]$ of a parameter $\varepsilon$ such that when $\varepsilon$ is sufficiently small, the generalized Tobin model has a continuous family of periodic solutions $(k(t, \varepsilon), x(t, \varepsilon), q(t, \varepsilon))^{\mathrm{T}}$, which collapse to the equilibrium point $(k_0, x_0, q_0)^{\mathrm{T}}$ as $\varepsilon \to 0$.

This theorem is very important since it proves the existence of regular oscillations in the system. Such oscillations will continue permanently if the stability of the cycles can be identified. Non-equilibrium economic development is no longer a short-term phenomenon. The generalized Tobin model can thus be applied to explain business cycles. We will improve the results of Benhabib and Miyao in the following aspects: (i) to find stability conditions of the cycle; (ii) to explicitly interpret the parameter $\varepsilon$; (iii) to find the explicit expression of the cycle; and (iv) to discuss whether the Hopf bifurcation is supercritical or subcritical. First, we show that if the Jacobian has a pair of purely imaginary eigenvalues, then the three eigenvalues are given by

$$\Theta_1 = -a_1 , \quad \Theta_{2,3} = \pm\sqrt{\mathrm{i}a_2} = \pm\mathrm{i}\Theta_0 . \tag{5.5.10}$$

As discussed in Benhabib and Miyao, we know that $a_i (i = 1, 2, 3)$ are all positive. The existence of purely imaginary eigenvalues means that (5.5.9) can be written in the following form:

$$\begin{aligned}\Theta^3 + a_1\Theta^2 + a_2\Theta + a_3 &= (\Theta^2 + a^*)(\Theta + a') \\ &= \Theta^3 + a'\Theta^2 + a^*\Theta + a'a^* = 0 .\end{aligned} \tag{5.5.11}$$

Thus (5.5.10) is true.

The following bifurcation analysis selects $\beta$ as a bifurcation parameter. The value of $\beta$ which satisfies (5.5.11) is denoted by $\beta_0$. A small shift of $\beta$ from $\beta_0$ is expressed by $v$, i.e., $v = \beta - \beta_0$. The eigenvalues are continuous functions of $\beta$. We denote $\Theta(v)$ the eigenvalue which equals $i\Theta_0$ at $v = 0$ (i.e., $\beta = \beta_0$). It can be shown that the assumption that $\Theta_v(0)$ is not equal to zero is quite acceptable (see Benhabib and Miyao 1981).

Introduce the following real numbers

$$G_1 = \alpha^2 g^* \left[(1-s)x_0 g_k + (n - sf')g_q\right](x_0 - ng_q) + \alpha g^* g_q \Theta_0^2 ,$$
$$G_2 = \alpha g^* \Theta_0 \left[g_q(\alpha x_0 - \alpha n g_q) - (1-s)x_0 g_k - (n - sf')g_q\right] , \tag{5.5.12}$$

where $g^* = 1/\left[(1-s)\{\Theta_0^2 + (\alpha x_0 - \alpha n g_q)^2\}\right]$. Then we can prove the following theorem.

**Theorem 5.5.2.** The bifurcating cycle in the generalized Tobin model has period $2\pi/S(\varepsilon)$, and can be approximately expressed as

$$k(t,\varepsilon) = k_0 + 2\varepsilon \cos\left[S(\varepsilon)t\right] + O\varepsilon^2) ,$$

$$x(t,\varepsilon) = x_0 + 2\varepsilon G_1 \cos\left[S(\varepsilon)t\right] - 2\varepsilon G_2 \sin\left[S(\varepsilon)t\right] + O(\varepsilon^2) ,$$

$$q(t,h) = q_0 + \varepsilon \left\{ \frac{n - sf'}{x_0}(1-s) + \frac{nG_1}{x_0} \right\} \cos\left[S(\varepsilon)t\right]$$
$$- \varepsilon \left\{ \frac{\Theta_0}{x_0}(1-s) + \frac{nG_2}{x_0} \right\} \sin\left[S(\varepsilon)t\right] + O(\varepsilon^2) , \tag{5.5.13}$$

in which $\varepsilon$ is the expansion amplitude parameter and

$$v(\varepsilon) = \beta - \beta_0 = \frac{\varepsilon^2 v_2}{2} + O(\varepsilon^4) ,$$

$$S(\varepsilon) = \Theta_0 + \frac{\varepsilon^2 S_2}{2} + O(\varepsilon^4) , \tag{5.5.14}$$

where $v_2$ and $S_2$ are constants to be given. When $\mathrm{Re}(\Theta_v) > 0$, if $v_2 > 0$ the cycle is supercritically stable; if $v_2 < 0$, the cycle is unstable. When $\mathrm{Re}(\Theta_v) < 0$, if $v_2$ is negative, the cycle is subcritically stable; if $v_2$ is positive, the cycle is unstable.

The theorem is proved in Zhang (1989).

This theorem shows that the loss of stability that occurs as expectations adjust is associated with the emergence of bounded, persistent oscillations in price, output and expectations. This holds no matter how quickly prices adjust since there always exists a value of $\beta$ at which the stability of the equilibrium is lost. The behavior of the system is illustrated in Fig. 5.7.

Whether the bifurcation is subcritical or supercritical depends on the higher order non-linear terms in the system. We now consider the subcritical case. For a left neighborhood of $\beta_0$, the economy will be locally stable around the equilibrium point. A large shock may throw the economy out or beyond the orbit, in which case it does not have a natural tendency to return to the stationary point.

For simplicity, we introduce: $K_0(t) = [k(t,\varepsilon) - k_0]/\varepsilon$, $X_0(t) = [x(t,\varepsilon) - x_0]/\varepsilon$, $Q_0(t) = [q(t,\varepsilon) - q_0]/\varepsilon$. Here, we note that $g_k > 0$, $g_q < 0$, and $g^* > 0$. If $n$ is approximately equal to $sf'(k_0)$, then one has $G_2 < 0$ and approximately

$$M_0(t) = a' G_1 K_0(t) + a'' \sin\left[d(\varepsilon)t\right] ,$$
$$Q_0(t) = b' G_1 K_0(t) + b'' \sin\left[d(\varepsilon)t\right] , \tag{5.5.15}$$

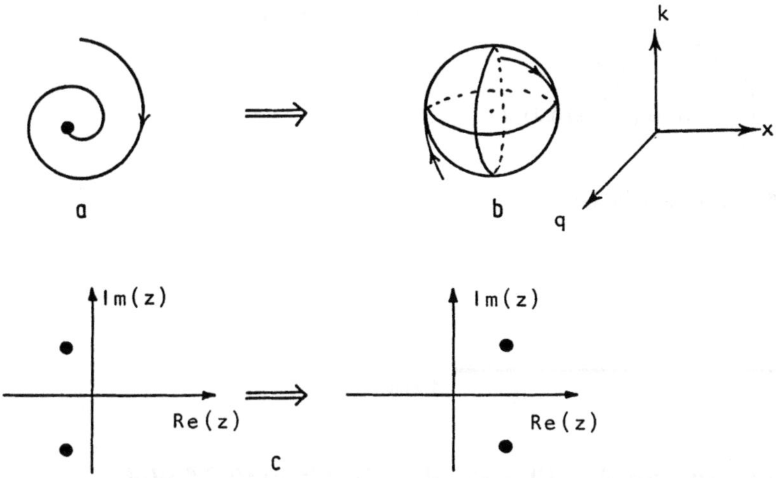


**Fig. 5.7.** Hopf Bifurcation from a fixed point (a) to a limit cycle (b), and behavior of $Z$ (c)

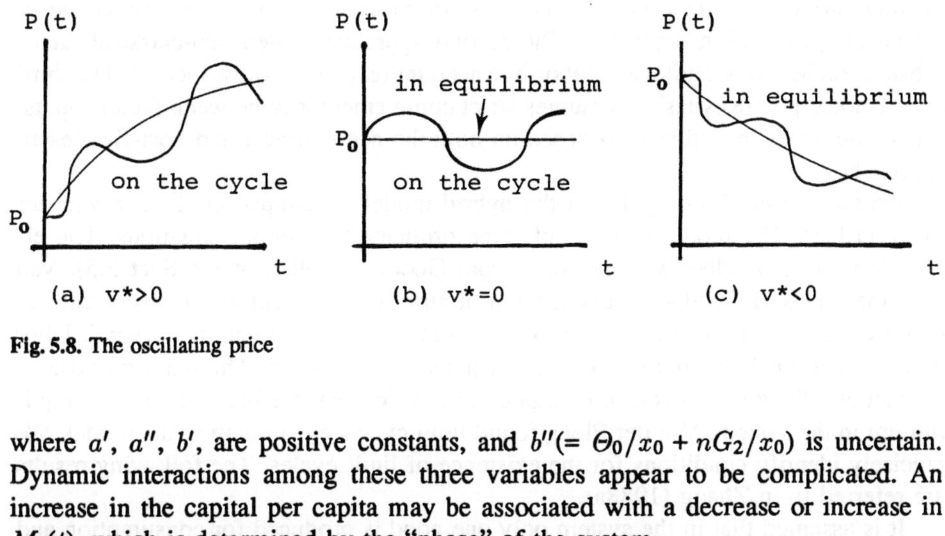


**Fig. 5.8.** The oscillating price

where $a'$, $a''$, $b'$, are positive constants, and $b''(=\Theta_0/x_0 + nG_2/x_0)$ is uncertain. Dynamic interactions among these three variables appear to be complicated. An increase in the capital per capita may be associated with a decrease or increase in $M_0(t)$, which is determined by the "phase" of the system.

It is interesting to study the behavior of the other variables on the cycle. The motion of price is given by

$$p(t) = \frac{L(t)x(t)}{M(t)} = \frac{L(0)x(t)}{M(0)} \exp\left[(n-z)t\right] = cx(t)\exp(v^* t) ,$$

where $c$ is constant and $v^* = n - z$. The behavior of price is shown in Fig. 5.8.

In the long run the price will go towards infinity or zero if the labor growth rate is not equal to the money growth rate. As $K(t) = k(t)L(t)$, the motion of the capital is shown in Fig. 5.9.

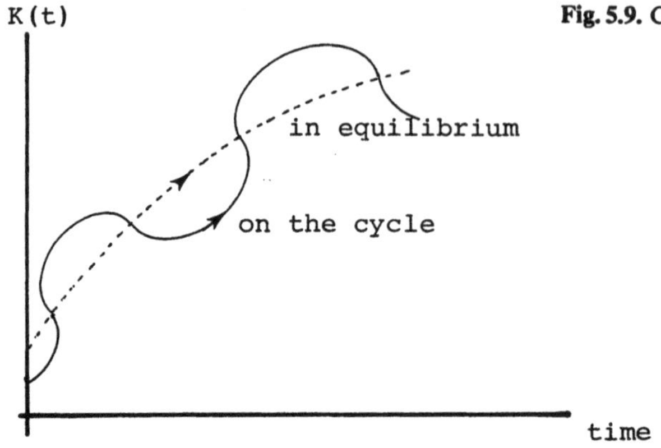


Fig. 5.9. Capital growth

## 5.6 Oscillations in van der Ploeg's Hybrid Growth Model

Three mechanisms for achieving balanced economic growth have been suggested in the literature on economic growth. The first approach is based on the mechanism of Malthusian population explosion. The second approach, called neo-classical, takes substitution between capital and labor into account (e. g., the Solow model). The third one, denoted post-Keynesian, assumes strict complementarity between factor inputs. The model to be investigated is based on both the neo-classical and post-Keynesian approaches.

We re-examine limit cycles in the hybrid model of damped conflict in van der Ploeg (1983). The model consists of three ordinary differential equations. The elementary ideas of Ploeg's work come from Goodwin (1967, or see Sect. 3.5). Van der Ploeg relaxed Goodwin's assumption of strict complementarity between factors by introducing technological progress and by allowing entrepreneurs to recruit labor until the marginal productivity of labor equals the real wage. The main purpose of the van der Ploeg model is to investigate limit cycles near the balanced growth equilibrium in the system. Van der Ploeg found limit cycles by simulating the model. We precisely identify conditions for the existence of limit cycles. The following results are referred to in Zhang (1988a).

It is assumed that in the system only one good is produced for consumption and investment purposes. The supply of labor, $L$, grows exogenously at a rate $n$. The demand for the product consists of goods for consumption, $C$, and gross investment, $I$. The receipts of production, $Q$, provide for net income, $Y$, in the form of wages, $WE$ (where $W$ is wage level and $E$ is the number of workers employed), profits $f$, and for depreciation, $D$, at a rate of $d_1$. Net income is either spent on consumption goods or saved, $S$. The balance sheet identity is

$$\frac{dK}{dt} = S = I - D \, .$$

The consumption function takes the form of

$$C = (1 - s)Y + rK \, , \quad s > 0 \, , \quad r > 0 \, ,$$

where $s$ and $r$ are the propensities to save out of net income and wealth, respectively. The production possibilities are characterized by the capital-output ratio $\alpha$, and the rate of labor-augmenting technical progress $w$. The warranted rate of economic growth is given by

$$g^w = bs - r \,, \tag{5.6.1}$$

where $b = \alpha^{-1} - d_1$ represents the ratio of net income to capital. The natural rate of growth supportable by a fully employed supply of labor and technical progress, is given by

$$g^n = w + n \,.$$

The warranted rate of growth increases when the propensity to save is raised and when the wealth effect, speed of depreciation or capital-output ratio are reduced, since these factors tend to increase accumulation and raise the output that can be produced with the existing capacity. In the long-term balanced growth the warranted growth rate has to equal the natural one. The productivity of labor is determined by a Cobb-Douglas production function with disembodied labor-augmenting technical progress and constant returns to scale

$$Q = dK^{w_2} E^{*1-w_2} \,, \quad 0 < w_2 < 1 \,,$$

$$\frac{1}{E}\left(\frac{dE}{dt}\right) = \frac{1}{E^*}\left(\frac{dE^*}{dt}\right) - w^* \,, \quad w^* > 0 \,, \tag{5.6.2}$$

where the constant $d$ depends on the initial state of the economy, $E^*$ denotes the efficiency units, and $w^*$ is a constant dependent on the disembodied labor-augmenting technical progress.

The change rate of the employment rate $\beta(= E/L)$ is given by

$$\frac{1}{\beta}\frac{d\beta}{dt} = g - g^n = \frac{s^*}{\alpha} - r - \frac{1}{\alpha(1-w_2)}\frac{d\alpha}{dt} - w^* - n \,, \tag{5.6.3}$$

where $g(= g^w - d\alpha/dt/\alpha)$ is the actual rate of growth in real output and $s^*$ is defined by:

$$s^* = s_1(1 - z - \alpha d_1) + s'z = s_2 - s_1 \alpha d_1 \,,$$

where $z$ is the share of labor in net income $[z = (1 - \alpha d_1)WE/Y]$, $s_2 = s_1(1 - z) + s'z'$, and $s_1$ and $s'$ are the saving rates of profit and wages, respectively. The growth in real wage $W$ is assumed to be dependent upon the bargaining strength of workers, proxied by the level of excess demand for labor. Its change rate consists of the real Phillips curve as

$$\frac{dW}{dt} = W(m_1\beta - m_2) \,, \quad m_1,\, m_2 > 0 \,, \tag{5.6.4}$$

where $m_1$ and $m_2$ are constant parameters.

Under the assumption that the capital-output ration always adjusts to the "desired" capital-output ratio determined by the entrepreneurs' optimal behavior, the dynamics of capital-output ratio and share of labor in net income are given by Zhang (1988a).

$$\frac{d\alpha}{dt} = v\alpha \left[ \left( \frac{z}{1 - w_2} \right)^{(1-w_2)/w_2} - 1 \right] , \tag{5.6.5}$$

$$\frac{dz}{dt} = z \left[ m_1 \beta - m_2 - \frac{w_2}{\alpha(1 - w_2)} \frac{d\alpha}{dt} - w^* \right] , \tag{5.6.6}$$

where $v(> 0)$ is an adjustment speed.

The complete hybrid model consists of three differential equations (5.6.3), (5.6.5) and (5.6.6). A unique equilibrium is equal to

$$\beta_0 = \frac{m_2 + w^*}{m_1} , \quad \alpha_0 = \frac{s^*}{r + w^* + n} , \quad z_0 = 1 - w_2 . \tag{5.6.7}$$

The equilibrium can be attained when the share of labor is appropriate, the employment is balanced and the capital-output ratio equals its desired value.

We note that there exists a meaningful value of $v$, denoted by $v_0$, such that

$$v_0 = r + w^* + n > 0 . \tag{5.6.8}$$

As shown in Zhang (1988a), at $v = v_0$ one of the three eigenvalues of the Jacobian is negative and the other two are purely imaginary. According to the Hopf bifurcation theorem, this is a sufficient condition for Hopf bifurcation to occur near the equilibrium. In this study, we choose $v$ as bifurcation parameter, as van der Ploeg (1987) did, although it is possible to select other parameters as bifurcation parameter. Let $x$ denote small perturbations of $v$ from $v_0$, i. e., $x = v - v_0$. If we introduce a positive number $u$ as

$$u = \left( \frac{m_1 v_0 \beta_0}{w_2} \right)^{1/2} , \tag{5.6.9}$$

then it can be shown that the two purely imaginary eigenvalues at $x = 0$ are equal to $\mathrm{i}u$ and $-\mathrm{i}u$, respectively.

To guarantee the loss of stability of the equilibrium, we need to calculate the real part of the derivative of the eigenvalue corresponding to $\mathrm{i}u$ with respect to $x$. It is shown that the requirement that the number is not equal to zero can be generally accepted, though it is not easy to interpret this condition economically.

**Theorem 5.6.1.** There exists a limit cycle near the equilibrium with small $x$. The cycle, bifurcating from the equilibrium with period $2\pi/S(\varepsilon)$, can be expressed as

$$\begin{aligned} \beta(\varepsilon, t) &= \beta_0 + 2\varepsilon \left\{ z_0 v_0 \cos\left[S(\varepsilon)t\right] - u \sin\left[S(\varepsilon)t\right] \right\} + O(\varepsilon^2) , \\ \alpha(\varepsilon, t) &= \alpha_0 + \frac{2\varepsilon\alpha_0 v_0 z_0 m_1}{w_2 u} \sin\left[S(\varepsilon)t\right] + O(\varepsilon^2) , \\ z(\varepsilon, t) &= z_0 + 2\varepsilon z_0 m_1 \cos\left[S(\varepsilon)t\right] + O(\varepsilon^2) , \end{aligned} \tag{5.6.10}$$

where $\varepsilon$ is an amplitude expansion parameter, and $x$ and $S$ are functions of the parameters in the system.

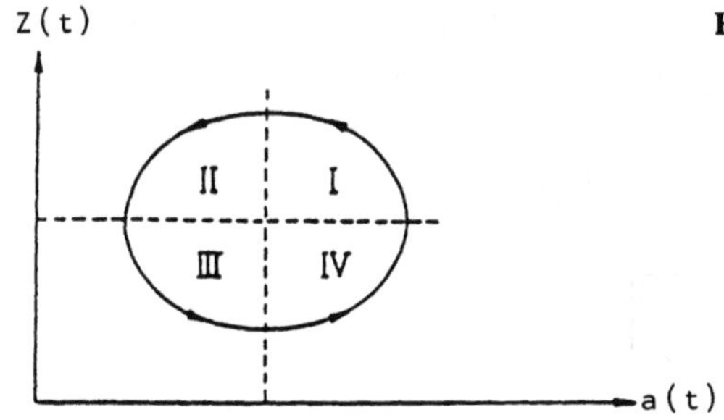


Fig. 5.10. Dynamics on the cycle in the $(a,z)$ plane

The proof of the theorem can be found in Zhang (1988a). It should be noted that we have calculated higher terms of the approximation and identified the stability conditions for the limit cycle. Moreover, all of the conditions for the existence of limit cycles can be verified by the simulation example given by van der Ploeg (1983).

Here, we try to interpret the economic implications of Theorem 5.6.1. First, we limit our discussion to the $(a, z)$ plane. If we neglect the higher terms, the motion of $a$ and $z$ on the cycle can be simply described by

$$\alpha(\varepsilon, t) = \alpha_0 + b_1 \sin\left[S(\varepsilon)t\right] ,$$
$$z(\varepsilon, t) = z_0 + b_2 \cos\left[S(\varepsilon)t\right] , \tag{5.6.11}$$

where $b_1$ and $b_2$ are appropriate parameters determined from (5.6.10). The motion is illustrated in Fig. 5.10.

Let us consider the case when the system begins to move in region I. The capital-output ratio is decreasing, while the share of labour in net income is increasing. This can occur if the labor skill is improved. After a certain period when the share of labor arrives at the maximal value the share decreases and the capital-output ratio keeps decreasing. This situation occurs when much labor is employed. The motion in region III implies that product technology is introduced into the system. The behavior of the system never becomes stationary.

From Theorem 5.6.1, we see that the dynamics of the employment rate is much more complicated than the behavior of the other two variables. The change in the employment rate can be expressed as a linear combination of the share of labor and the capital-output ratio. Let $\delta$ denote the difference between the actual value of a variable in the cycle and its equilibrium value. Then we have

$$\delta\beta(\varepsilon, t) = a_1 \delta z(\varepsilon, t) - a_2 \delta\alpha(\varepsilon, t) , \tag{5.6.12}$$

where $a_1$ and $a_2$ are positive parameters determined by (5.6.10). When the share of labor in net income increases over its equilibrium value, the current employment rate tends to be larger than its equilibrium value. If the capital-output ratio increases over its equilibrium value, the employment rate tends to be less than its equilibrium. Change of the employment rate is "related" to change in the share of labor and the capital-output ratio.

We may also interpret the economic cycle as a result of structural change in the system. The cycle bifurcates from a stationary point when the system goes

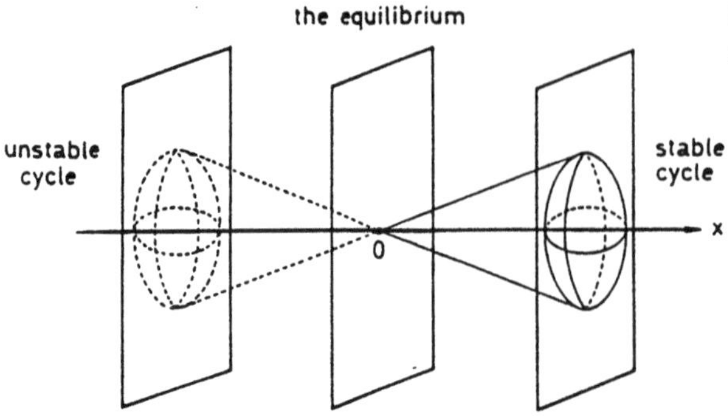


**Fig. 5.11.** Hopf bifurcation in the van der Ploeg model

from stability into instability. The new limit cycle pattern of behavior results from technological change, as the bifurcation parameter is the adjustment speed of the output ratio to its equilibrium value. Perturbations in the parameter imply that the technological characteristics of the system are changed. As a result of the structural change, the behavior of the system changes from steady-state equilibrium to non-equilibrium.

Finally, we illustrate supercritical limit cycles with variables of $x$ in Fig. 5.11.

## 5.7 Periodic Optimal Employment Policy

We will provide an example of applying the Hopf bifurcation theorem to a four-dimensional economic problem. This section is concerned with periodic behavior of firms in a well-informed environment.

We are concerned with the behavior of a firm which is influenced by the government. The model of this section is suggested by Long and Siebert (1985) and re-examined by Steindal et al. (1986) and Zhang (1988f).

The model deals with the behavior of the firm under influences of financial penalties or subsidies imposed by the government. The product of the firm is sold in a competitive market. The price of the product, $p$, is exogenously fixed as the firm cannot affect the market. Since the labor market is also competitive, the wage of labor, $w$, is also exogenously given. The capital is not changed during the study period, and the product is only dependent on the amount of labor employed, $L$. The production function is denoted by $F(L)(F'(0) = \infty$, $F'(\infty) = 0$, $F' > 0$, and $F''(L) < 0$ for $0 < L < \infty)$.

It is assumed that the firm can control its product scale by choosing the hiring/firing rate $v$ of the labor, and that it costs the firm for training new workers and for deciding to layoff. We denote labor adjustment (training, integration and layoff) cost function by $k(v)$ which is specified by $k(v) = v^2/2$. The firm's objective is to maximize the present value of the profit stream

$$\max \int_0^\infty \exp(-rt)\left[pF(L) - wL + f(L - A) - k(v)\right] dt \, ,$$

subject to

$$\frac{dA}{dt} = m(L - A) ,$$
$$\frac{dL}{dt} = v - qL , \tag{5.7.1}$$

where $A$ is the firm's normal employment defined by the government, $r$ the discount rate, $q$ the voluntary quit rate. It is assumed that the government determines the normal employment level by taking the historical employment level of the firm into account. The average value of the historical employment level is given by

$$A(t) = m \int_{-\infty}^{t} L(s) \exp\left[-(t - s)\right] ds , \tag{5.7.2}$$

where $m$ is a positive number. Taking derivatives of (5.7.2) with respect to $t$ yields $dA/dt = m(L - A)$.

The behavior of the government is described by the function $f(L - A)$ which satisfies

$$f(0) = 0 , \quad f(L - A) > 0 \quad \text{if} \quad A < L ,$$
$$f(L - A) < 0 \quad \text{if} \quad A > L , \tag{5.7.3}$$

in which $f(L - A)$ has sufficient differentials. As there are always unemployed people in the labor market, the government likes the firm to employ more people. Equation (5.7.3) states that if the firm decides on a higher employment level than the normal one determined by the government, then the firm can get financial reward (subsidy) from the government; if the firm decides on a lower level, it pays a penalty (tax) to the government.

The Hamiltonian associated with the problem is defined by

$$H = pF - wL + d - k - \alpha m(L - A) + \beta(v - qL) , \tag{5.7.4}$$

where $\alpha$ and $\beta$ are the costate variables. Applying the maximum principle yields

$$\beta = v ,$$
$$\frac{dA}{dt} = m(L - A) ,$$
$$\frac{dL}{dt} = \beta - qL ,$$
$$\frac{d\alpha}{dt} = f' + \alpha(m - r) ,$$
$$\frac{d\beta}{dt} = -pF' + w - f' - \alpha m + \beta(q - r) . \tag{5.7.5}$$

The transversality conditions are given by

$$\lim_{t \to +\infty} \{\alpha(L - A) \exp(-rt)\} = 0 ,$$
$$\lim_{t \to +\infty} \{\beta(\beta - qL) \exp(-rt)\} = 0 . \tag{5.7.6}$$

We show that the Hopf bifurcation theorem holds for (5.7.5) and (5.7.6).

Zhang (1988f) proves that for appropriate values of the parameters there is a positive unique equilibrium in the system, denoted by $(A_0, L_0, a_0, b_0)$.

**Assumption 5.7.1.** Let there be meaningful values of the parameters such that two eigenvalues have negative real parts, and the other two are purely imaginary.

We can verify this assumption by specifying $f$ as

$$f(z) = c_1 \ln\left(\frac{1 + c_2 z}{1 - c_3 z}\right) , \tag{5.7.7}$$

where $z = L - A$, and $c_i$ are positive constants. The function $f(z)$ satisfies the requirements concerning the financial function. If we take: $m = r = q = -pF''(L_0)/f''(0)$, the eigenvalues are equal to $-m$, $-q$, $\pm i(f'')^{1/2}$, respectively.

We select the voluntary quit rate as the bifurcation parameter, and denote by $q_0$ the value of $q$ which satisfies Assumption 1. Let $x = q - q_0$. At $x = 0$ there is one purely imaginary eigenvalue which is equal to $iv$, where $v$ is a positive number.

**Assumption 5.7.2.** Let $(2q_0 - r)(mg_1 - vg_2)$ be not equal to $2v(mg_2 + vg_1)$, where $g_i$ are determined by

$$\begin{aligned}
&\left[mf''(iv + r)(r - m + iv)(m + iv) + r(iv + m)^2\right.\\
&\left.\times(r - m + iv)^2 - if''vm(iv + r)\right](g_1 - ig_2)\\
&= (iv + m)(r - m + iv)^2 \,.
\end{aligned} \tag{5.7.8}$$

This assumption guarantees the loss of stability of the equilibrium when $q$ is shifted from $q_0$.

**Theorem 5.7.1.** Let Assumptions 1 and 2 hold. There then exists a limit cycle of period $2\pi/s(\varepsilon)$ near the equilibrium. The cycle is approximately given by

$$\begin{aligned}
A(\varepsilon, t) &= A_0 + 2\varepsilon m \cos\left[s(\varepsilon)t\right] + O(\varepsilon^2) \,,\\
L(\varepsilon, t) &= L_0 + 2\varepsilon \left\{m \cos\left[s(\varepsilon)t\right] - v \sin\left[s(\varepsilon)t\right]\right\} + O(\varepsilon^2) \,,\\
\alpha(\varepsilon, t) &= \alpha_0 + 2\varepsilon \left\{\frac{v^2 f'' \cos\left[s(\varepsilon)t\right]}{(r - m)^2 + v^2} - \frac{vf''(r - m)\sin\left[s(\varepsilon)t\right]}{(r - m)^2 + v^2}\right\} + O(\varepsilon^2) \,,\\
\beta(\varepsilon, t) &= \beta_0 + 2\varepsilon\left\{(mq_0 - v^2)\cos\left[s(\varepsilon)t\right]\right.\\
&\qquad \left.-v(m + q_0)\sin\left[s(\varepsilon)t\right]\right\} + O(\varepsilon^2) \,,
\end{aligned} \tag{5.7.9}$$

where $\varepsilon$ is the expansion amplitude parameter, and $x(\varepsilon)$ and $s(\varepsilon)$ are given by

$$\begin{aligned}
x(\varepsilon) &= \frac{x_2 \varepsilon^2}{2} + O(\varepsilon^4) \,,\\
s(\varepsilon) &= v + \frac{s_2 \varepsilon^2}{2} + O(\varepsilon^4) \,,
\end{aligned} \tag{5.7.10}$$

in which $x_2$ and $s_2$ are real numbers dependent on the parameters of the system. Moreover, if $N_1$ is positive, the cycle is stable, while if $N_1$ is negative it is unstable. The three numbers $x_2$, $s_2$ and $N_1$ are explicitly given in Zhang (1988f).

Even from (5.7.2), we know that if $L(t)$ is periodic, $A(t)$ is oscillatory with the same period as that of $L(t)$. An increase in the actual employment level $L(t)$ does not imply that the normal employment level $A(t)$ becomes higher. This can be seen easily from the following relation

$$\delta A(t) = \delta L(t) + v(t) \sin\{s(\varepsilon)t\} + O(\varepsilon) , \tag{5.7.11}$$

obtaines from (5.7.9). In (5.7.11), $\delta A = (A - A_0)/\varepsilon$, $\delta L = (L - L_0)/\varepsilon$. The existence of limit cycles is due to the behavior of the government which determines the normal employment level by taking the historical employment level of the firm into account.

When the hiring/firing rate $v$ is equal to zero, then $\delta A = \delta L$. That is, the change in the normal level forecast by the government is equal to the actual change of the employment level.

## 5.8 Optimal Economic Growth Associated with Endogenous Fluctuations

A very interesting example for the existence of business cycles is related to multi-sector models of optimal growth. It is well known that much effort has been devoted to these types of model (e. g., Cass and Shell 1976, Brock and Scheinkman 1976, Araujo and Scheinkman 1977). These types of models are potentially unstable. Recently, Benhabib and Nishimura (1979) and Zhang (1988b) re-examined the dynamic behavior of these models, using the Hopf bifurcation theorem.

Consider the following optimal growth problem

$$\max \int_0^{\infty} U\left[T(y, k)\right] \exp\{-(r - g)t\} dt , \tag{5.8.1}$$

subject to

$$\frac{dk_i}{dt} = y_i - gk_i , \quad i = 1, \dots , n , \tag{5.8.2}$$

where the vectors $y$ and $k$ represent output per capita and stocks of capital goods per capita, respectively, consumption is given by $c = T(y, k)$, and $U(T)$ is the utility derived from consumption. $g(\geqslant 0)$ and $r(\geqslant 0)$ are the rate of population growth and the rate of interest, respectively.

In order to prove the existence of Hopf bifurcation, the following six assumptions (also accepted here) were made by Benhabib and Nishimura (1979):

A1) All goods are produced nonjointly with production functions homogeneous to degree one, strictly quasi-concave for non-negative inputs, and twice differentiable for positive inputs;

A2) If we denote by $(K_{ij})$ the set of inputs used in the production of good $j$, then the $j$th good cannot be produced without $(K_{ij})$. Applying the maximum principle to the problem (5.8.1–5.8.2) yields

$$\frac{dk_i}{dt} = y_i - gk_i ,$$
$$\frac{dq_i}{dt} = -U'w_i + rq_i ,$$
$$q_i = U'p_i , \quad p_i = -\frac{\partial T}{\partial y_1} ,$$
$$w_i = \frac{\partial T}{\partial k_i} , \quad i = 1, \dots , n , \tag{5.8.3}$$

where $p_i$ and $w_i$ are the price and rental of the $i$th good in terms of the price of the consumption good. They are uniquely determined for $r \in (g, r^*)$, where $r^*$ is given and may be positively infinite. There exists a unique equilibrium.

A3) At the steady state, the capital coefficient matrix is indecomposable;

A4) At the steady state, direct labor and at least one capital input are required in production of the consumption good;

A5) Near the steady state, marginal utility of consumption is constant, i. e., $U'' = 0$, and $U' = 1$; and

A6) The input coefficient matrix is nonsingular near the steady state.

These assumptions are explained by Benhabib and Nishimura (1979). (A3–A6) are assumed to describe the system in the local form.

It can be proved that if (A1–A6) hold, then we have: (i) $T(y, k)$ is twice differentiable; (ii) the dynamics near the steady state is given by

$$\frac{dk_i}{dt} = y_i(k,p) - gk_i ,$$
$$\frac{dp_i}{dt} = -w_i(k,p) + rp_i , \tag{5.8.4}$$

where $y$ and $w$ are differentiable; (iii) the steady state $(k_0, p_0)$ of (5.8.4) is uniquely determined for $r \in (g, r^*)$ where $g < r^* < +\infty$; (iv) the functions $c(r)$, $p(r)$, $k(r)$ and $y(r)$ of $r$ at the steady state are all positive and continuous for $r \in (g, r^*)$; and (v) $T(k, y)$ is strictly concave in $y$ for fixed $k$ near the steady state.

**Assumption 5.8.7.** Let there exist a value of $r$ denoted by $r_0$, such that the Jacobian at the equilibrium has one pair of conjugate eigenvalues $z_{1,2} = \alpha(r) \pm \mathrm{i}\beta(r)$ which, at $r = r_0$, satisfy $\alpha(r_0) = 0$, $\beta(r_0) > 0$, and $d\alpha(r_0)/dr$ is not equal to zero.

Clearly, Assumption 5.8.7 is required in Theorem 2 by Benhabib and Nishimura (1979). We select $r$ as bifurcation parameter with the critical value $r_0$. Introduce $x = r - r_0$ and denote $z(x) = \alpha(x) + \mathrm{i}\beta(x)$. Let $L(x)$ stand for the Jacobian of the system (5.8.4) evaluated at the equilibrium.

**Assumption 5.8.8.** Let $\pm\mathrm{i}\beta_0$ be simple, isolated eigenvalues of $L(0)$.

**Assumption 5.8.9.** All real parts of the other eigenvalues except $z_{1,2}(r)$ of $L(0)$ are negative.

**Assumption 5.8.10.** The strict loss of stability condition can be guaranteed, i. e., $R_1$ is not equal to zero where $R_1 = \text{Re}(R)$

$$R = \sum_{i=n+1}^{2n} Y_i \bar{Y}_i^* ,$$

and $Y$ and $Y^*$ are solutions of

$$\begin{aligned} &L(0)Y = \mathrm{i}\beta_0 Y , \quad L^{\mathrm{T}}(0)Y^* = -\mathrm{i}\beta_0 Y^* , \\ &\langle Y, Y^* \rangle = 1 , \quad \langle Y, \bar{Y}^* \rangle = 0 , \end{aligned} \tag{5.8.5}$$

where $\langle , \rangle$ is the product operator in $C^{2n}$.

If Assumption 5.8.8 holds, we can always find a pair of $Y$ and $Y^*$ which satisfy (5.8.5). $Y$ and $Y^*$ can simply be determined by solving algebraic equations, although the calculation may be tedious. Assumption 5.8.9 implies that all of the eigenvalues are imaginary since according to Proposition 3 in Benhabib and Nishimura (1979) the real eigenvalues of $L(0)$ come in non-zero pairs of opposite sign.

Introducing real numbers: $R_2 = \text{Im}(R)$, $N_1 = \text{Re}\{[N(U^1, U^2), X^*]\}$, $N_2 = \text{Im}\{[N(U^1, U^2), X^*]\}$, where $[N(U^1, U^2), X^*]$ is given in Zhang (1988b), we have

**Theorem 5.8.1.** Let the optimization problem satisfy Assumptions 1–10, $y(k,p)$ and $w(k,p)$ be $C^\Theta$, $\Theta \geqslant 3$. Then there exist limit cycles bifurcating from the equilibrium $(k_0, p_0)$ with the bifurcation parameter $r$ of the critical value $r_0$. The bifurcating cycles, with period $2\pi/s(\varepsilon)$, are explicitly expressed as

$$\begin{pmatrix} k(\varepsilon,t) \\ p(\varepsilon,t) \end{pmatrix} = \begin{pmatrix} k_o \\ p_0 \end{pmatrix} + 2\varepsilon \left[\cos\{s(\varepsilon)t\} \,\text{Re}\,(Y) - \sin\{s(\varepsilon)t\} \,\text{Im}\,(Y)\right] + \mathrm{O}(\varepsilon^2) , \tag{5.8.6}$$

where $\varepsilon$ is the expansion amplitude parameter and

$$\begin{aligned} x(\varepsilon) &= -\frac{\varepsilon^2 N_1}{2R_1} + \mathrm{O}(\varepsilon^4) , \\ s(\varepsilon) &= \beta_0 - \frac{\varepsilon^2 R_2 N_1}{2R_1} + \frac{\varepsilon^2 N_2}{2} + \mathrm{O}(\varepsilon^4) . \end{aligned} \tag{5.8.7}$$

Moreover, if $N_1$ is positive, the bifurcating cycle is unstable. If $N_1$ is negative, the cycle is supercritical or subcritically asymptotically stable according to whether $R_1$ is negative or positive.

Perfect competition and utility maximization are still held for these results. Instability does not mean that the economic system must destroy itself due to unstable growth. However, economic phenomena in such unstable systems may be very complicated.

## 5.9 Remarks on Possible Further Bifurcations from Limit Cycles

We provided some examples to show that business cycles can occur from various economic mechanisms. The examples cited above belong to different economic theories. Evolution with instabilities is not limited to a special market or economic mechanism. These results also shed light on observed economic oscillations. They prove that such fluctuations can be created endogenously from purely economic mechanisms.

We will mention some analytical problems in improving the results in this chapter. Consider a nonlinear dynamic system of high dimensions

$$\frac{dx}{dt} = f(x, r) ,$$

where $r$ are parameters. Assume that at a value $r_0$ of $r$ the equilibrium exists and the corresponding Jacobian has $n$ pairs of purely imaginary eigenvalues and $m$ zero eigenvalues. We are interested in the behavior of the dynamic system when $r$ shifts from $r_0$. Although mathematicians have investigated the problem under different conditions, the problem is not yet completely solved. Hopf bifurcations are a special case of the question. In the next chapter, we consider the case of two pairs of pure imaginary eigenvalues at a critical point. Moreover, it was shown by Huseyin (1986) that if the system is located at a critical point where the Jacobian of the system exhibits a double zero eigenvalue of index one and a pair of pure imaginary eigenvalues, it may exhibit static bifurcations, Hopf bifurcations, secondary Hopf bifurcations, and bifurcations into two- or three-dimensional tori.

Another question is as follows. Suppose that we have found stable limit cycles as in the preceding examples. As it is impossible to fix parameters permanently, it is important to know what will happen if the system becomes unstable because of shifts in parameters. Secondary bifurcation may occur. Much work has been carried out on this problem (e. g., Haken 1983, Chow and Hale 1982). In what follows, we mention some cases of further bifurcations. The discussion is referred to in detail in Iooss and Joseph (1980).

Consider the following autonomous differential equation

$$\frac{dx}{dt} = f(r, x) , \quad x \in C^n , \tag{5.9.1}$$

where $f$ is assumed to be sufficiently smooth and $f(r, 0)$ need not vanish. We look for the conditions under which subharmonic solutions – $nT$-periodic solution with integer $n \geqslant 1$ – can bifurcate from $T$-periodic solutions. All of the examples provided above can be re-examined to find more complicated behavior in this direction. However, it is sufficient for us to illustrate how further analysis can be carried out.

The given periodic solutions are of the form

$$x = X\big(w(r)t, r\big) = X(s, r) = X(s + 2\pi, r) ,$$

where $w(r)$ is the frequency. For instance, $X(s, t) = x_0(r) + U(s, r)$ may be the Hopf solution in the preceding examples, where $x_0$ is the equilibrium point and $U$

a periodic function. $X$ satisfies

$$w(r)\frac{dX}{ds} = f(r, X) . \tag{5.9.2}$$

Introducing $x = X(s,r) + v(t)$, $s = w(r)t$, we have the linearized problem

$$\frac{dv}{dt} = f_v\big(r, X(s,r) \mid v\big) , \tag{5.9.3}$$

where $f_v(r, X(s,r) \mid v)$ is the first derivative of $f(r,x)$, evaluated at $x = X(s,r)$, acting on $v$. Floquet theory implies that we may ascertain the stability of the cycle $X(s,r)$ by study of the exponents $z(r) = a(r) + ib(r)$ in the representation: $v(t) = L(s)\exp(zt)$, $L(s) = L(s+2\pi)$. The exponents are eigenvalues of the spectral problem

$$zL = w(r)\frac{dL}{ds} + f_v\big(r, X(s,r) \mid L\big) . \tag{5.9.4}$$

The adjoint problem is defined by

$$zL^* = w(r)\frac{dL^*}{ds} + F_v^*\big(r, X(s,r) \mid L^*\big) ,$$

where $F_v^*\big(r, X(s,r) \mid .\big)$ is the unique linear operator satisfying

$$\big\langle F_v^*\big(r, X(s,r) \mid A\big), B\big\rangle = \big\langle A, F_v^*\big(r, X(s,r)\big) \mid B\big\rangle ,$$

where $\langle , \rangle$ is the product operator in $C^n$ and $A$ and $B \in C^n$. Suppose that the periodic solution is at criticality at $r = r_0$, i.e., $a(r_0) = 0$, $b(r_0) = b_0$. Introduce $w(r_0) = w_0$, $X(s, r_0) = X_0(s)$ and

$$J_0 = -w_0\frac{d}{ds} + f_v\big(r_0, X_0(s) \mid .\big) ,$$

$$J_0^* = w_0\frac{d}{ds} + f_v^*\big(r_0, X_0(s) \mid .\big) . \tag{5.9.5}$$

The spectral problems at criticality are

$$\mathrm{i}b_0 L_0 = J_0 L_0 , \quad -\mathrm{i}b_0 L_0^* = J_0^* L_0^* . \tag{5.9.6}$$

If the Floquet exponent $\mathrm{i}b_0$ is an eigenvalue of $J_0$ at criticality, then $\mathrm{i}(b_0 + jw_0)$ where $j$ is an integer, is also an eigenvalue with eigenvector $Y(s) = \exp(-\mathrm{i}js)L_0(s) = Y(s+2\pi)$. The Floquet multiplier at criticality

$$k_0 = \exp\big[z(r_0)T(r_0)\big] = \exp\left[\mathrm{i}(b_0 + nw_0)\frac{2\pi}{w_0}\right] = \exp\left(\frac{2\pi \mathrm{i} b_0}{w_0}\right) , \tag{5.9.7}$$

maps repeated points on the imaginary axis of the complex $z$-plane into unique points of the complex $k$-plane. We may cover the unit circle of the $k$-plane by restricting our considerations to the principal branch

$$0 \leqslant \frac{b_0}{w_0} < 1 , \tag{5.9.8}$$

where $b_0$ and $w_0$ are defined as above. We say that

$$\frac{b_0}{w_0} = \frac{m}{n}, \tag{5.9.9}$$

satisfying (5.9.8) is in the set of rational points if $m$ and $n$ are integers and $m = 0$ when $n = 1$; otherwise $m$ is not equal to zero. Here, we are only concerned with the case in which (5.9.9) holds.

**Assumption 5.9.1.** We assume that if $b_0 = 0$, then 0 is an isolated double eigenvalue of $J_0$; while if $b_0$ is not equal to 0, $ib_0$ is an isolated simple eigenvalue of $J_0$.

As shown by Iooss and Joseph, the strict crossing condition will suffice to guarantee the existence of subharmonic bifurcation of $2\pi/w(r)$ periodic solutions. By a strict crossing we mean that $a_r(r_0) > 0$. The conditions for strict loss of stability are investigated in Iooss and Joseph (1980).

Suppose that Assumption 5.9.1 holds with $b_0/w_0 = m/n$ along with the strict crossing conditions. We then have the following results.

i) When $n = 1$ a single, one-parameter $h$ family of $2\pi/s^*(h)$-periodic solutions of (5.9.1) bifurcates on both sides of criticality. When $n = 2$ a single, one-parameter $h$ family of $4\pi/s^*$-periodic solutions bifurcates on one side of criticality. Supercritical ($r(h) > 0$) bifurcating solutions are stable; subcritical ($r(h) < 0$) bifurcating solutions are unstable;
ii) When $n = 3$ a single, one-parameter family of $6\pi/s^*(h)$-periodic solutions bifurcates and is unstable on both sides of criticality;
iii) When $n = 4$, $m = 1$ or 3, then two one-parameter families of $8\pi/s^*(h)$-periodic solutions bifurcate. The stability depends on details of the problem.
iv) When $n \geqslant 5$, there is in general no small-amplitude $2n\pi/s^*(h)$-periodic solution near criticality.

It should be noted that the conclusions (iii) and (iv) are true when some additional conditions are satisfied. In all of these cases, $s^*(h)$ is such that $s^*(0) = w_0$, so bifurcating solutions have periods close to a multiple of $2\pi/w(h)$.

The reader may try to investigate secondary bifurcation for the economic problems provided in this chapter.

## 5.10 Competitive Business Cycles in an Overlapping Generations Economy – A Discrete Model

Since Gale (1973) pointed out that equilibrium cycles may arise in overlapping generations models, it has been known that a perfectly competitive economy can exhibit persistent fluctuations under "laissez-faire" (see, for instance, Jullien 1988, Grandmont 1985, 1986). In what follows, we describe a discrete dynamic economic model which exhibits complicated behavior. The following model is due to Jullien (1988).

It is assumed that the economy produces a single good which can be consumed during the period or stored as an input (capital good) for future generations. Each generation lives two periods and reproduces identically. The young generation sells one unit of labor inelastically at a real wage $w_t$, consumes the quantity $C_{1t}$, and saves the real quantity $S_t$ for next period consumption by holding money and capital. The old generation spends all its savings from the previous period. A typical consumer solves the following optimization problem

$$\begin{aligned} &\max U(C_{1t}, C_{2t}) \\ &\text{s.b. } C_{1t} + S_t \leqslant w_t , \\ &\qquad C_{2t} \leqslant R_{t+1} S_t , \quad C_{1t},\ C_{2t} \geqslant 0 , \end{aligned} \tag{5.10.1}$$

where $U$ is the utility function, $R_{t+1}$ is referred to as the real rate of interest between $t$ and $t+1$. Under certain assumptions, the problem has a unique solution characterized by the savings function $S(w_t, R_{t+1})$.

The economy is initially endowed with a fixed quantity of capital $k_0$ and a quantity of money $M$. Production is made through a neoclassical constant return to scale technology. Output per capital is denoted by $y_t = f(k_t)$. We make the following hypotheses on properties of $S$ and $f$.

H1) $0 < S(w,R) < w$, $S$ is continuously differentiable, increasing with $w$, and $RS$ is increasing with $R$,

$$\lim_{w\to\infty} S(w,R) = \infty , \qquad \lim_{R\to\infty} RS(w,R) = \infty .$$

H2) $f$ is increasing, strictly concave on $R_+$, and $C^2$ on $R_+^*$,

$$\begin{aligned} &\lim_{k\to\infty} f'(k) \in (0,1) , \qquad \lim_{k\to 0} f'(k) = \infty , \\ &\lim_{k\to\infty} f(k) - kf'(k) = \infty , \qquad \lim_{k\to 0} f(k) - kf'(k) = 0 \end{aligned}$$

and $kf'(k)$ is non-decreasing.

The competitive behavior of firms leads to the equalization of the marginal productivity of each sector to its cost

$$R_t = f'(k_t) , \qquad w_t = f(k_t) - k_t f'(k_t) = W(k_t) . \tag{5.10.2}$$

**Definition 5.10.1.** A perfect foresight equilibrium is a sequence of prices $(p_t)_{t\geqslant 0}$, capital stocks $(k_t)_{t\geqslant 0}$, interest rates $(R_t)_{t\geqslant 0}$, and real wages $(w_t)_{t\geqslant 0}$ which achieves a competitive temporary equilibrium with perfect foresight at each date.

The conditions for a perfect foresight equilibrium are

$$\begin{aligned} &m_t + k_{t+1} = S\big(W(k_t), f'(k_{t+1})\big) , \\ &m_{t+1} = f'(k_{t+1}) m_t , \quad k_t > 0, \quad m_t \geqslant 0, \quad \text{at each } t \geqslant 0 , \end{aligned} \tag{5.10.3}$$

where $m_t = M/p_t$. We call an equilibrium non-monetary or monetary according to whether $m_0 = 0$ or $m_0 > 0$. Combining the two equations, we may rewrite the first equation of (5.10.3) as $m_{t+1} + k_{t+1}f'(k_{t+1}) = S(W(k_t), f'(k_{t+1}))f'(k_{t+1})$, from which we can obtain $k_t = g(k_{t+1}, m_{t+1})$. It can be easily shown that $g$ is $C^1$, increasing in each of its arguments, and $g(k,m)$ tends to 0 (infinity) when $k$ goes to 0 (infinity) with $m$ fixed. Now, the perfect foresight equilibrium can be rewritten as

$$(k_t, m_t) = F(k_{t+1}, m_{t+1}) ,$$
$$k_t > 0 , \quad k_0 \quad \text{given} \tag{5.10.4}$$

where $F(k,m) = (g(k,m), m/f'(k))$ for $k > 0$ and $m \geqslant 0$.

We introduce $X_t = (k_t, m_t)$. A monetary steady state, $X^* = (k^*, m^*)$, is defined by a solution of $f'(k^*) = 1$, $m^* = S(W(k^*), 1) - k^*$. From $f'(k^*) = 1$ we have a well defined solution $k^*$. We can prove that if $S(W(k^*), 1) > k^*$, then under H1 and H2 there exist a unique monetary steady state $X^*$ and at least one inefficient non-monetary steady state ($k > k^*$). Only one point should be noted, i. e., a non-monetary steady state is a point where $S(W(k), f'(k)) = k$.

In order to demonstrate the existence of periodic solutions to the map (5.10.4), Jullien first reduced the two-dimensional problem to a one-dimensional problem and then proceeded to identify the existence of these solutions. We now reduce the dimension.

Define $F^n(X) = (k_n(X), m_n(X))$, which is bounded since $W(g(k,m)) > m/f'(k) + k$.

**Theorem 5.10.1.** There exists a compact set $K \subseteq R_+^{*2}$ and a function $h$, decreasing and $C^1$, from $R_+^*$ to $R_+^{*2}$ such that if we define the sets

$$\Gamma = \left\{(k,m) \in R_+^{*2} \mid k = h(m)\right\} ,$$
$$\Gamma_+ = \left\{(k,m) \in R_+^{*2} \mid k > h(m)\right\} ,$$
$$\Gamma_- = \left\{(k,m) \in R_+^{*2} \mid k < h(m)\right\} ,$$

then $\{\Gamma, \Gamma_+, \Gamma_-\}$ is an $F$-invariant partition of $R_+^{*2}$ and

$$X \in \Gamma_+ \iff \lim_{n\to\infty} m_n(X) = \infty , \quad X \in \Gamma_- \iff \lim_{n\to\infty} m_n(X) = 0 ,$$
$$X \in \Gamma \iff F^n(X) \in K \quad \text{for large enough} \quad n .$$

**Theorem 5.10.2.** If $(X_t)_{t\geqslant 0}$ is a competitive perfect foresight equilibrium, then at least one of the following two situations must prevail: (i) $X_t$ converges to $\Gamma$ as $t \to \infty$; or (ii) $X_t$ converges to an inefficient non-monetary steady state as $t \to \infty$.

The proofs are given in Jullien (1988). Theorem 5.10.2 guarantees that all cycles must belong to the curve $\Gamma$. This allows us to restrict the dimension of the map to one by focusing on the dynamics on $\Gamma$. We define a new function

$$\Phi : R_+^* \to R_+^*$$
$$m \to \frac{m}{f'\big(h(m)\big)} .$$

The function $\Phi$ has the following properties: $\Phi$ is $C^1$, $\Phi(m) - m > 0$ for $m < m^*, \Phi(m) - m < 0$ for $m > m^*$. Let $K_\Phi$ be the projection of $K$ on the second axis. The set $K_\Phi$ is a $\Phi$-invariant compact set belonging to $R_+^*$.

From Theorem 3.7.1, it can be proved that if $X = (k, m)$ is a cyclical point of order $p$, then $X$ belongs to $\Gamma$ and $V((1, h'(m))$ is the eigenvector of $DF^n(X)$ associated with its smallest eigenvalue (it has two distinct eigenvalues). According to the definition of a cyclical solution, $F^p(X) = X$ with $p$ the smallest integer for it to hold. Then $X$ belongs to $\Gamma$ because its orbit is bounded. Since $V$ is tangent to $\Gamma$ at $X$ and $\Gamma$ is $F$-invariant, $DF^p(X)V$ is also tangent to $\Gamma$ at $X$. Hence, there exists $\alpha$ such that $DF^p(X)V = \alpha V$. $V$ is an eigenvector of $DF^p(X)$. But $DF^p(X)$ has all the elements positive and two distinct real eigenvalues. The coordinates of the eigenvector associated with the largest eigenvalue have the same sign. The coordinates of the eigenvector associated with the smallest eigenvalue are of opposite signs. $h'(m)$ is negative, so $\alpha$ must be the smallest eigenvalue.

The cycles of $\Phi$ are equivalent to the cycles of $F$ in the sense that

$$m \quad \text{is not equal to} \quad 0\,,$$

$$F^n(k,m) = (k,m) \iff k = h(m) \quad \text{and} \quad \Phi^n(m) = m\,.$$

**Theorem 5.10.3.** A sufficient condition for the existence of a cycle of order 2 is that

$$\begin{aligned} S\big(W(k^*),1\big) - k^* + 2S_R\big(W(k^*),1\big) + 2k^*S_w\big(W(k^*),1\big) \\ - 2/f''(k^*) < 0\,. \end{aligned} \tag{5.10.5}$$

*Proof.* A direct calculation shows that equation (5.10.5) is equivalent to $\Phi'(m^*) < -1$. It can be shown that for $m$ sufficiently small, $\Phi^2(m)$ greater than $m$ and equivalent to $m/f'(k_s)$, where $k_s = \inf\{k > k^* \mid g(k,0) = k\}$. $\Phi'(m^*)^2 > 0$ so that for $m$ smaller than but close to $m^*$, $\Phi^2(m) < m$. By continuity there exists a point $m$ between 0 and $m^*$ such that $\Phi^2(m) = m$. □

It should be noted that Jullien (1988) gives conditions for the existence of cycles of order 3 and provides some examples for 2-period and 3-period cases.

# 6. Economic Chaos in Deterministic Systems

All depends, then on finding out these easier problems and on solving them by means of devices as perfect as possible and concepts capable of generalization.

*D. Hilbert*

In the previous chapter, we investigated business cycles which are created by different economic mechanisms. However, the practically observed economic data hardly exhibit such regular oscillations. Economic variables often appear to fluctuate irregularly. This chapter will explain such endogenous "chaotic" economic phenomena. We show that there is order in chaos; the "random" economic behavior may have an underlying geometric form. Such phenomena are deterministic, generated by fixed rules that do not involve any elements of chance. In principle the future is completely determined by the past; in practice it is almost impossible to forecast the future precisely in a chaotic world.

## 6.1 Chaos in Deterministic Systems

According to the *Encyclopedia Britannica* the word "chaos" is derived from the Greek and originally meant the infinite empty space which existed before all things. The later Roman conception interpreted chaos as the original crude shapless mass into which the Architect of the world introduces order and harmony.

In this study the word in its technical sense refers to irregular motion that is generated by nonlinear systems whose dynamic laws uniquely determine the time evolution of a state of the system from a knowledge of its previous history.

By "deterministic motion" we mean that there exists a prescription, either in terms of differential or difference equations, for calculating the future behavior of a system from given initial conditions. Although we are used to the assumption that deterministic motion is rather regular and far from being chaotic, it was already discovered at the turn of the century by H. Poincaré that certain mechanical systems whose time evolution is governed by Hamilton's equations could display chaotic motion. Unfortunately, this was considered as a mere curiosity, and it took another 70 years until, in 1963, E.N. Lorenz found that even a simple set of three coupled, first-order, nonlinear differential equations can lead to completely chaotic trajectories. Lorenz discovered one of the first examples of deterministic chaos in dissipative systems.

"Deterministic chaos" is now a very active field of research. Many methods have been developed to classify different types of chaos. It should be emphasized that there is no generally accepted definition of chaos, and in the literature chaos is often referred to in the context of a dissipative system as the phenomenon related to the occurrence of randomness and unpredictability in completely deterministic systems, which has been called "dynamical stochasticity", "deterministic chaos", "self-generated noise", "intrinsic stochasiticity" and "Hamiltonian stochasticity" (see Hao 1984, Guckenheilmer and Holmes 1983, Wiggins 1988). In the appendix to this chapter, we provide some criteria to distinguish chaos from regular motion such as limit cycles and aperiodic solutions.

Up to now, there are at least three routes or transitions by which nonlinear systems can become chaotic if an external control parameter is varied (Schuster 1988). All of these routes can be realized experimentally, and they show a fascinating universal behavior which is reminiscent of the universality found in second-order equilibrium transitions. The first route to chaos has recently been found by Grossmann and Thomae (1977), Feigenbaum (1978), and Coullet and Tresser (1978). They considered a simple difference equation which, for example, has been used to describe the time dependence of populations in biology, and found that the population oscillated in time between stable values (fixed points) whose numbers also serve as distinct values of an external parameter. This continues until the number of fixed points becomes infinite at a finite parameter value, where the variation of population in time becomes irregular. The second approach, known as the intermittency route, has been discovered by Manneville and Pomeau (1979). Intermittency means that a signal which behaves regularly in time becomes interrupted by statistically distributed periods of irregular motion (intermittent bursts). The average number of these bursts increases with the variation of an external control parameter until the motion becomes completely chaotic.

The third route was found by Ruelle and Takens (1971) and Newhouse et al. (1978). They suggested a transition to turbulent motion which was different from that proposed much earlier by Landau (1944) and Landau and Lifshitz (1959). Landau considered turbulence in time as the limit of an infinite sequence of instabilities (Hopf bifurcations), each of which creates a new basic frequency. However, Ruelle, Takens and Newhouse showed that after only two instabilities in the third step the trajectory becomes attracted to a bounded region of phase space in which initially close trajectories separate exponentially such that the motion becomes chaotic. These particular regions of phase space are called strange attractors. Figure 6.1a shows Landau's route to chaos and reveals that as the parameter $r$ increases, more and more fundamental frequencies are generated by Hopf bifurcations. In Fig. 6.1b, we describe the Ruelle-Takens-Newhouse route to chaos.

Time series of many economic variables exhibit noisy fluctuations. A traditional explanation of such fluctuating motion is that the economy is subject to random shocks. There are storms, earthquakes and similar exogenous phenomena which influence economic (e.g., agricultural) activities. The studies of chaos by mathematicians influenced the development of economics. Recently, economists have tried to interpret chaotic phenomena in terms of deterministic systems. Economic chaos is not necessarily created by exogenous shocks. Economic chaos can be created endogenously in a relatively simple nonlinear system.

We should like to mention some studies here about economic chaos. Benhabib and Day (1981, 1982) and Grandmont (1985) constructed overlapping generation

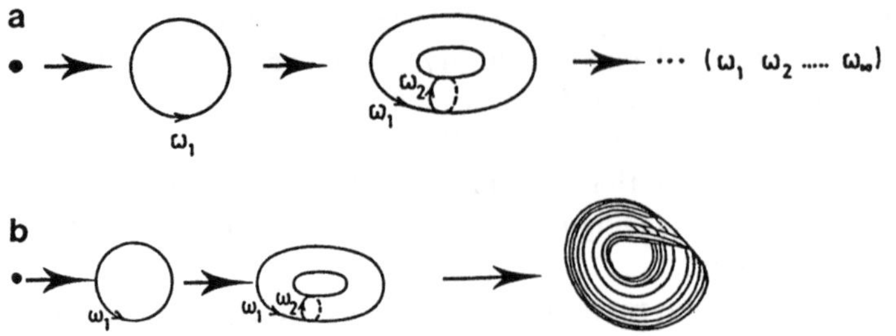


**Fig. 6.1.** (a) Landau's route to chaos. (b) The Ruelle-Takens-Newhouse route to chaos

models of monetary phenomena and quasi-dynamic models of consumer choice with endogenous preferences, which exhibit chaotic dynamics. Day (1983) considers classical economic growth which is a "Malthusian" type model where the level of output determines the rate of population growth. Output in turn depends on the size of the labor force. The interaction of these two effects may result in cycles or chaos. Dana and Montrucchio (1986) discuss the occurrence of periodic and chaotic phenomena in infinite horizon duopoly games where firms maximize their discounted sum of profits and use Markov-perfect equilibrium strategies. Chaos has also been established in neoclassical optimal growth models (e.g., Boldrin and Montrucchio 1986, Deneckere and Pelikan 1986). In what follows, we examine the way in which chaos can occur in deterministic dynamic systems.

## 6.2 Economic Chaos in a Discrete System

This section shows how some very simple equations may yield rather complicated dynamic behavior, We are concerned with one-dimensional discrete maps

$$x_{n+1} = f(x_n) \; .$$

It is well known that chaos may occur even if the discrete maps take on very simple forms.

The model to be discussed is proposed by Stutzer (1980). First, consider a macroeconomic growth model proposed by Haavelmo (1954)

$$\frac{dN}{dt} = N\left(a - \frac{\beta N}{Y}\right) \; , \quad a, \beta > 0 \; ,$$

$$Y = AN^{\alpha} \; , \quad A > 0 \; , \quad 0 < \alpha < 1 \; , \tag{6.2.1}$$

where $N$ is population, $Y$ is real output, and $a$, $\beta$, $\alpha$ and $A$ are constant parameters. Substituting the second equation into the first yields

$$\frac{dN}{dt} = N\left(a - \frac{\beta N^{1-\alpha}}{A}\right) \; . \tag{6.2.2}$$

We see that the growth law is a generalization of the familiar logistic form widely used in biological population and economic analysis. It is not difficult to see that the dynamics of this system are very simple. If the initial condition $N(0) > (<)$ $(aA/\beta)^{1/(2-\alpha)}$, then both $N$ and $Y$ will decrease (increase) monotonically until they approach their respective unique equilibria.

If we replace time derivatives by first differences and accept discrete time, then (6.2.2) may be rewritten as

$$N_{t+1} = N_t \left[(1+a) - \frac{\beta N_t^{1-\alpha}}{A}\right] ,$$

which can be further simplified as

$$x_{t+1} = (1+a)x_t \left(1 - x_t^{1-\alpha}\right) = F(x_t;\, a, \alpha) , \qquad (6.2.3)$$

in which the new variable is defined by the transformation

$$N_t = \left[\frac{A(1+a)}{\beta}\right]^{1/(1-\alpha)} x_t .$$

We will analyze the dynamics of (6.2.3).

Before analyzing the dynamics of the system, we define some basic concepts for a first-order difference equation: $x_{t+1} = F(x_t)$, where $F : J \to J$ is continuous, and $J$ is a closed and bounded interval of the real line. The $n$-fold composition of $F$ with itself is denoted by $F^n(x)$, with $F^0(x) = x$ denoting the identity map.

**Definition 6.2.1.** (*N-period Point*). A point $p \in J$ is termed a non-degenerate [degenerate] periodic point with period $n$, or an $n$-periodic point, if and only if $F^n(p) = p$ and $p \neq [=] F^k(p)$, for all [some] $1 \leq k < n$. A point $p \in J$ is termed periodic if it is an $n$-periodic point for some $n \geq 1$. A 1-periodic point is called a steady-state, or equilibrium, or fixed point of $F$.

**Definition 6.2.2.** (*Cycle, Period*). If $p$ is an $n$-periodic point, then each point in the collection of points $\{p, F(p), \ldots, F^{n-1}(p)\}$ is also an $n$-periodic point, and the collection is defined as the periodic orbit, or cycle, of $p$. If $p$ is non-degenerate, then each point in the periodic orbit is distinct, and the orbit is said to have length, or period, $n$.

For example, consider the simplest difference equation

$$c_{t+1} = hc_t ,$$

to which the solution is $c_t = c_0 h^t$. This grows exponentially if $|h| > 1$. When $0 < h < 1$, the system converges to a stationary state. In the case where $h = -1$ there is a 2-cycle and this occurs for a single value of the parameter.

**Definition 6.2.3.** (*Asymptotical Period*). A point $p$ is asymptotically periodic if there is a periodic point $q \neq p$ for which

$$\lim_{n \to +\infty} \left[F^n(q) - F^n(p)\right] = 0 .$$

**Definition 6.2.4.** (*Local Stability*). A $k$-period point $p$ and its corresponding periodic orbit are said to be locally asymptotically stable, if for some open interval $I$ about $p$,

$$|F^k(p) - x| < |p - x| \quad \text{for all } x \in I\,.$$

**Definition 6.2.5.** (*Chaotic Dynamics*). The term chaotic dynamics refers to the dynamic behavior of certain equations $F$ which possess: (a) a non-degenerate $n$-period point for each $n \geq 1$, and (b) an uncountable set $S \in J$, containing no periodic points and no asymptotically periodic points. The trajectories of such points wander around in $J$ "randomly".

For the rest of this section, we take $J = [0, 1]$. For simplicity, we specify $\alpha = 1/2$ in (6.2.3). In this case $F$ maps $J$ into itself. It should be mentioned that none of the qualitative properties are affected by the particular choice of $0 < \alpha < 1$. Thus the model is written as

$$x_{t+1} = (1 + a)x_t \left(1 - x_t^{1/2}\right) = F\left(x_t;\, a, \tfrac{1}{2}\right)\,. \tag{6.2.4}$$

The geometry of $F$ for different values of $a$ ($0 \leq a \leq 5.75$, and $x(0) \in [0, 1]$) is depicted in Fig. 6.2.

For each value of $a$, equilibrium points are given by the intersection of the graph of $F(x_t;\, a)$ with the 45-degree line in Fig. 6.2. For each value of $a$, there are two equilibria: $x_0 = 0$ and $x_0 = [a/(1 + a)]^2$. The point $x_0 = 0$ is unstable and repels nearby points. The local stability of the other can be determined by linearization at the equilibrium. We have

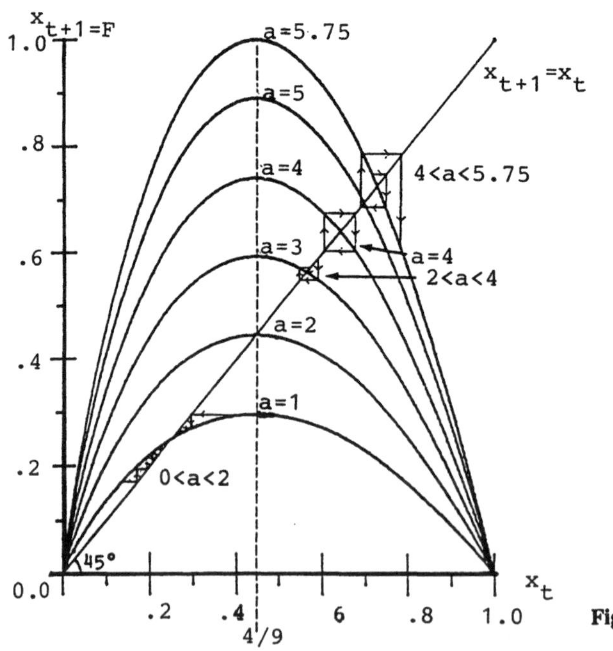


Fig. 6.2. Equilibrium and stability

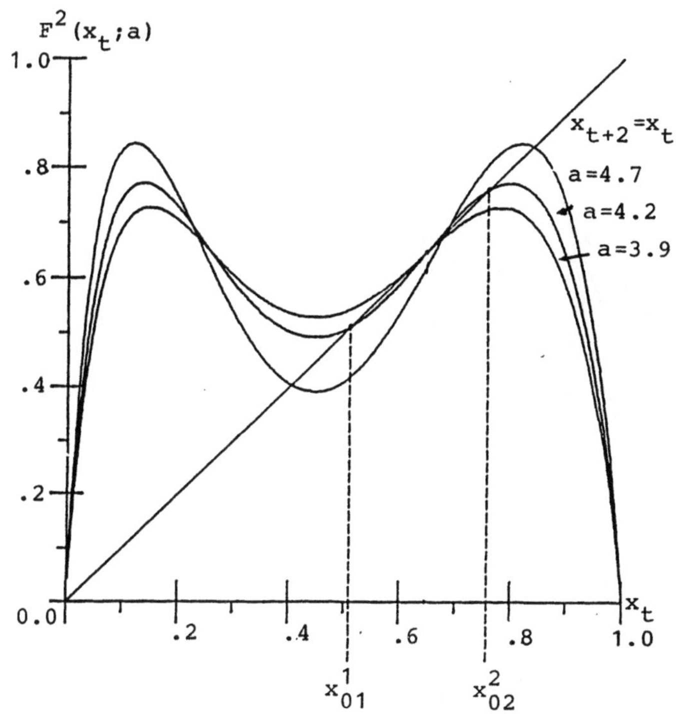


**Fig. 6.3.** Bifurcation of $x_0$ into a 2-period orbit

$$F'(x_0; a) = 1 - \frac{a}{2} = \Theta(a) . \tag{6.2.5}$$

The eigenvalue $\Theta(a)$ determines the local stability of $x_0$. When $0 < \Theta < 1$, $x_0$ attracts nearby points in an exponential, montonic fashion. When $0 > \Theta > -1$, $x_0$ attracts nearby points in a damped oscillatory manner. When $\Theta = -1$, $x_0$ is neither stable nor unstable. Finally, if $|\Theta| > 1$, $x_0$ is unstable. These behaviors occur when $0 < a < 2$, $2 < a < 4$, $a = 4$, and $4 < a < 5.57$, respectively. This is illustrated in Fig. 6.2.

When the equilibrium is stable, i.e., $a < 4$, the trajectory starting at any point always approaches it. In this region a traditional comparative statics analysis shows that an increase in the parameter $a$ will increase $x_t$ for sufficiently large $t$. If $4 < a < 5.75$, trajectories do not approach the equilibrium, but are bounded by 0 and 1. In fact, as the parameter $a$ exceeds 4, the unstable equilibrium point bifurcates into two stable points of period two, i.e., into a stable periodic orbit of length 2. For $a = 4.2$, Fig. 6.3 shows the two non-degenerate fixed points of $F^2(x; 4.2)$, labeled $x_{01}^2$ and $x_{02}^2$, respectively.

As shown by Stutzer, the 2-period cycle becomes unstable for values of $a$ in excess of about 4.8, and each 2-period point bifurcates into two 4-period points, producing a stable cycle of length four denoted by $\{x_{01}^4, x_{02}^4, x_{03}^4, x_{04}^4\}$. Figure 6.4 illustrates the phenomenon.

This pitchfork bifurcation process continues as the parameter $a$ increases, producing non-degenerate orbits of length $2k$ ($k = 2, \ldots$). These orbits are called

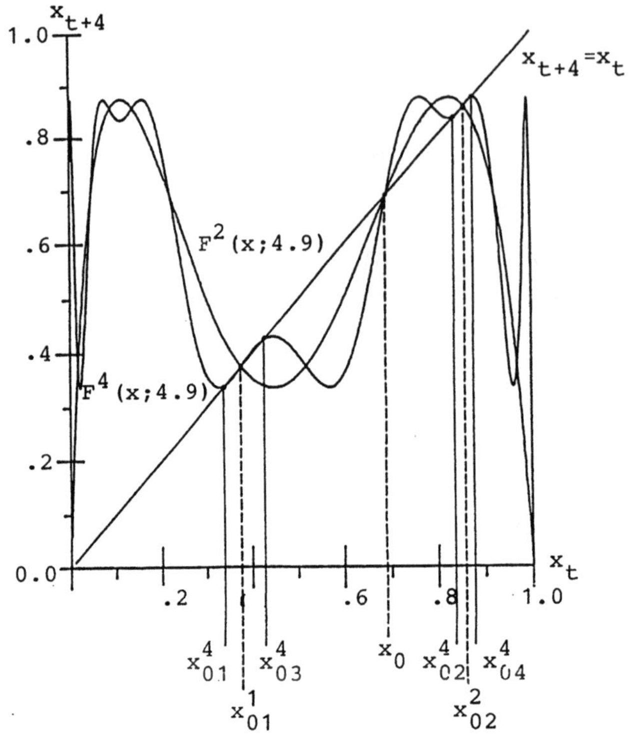


**Fig. 6.4.** The 4-period orbit for $a = 4.9$

harmonics of the 2-period orbit. It can be shown that all the harmonics occur prior to the parameter $a$ reaching 5.54, although how much prior to this value is not known. Thus, the range of $a$, within which a stable orbit of length $k$ first appears and later becomes unstable and bifurcates to a $2k$-period orbit, decreases in length as the parameter $a$ increases to a limiting value $a_c < 5.54$.

The range of $a_c < a \leq 5.75$ is termed the chaotic region. As the parameter $a$ enters this region, even stranger behavior can occur. For example, a 3-period orbit exists at values of $a$ near 5.540. This then gives rise to orbits of periods $3k$ ($k = 2, \ldots$) via the pitchfork process just described. In fact, if we can locate the 3-period orbit, a remarkable theorem of Li and Yorke (1975) demonstrates that for any $F(x_t;\, a)$ in which a non-degenerate 3-period orbit arises, there must also exist non-degenerate points of all periods, as well as an uncountable set of periodic (not asymptotically periodic) points whose trajectories wander "randomly" throughout the domain of $F$.

**Theorem** (*Li and Yorke*). Let $J$ be an interval and let $F : J \to J$ be continuous. Assume that there is a point $a_1 \in J$ for which the points $a_2 = F(a_1)$, $a_3 = F^2(a_1)$ and $a_4 = F^3(a_1)$, satisfy

$$a_4 \leq a_1 < a_2 < a_3 \quad (\text{or} \quad a_4 \geq a_1 > a_2 > a_3) \; .$$

Then: (i) For every $k = 1, 2, \ldots$ , there is a periodic point in $J$ having period $k$; (ii) there is an uncountable set $S \in J$ (containing no periodic points), which satisfies the following conditions:

A) For every $p, q \in S$ with $p$ not equal to $q$,

$$\lim_{n \to \infty} \sup |F^n(p) - F^n(q)| > 0 \text{ and}$$

$$\lim_{n \to \infty} \inf |F^n(p) - F^n(q)| = 0 .$$

B) For every $p \in S$ and periodic point $q \in J$,

$$\lim_{n \to \infty} \sup |F^n(p) - F^n(q)| > 0 .$$

Our dynamic economic system satisfies the requirements in the theorem for some values of $a$. The existence of chaotic behavior is illustrated in Fig. 6.5.

In summary, as the autonomous growth rate $a$ exceeds a certain value, the steady state ceases to be approached monotonically, and an oscillatory approach occurs. If $a$ is increased further, the steady state becomes unstable and repels nearby points. As $a$ increases, one can find a value of $a$ where the system possesses a cycle of period $k$ for arbitrary $k$. Also, there exist an uncountable number of initial conditions from which emanate trajectories that fluctuate in a bounded and aperiodic fashion and are indistinguishable from a realization of some stochastic (chaotic) process.

Relatively small changes in structural parameters can lead to large, qualitative changes in system behavior. Moreover, the evolution of nonlinear low-order sys-

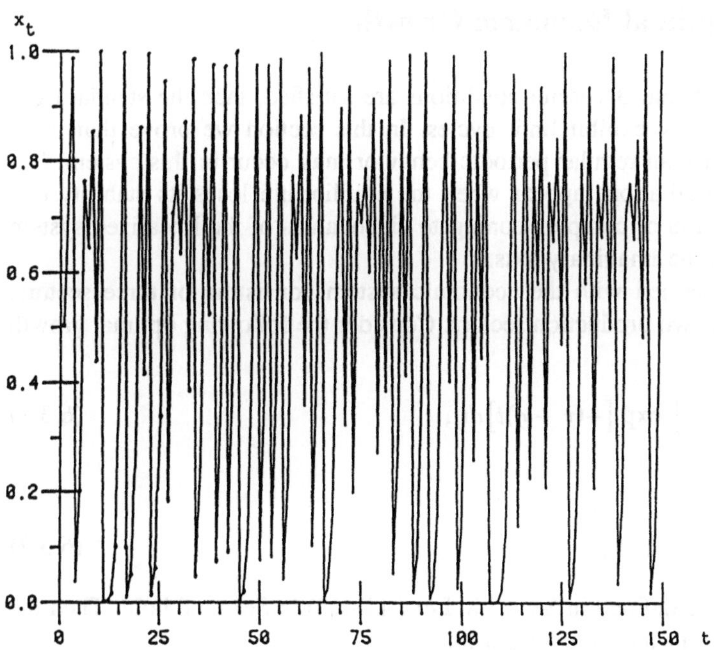


Fig. 6.5. The existence of chaos ($a = 5.75$, $\alpha = 1/2$)

tems can also be drastically affected by the initial conditions of the system. In the construction of models, this dependence is often overlooked. It can be concluded that qualitative changes caused by small changes in structural parameters and initial conditions, along with the possibility of measurement error in these variables, cast doubt on the ability to predict and control such nonlinear systems. Thus, even if the model specification is exact, prediction and control may be impossible in practice, due to unavoidable measurement error.

This example shows that a simple nonlinear first-order deterministic difference equation may exhibit chaotic, seemingly random fluctuations which might mistakenly be attributed to the influence of excluded variables or the influence of included, but assumed, random variables. Such phenomena cannot be observed in deterministic low-order linear difference equations. Chaos results from nonlinearity. Moreover, the results also mean that in the context of linear difference equation models of macroeconomic phenomena, the introduction of plausible, theoretically justifiable nonlinearities into the structural equations might explain observed economic fluctuations just as well, or better than, the addition of random variables.

The discrete time version of the original Haavelmo model has vastly different qualitative properties. No longer does the system always monotonically approach a steady state. This implies that the discrete time analog of a continuous time system cannot reliably be assumed to be found by replacing derivatives with first differences. Alternatively, if one is not sure which representation is the real system, these results stress the fundamental importance that choice of time domain and unit time length can have significant effects on the qualitative properties of models.

## 6.3 Aperiodic Optimal Economic Growth

In Chap. 5, we proved that if certain conditions are satisfied, then the standard optimal growth model may exhibit limit cycles. In this section we prove that more complicated behavior than regular periodic behavior may occur in this system. Endogenous irregular oscillations appear when the equilibrium loses its stability due to the fact that two pairs of complex conjugate eigenvalues of the linearized system simultaneously cross the imaginary axis.

We are only concerned with the economic system consisting of three sectors: one consumption and two production sectors. Consider the following optimal growth problem

$$\max \int_0^\infty U\left[T(y,k)\right] \exp\left[-(r-g)t\right] dt \,, \tag{6.3.1}$$

subject to

$$\frac{dk_i}{dt} = y_i - gk_i \,, \quad i = 1,2 \,, \tag{6.3.2}$$

where the variables are defined in Sect. 5.8. Let ($A1$–$A6$) in Sect. 5.8 hold. Then the system can be written in the following from

$$\frac{dk_i}{dt} = y_i(k, p) - gk_i \ ,$$

$$\frac{dp_i}{dt} = -w_i(k, p) + rp_i \ . \tag{6.3.3}$$

To guarantee the existence of aperiodic oscillations, we make the following assumption.

**Assumption 6.3.7.** Let system (6.3.3) possess two pairs of simple complex conjugate eigenvalues denoted by $z_{1'2}(r)$ and $z_{3'4}(r)$, respectively

$$z_{1'2}(r) = \alpha_1(r) \pm \beta_1(r) \ ,$$

$$z_{3'4}(r) = \alpha_2(r) \pm \beta_2(r) \ , \tag{6.3.4}$$

where $\alpha_i$ and $\beta_i$ are real numbers. It is assumed that there exists a value of $r$, denoted by $r_0$, such that

$$\alpha_1(r_0) = \alpha_2(r_0) = 0 \ , \quad \alpha_1(r)\alpha_2(r) > 0 \ , \quad r \neq r_0 \ ,$$

$$\frac{d\alpha_1(r)}{dr} \neq 0 \ , \quad \frac{d\alpha_2(r)}{dr} \neq 0 \ . \tag{6.3.5}$$

In this assumption, we require that $|r - r_0|$ is sufficiently small. It states that the Jacobian has two pairs of purely imaginary eigenvalues and all the eigenvalues lose stability simultaneously if there are perturbations which drive $r$ away from its critical value $r_0$. From the examples in Benhabib and Nishimura (1979), we see that this assumption is acceptable.

Let $x = r - r_0$ and write $z_i(x)$ instead of $z_i(r)$. From (6.3.5), we see that the real parts of the eigenvalues are always the same sign. If $x$ is varied in such a way that $\alpha_1(x)$ and $\alpha_2(x)$ change from negative to positive values, then the steady state loses its stability. According to bifurcation theory, new (possibly quite complicated) time-dependent solutions may be bifurcated from the branch of $(k_0, p_0)$. As $x$ crosses the stability boundary ($x = 0$) from the stable to unstable region, linearized stability theory predicts that the steady state loses its stability via an exponentially growing function of time $t$. This linearized exponentially growing function cannot represent the real solution of the system for very long as the nonlinear terms will become important as time passes. This is the reason that we have to take nonlinear terms of an unstable dynamic system into account.

Introduce an expansion amplitude parameter as follows:

$$\varepsilon^2 = \begin{cases} x \ , & \text{if } d\alpha_1(0)/dx > 0 \ , \\ -x \ , & \text{if } d\alpha_1(0)/dx < 0 \ . \end{cases} \tag{6.3.6}$$

Zhang (1989d) proved the following theorem

**Theorem 6.3.1.** Let the optimization problem satisfy (A1–A7). If $|\beta_1 - 2\beta_2|$, $|\beta_1 - \beta_2|$, $|2\beta_1 - \beta_2|$ are all O(1) with regard to $\varepsilon$, then

$$\begin{bmatrix} k(\varepsilon,t) \\ p(\varepsilon,t) \end{bmatrix} = \begin{bmatrix} k_0 \\ p_0 \end{bmatrix} + \varepsilon\left[C_1 R(m) \sin F\right.$$
$$\left. + C_2 R(m) \cos F + D_1 S(m) \sin G + D_2 S(m) \cos G\right] + \mathrm{O}\left(\varepsilon^2\right) \; , \qquad (6.3.7)$$

where $C_i$ and $D_i$ are constant 4-dimensional vectors and

$$A = \left(1 + w_2\varepsilon^2\right)\beta_1 t + A^*(m) \; ,$$
$$B = \left(1 + v_2\varepsilon^2\right)\beta_2 t + B^*(m) \; ,$$
$$m = \varepsilon^2 t \; , \qquad (6.3.8)$$

where $R(m)$, $S(m)$, $A^*(m)$ and $B^*(m)$ are scalar functions determined from

$$\begin{pmatrix} R' \\ RA^* + \beta_1 w_2 R \\ S' \\ SB^* + \beta_2 v_2 S \end{pmatrix} = \begin{pmatrix} R \\ S \\ R^3 \\ Y \\ R^2 S \\ RS^2 \\ S^3 \end{pmatrix} \qquad (6.3.9)$$

where $N = (n_{ij})_{4\times 6}$. $C_i$, $D_i$, $w_2$, $v_2$, $N$, are given in Zhang (1989d).

The proof of the theorem is given in Zhang (1989d).

The approximation (6.3.7) is valid at least for the time period of order $\mathrm{O}(1/\varepsilon^2)$. Stability of the solution is determined by the asymptotic behavior of $R(m)$ and $S(m)$ as $m \to \infty$. If they approach constant values or oscillate, then the bifurcating solution is stable. Calculating the parameters in the theorem is simple, though tedious.

The theorem describes possible irregular oscillations bifurcating from the equilibrium. In sharp contrast to the case of Hopf bifurcation at a pair of simple complex eigenvalues, it is possible for the time-dependent solution not to be periodic. The superposition of harmonics in $A$ and $B$ is not periodic if $\beta_1$ and $\beta_2$ are incommensurate. It is almost impossible to forecast the behavior of the system precisely. We illustrate the behavior of $p_i(\varepsilon, t)$ in Fig. 6.6 in which $\beta_1$ and $\beta_2$ are incommensurate.

The difference between the actual capital stock per capita and its equilibrium value consists of the sum of two parts: $C_{11}R(m) \sin A + C_{21}R(m) \cos A$ and $D_{11}S(m) \sin B + D_{21}S(m) \cos B$. Hence, if $\beta_1$ and $\beta_2$ are commensurate, we can observe regular periodic behavior, although during each period the behavior appears irregular.

It is interesting to study the behavior of the other variables in the system. If the theorem holds, then the output and consumption appear to be oscillatory, though their motion is bounded near the corresponding equilibrium values. Look at the dynamics of the capital stock $K_1(t)$. As $K_1(t) = k_1(\varepsilon, t)L_0 \exp(gt)$, where $L_0$ is the initial population, the capital stock grows oscillatorily as illustrated in Fig. 6.7. The capital stock may be far away from the equilibrium value if time $t$ becomes very large.

From the discussion above, we see that the variables move "randomly" near the equilibrium. They may be either periodic or aperiodic depending on the given initial conditions.

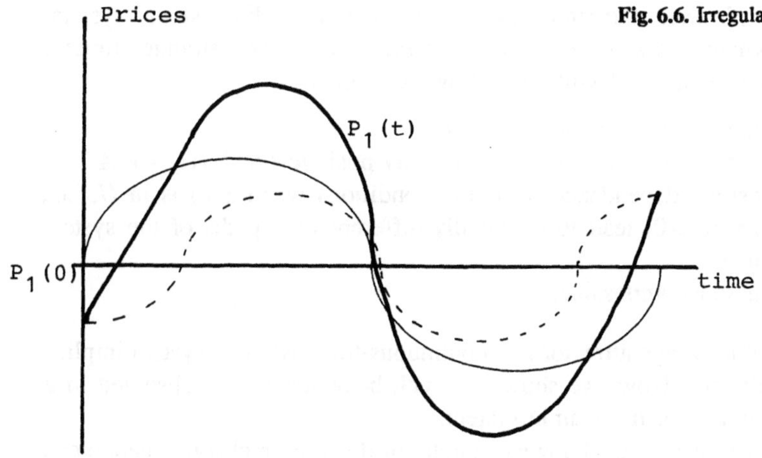


Fig. 6.6. Irregular oscillations in prices

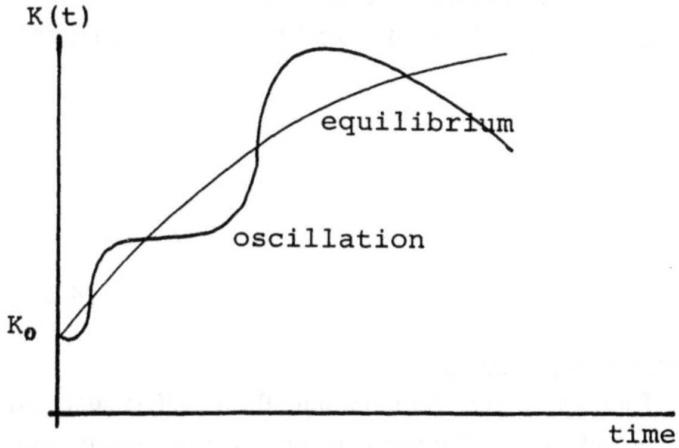


Fig. 6.7. Capital growth occurring far away from the equilibrium

## 6.4 Urban Dynamics – The Lorenz System

The previous examples showed a very important aspect of the evolutionary system. As time passes, new patterns of behavior are created as bifurcations. A sequence of bifurcations may drive the system from a equilibrium point to chaotic behavior. The Landau-Hopf route is an example for creating chaos through sequential bifurcations (see Sect. 3.7). Ruelle and Takens argued that the Landau-Hopf route is unlikely to occur in nature. It is enough to have four consecutive bifurcations to get a low-dimensional manifold in the phase space referred to as a strange attractor. Their schedule may be summarized as follows:

fixed point $\rightarrow$ limit cycle $\rightarrow$ 2-torus
$\rightarrow$ 3-torus $\rightarrow$ strange attractor.

This provides another way to search for chaos in dynamic systems. Strange attractors are precisely defined as follows.

**Definition 6.4.1.** (*Strange Attractor*). Let us consider an $n$-dimensional system: $dx/dt = f(x,r)$, where $r$ is a scalar. A bounded set $A$ in $R^n$ is a strange attractor of the system if there is a set $U$ with the following properties:

i) $U$ is an $n$-dimensional neighborhood of $A$.
ii) If $x(0)$ belongs to $U$, then $x(t)$ does too for any positive $t$ and $x(t) \to A$.
iii) There is a sensitive dependence on initial conditions when $x(0)$ is in $U$, i.e., small variations of $x(0)$ lead to essentially different time paths of the system after a short time.
iv) The attractor is indecomposable.

The existence of a strange attractor in a continuous-time dynamic system implies highly irregular behavior. Now, we show that such behavior can be observed in a very simple three-dimensional dynamic system.

The Lorenz system is the best-known example for illustrating chaotic phenomena in differential equations. Independent of the main interest in bifurcations, Lorenz, in 1963, published an article concerning turbulence. The Lorenz system consists of three ordinary differential equations that depend on three real positive parameters

$$\frac{dx}{dt} = \sigma(y - x) ,$$

$$\frac{dy}{dt} = rx - y - xz ,$$

$$\frac{dz}{dt} = xy - bz , \tag{6.4.1}$$

where $\sigma$, $r$, and $b$ are the real positive parameters.

Lorenz derived the system as follows. A two-dimensional fluid cell is warmed from below and cooled from above and the fluid moves in the form of convection. The resulting convective motion is modeled by a partial differential equation. The variables in the equation are expanded into an infinite number of modes, all but the three in (6.4.1) of which are set identically to zero. In (6.4.1), $x$ represents the rate of convective overturning. The variables $y$ and $z$ are the horizontal and vertical temperature variations, respectively. The three parametes $\sigma$, $r$ and $b$ are, respectively, proportional to the Prandtl number, the Rayleigh number, and some physical proportions of the region under consideration.

As the parameters vary, behavior of the flow also changes. It has been numerically shown that for some parameter values, solutions of the equations oscillate, apparently forever, in a pseudo-random (chaotic) way. In addition, there are some parameter values for which "perturbulence" – a phenomenon whereby trajectories oscillate chaotically for long periods of time before finally settling down to stable stationary or stable periodic behavior – can be observed. Also, we can identify "intermittent chaos" where trajectories alternate between chaotic and apparently stable behaviors. The system can also produce a kind of behavior, called "noisy periodicity", where trajectories appear chaotic enough though they stay very close to a non-stable periodic orbit. A plot of the Lorenz attractor is provided by Fig. 6.8, where $\sigma = 4$, $r = 80$, $b = 8/3$ (see Haken 1983, p. 31).

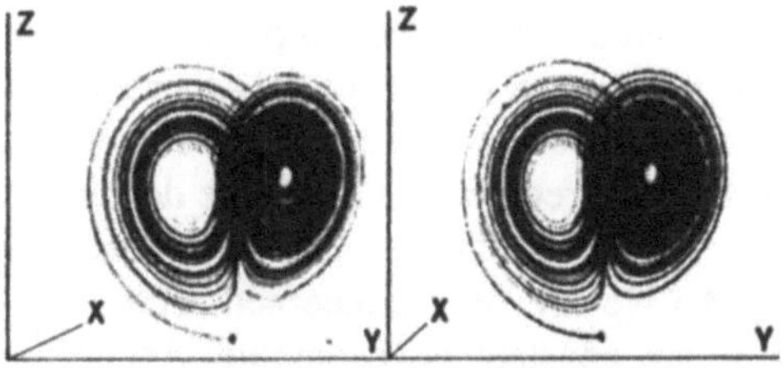


**Fig. 6.8.** A plot of the Lorenz attractor

It may be interesting to note that many other problems in science can be modeled by the Lorenz equations. For instance, Haken (1975) derives the Lorenz equations from a problem of irregular spiking in lasers, while York and York (1979) derive them from a problem concerning convection in a tordoidal region. Knobloch (1981) found that the problems of a disc dynamo can be reduced to the Lorenz system. Pedlosky and Frenten (1980) used the Lorenz equations to describe the dynamics of a weakly unstable, finite amplitude, baroclinic wave. There are also other problems which can be modeled by the equations (see Sparrow 1982). In what follows, we show that the Lorenz equations can be used to describe the dynamics of a small urban system within a metropolitan area, at least over the short term.

Consider an urban system within a metropolitan area. It is assumed that the urban system is very "small" in comparison to the metropolitan area in the sense of economic activities. This means that any change in economic conditions in the urban system will not affect the whole metropolitan area, which is structurally stable during the study period. We are concerned with the short-term dynamics of the urban system. Hence, we may treat the metropolitan area as a stationary environment. Evidently, this assumption cannot be valid in the long term.

It is assumed that firms and residents are free to choose their location sites either in the urban area or in the "outside world". As the urban area is very small, location and allocation behavior of the firms and households in the urban area cannot affect the locational conditions of the other parts of the metropolitan area.

It is assumed that locational charateristics of the urban area are described by the following three variables

$X$ = the output of the urban system;
$Y$ = the number of residents;
$Z$ = the land rent.

Produce of the urban area can be consumed by the residents or exported to the outside world. We suggest the following possible dynamics for the urban area

$$\frac{dX}{dt} = a_1 (a_2 Y - a_3 X) \,, \tag{6.4.2}$$

$$\frac{dY}{dt} = c_1 (c_2 X - c_3 Y) - c_4 X Z \,, \tag{6.4.3}$$

$$\frac{dZ}{dt} = d_1 XY - d_2 Z \,, \tag{6.4.4}$$

where $a_i$, $c_i$ and $d_i$ are positive parameters.

We define the parameter $a_2$ as per capita demand of the urban output of the urban residents. The parameter $a_3$ is interpreted as the rate at which the urban product is supplied to the urban area. As the demand of the urban product from the residents and the supply of the urban product to the urban market are assumed to be dependent on the scale of production and the number of residents, the two parameters may be influenced by the variables of the system. As we are only concerned with short-run behavior, the assumption that $a_2$ and $a_3$ are constant is acceptable. According to these definitions, we see that $a_2Y$ is the total demand of the residents for the urban product and $a_3X$ is the total supply of the urban product to the urban market. Consequently, the equation (6.4.2) means that rate of change in the urban product is proportional to the excess demand. If the demand is larger than the supply, the product tends to increase, and vice versa. The parameter $a_1$ is an adjustment speed coefficient. For simplicity, we assume that production is not affected by land rent. The change rate is only dependent on the excess demand for the urban product.

We assume that changes in the number of urban residents are given by two parts: $c_1(c_2X - c_3Y)$ and $-c_4XZ$. We interpret $c_2$ as the demand for labor from the firms to produce unit product. Hence, $c_2X$ is the total demand for labor from the urban labor market. The parameter $c_3$ is defined as the rate of the urban residents choosing to work in the urban area to the number of urban residents. The number $c_3Y$ gives the total supply of labor to the urban labor market. The term $(c_2X - c_3Y)$ is the excess demand for urban labor. The direction of migration is affected by the excess demand. The migration is also influenced by land rent. The people would choose to live in the place where land rent is low. The term $-c_4XZ$ takes this hypothesis into account.

In (6.4.4), we assume that any change in the rate of the land rent is negatively related to the current land rent level. This is based on the assumption that if the land rent is already very high, it is difficult to increase the rent further. The term $d_1XY$ means that changes in the rate of the land rent are positively related to $X$ and $Y$.

To show that the system (6.4.2–4) is identical to the Lorenz system, we make the following transformation

$$t = \frac{t^*}{c_1c_3}, \quad \sigma = \frac{a_{13}}{c_1c_3}, \quad r = \frac{a_2c_2}{a_3c_3}, \quad b = \frac{d_2}{c_1c_3},$$

$$x = \left(\frac{c_4}{d_1}\right)^{1/2}\frac{d_1X}{c_1c_3}, \quad y = \left(\frac{c_4}{d_1}\right)^{1/2}\frac{d_1a_2Y}{a_3c_1c_3},$$

$$z = \frac{c_4a_2Z}{a_3c_1c_3}. \tag{6.4.5}$$

It can easily be identified that (6.4.5) transforms (6.4.2–6.4.4) into (6.4.1). We have thus interpreted the Lorenz equations in the context of the urban problem. It is possible to apply the results obtained for the Lorenz equations to explain urban development.

There are many studies on the behavior of the Lorenz system (e.g., Sparrow 1982). Different analytic and simulation methods are combined to study the system.

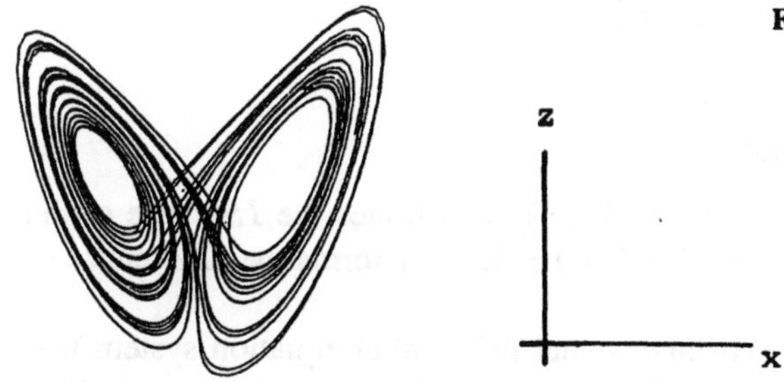


**Fig. 6.9.** Urban chaotic dynamics

We take the following values of the scaled parameters: $\sigma = 10$, $b = 8/3$, and $r = 28$. According to the simulation results in Sparrow (1982), the behavior is displayed in Fig. 6.9.

From the figure, we see some of the properties of the solutions: (i) the trajectory is not periodic; (ii) the figure does not appear to show a transient phenomenon since, regardless of how long the numerical integration is continued, the trajectory is going to continue to wind around and around without settling down to either periodic or stationary behavior; (iii) the topology of the figure is not dependent on the choice of initial conditions or integrating route; and (iv) it is impossible to predict the details of how the trajectory will develop over any period other than a very short time interval.

## 6.5 Chaos in an International Economic Model

This chapter will show that international trade among the economies which exhibit limit cycles may involve the occurrence of a strange attractor and hence chaos. In some sense, international trade activities can be considered as perturbations to the isolated economies. The following model is suggested by Lorenz (1987).

Consider three economies (nations, regions or cities), each of which is described by the simplified deterministic Keynesian equations (see Chap. 5)

$$\frac{dY_i}{dt} = \alpha_i \left[ I_i(Y_i, r_i) - S_i(Y_i, r_i) \right] ,$$

$$\frac{dr_i}{dt} = \beta_i \left[ L_i(Y_i, r_i) - \frac{M_i}{p_i} \right] , \qquad (6.5.1)$$

where the subscript $i$ denotes the $i$th economy, and the other variables are defined as

$Y$ = income ;

$r$ = interest rate ;

$M$ = constant nominal money supply ;

$p$ = fixed goods price ;

$I(Y, r)$ = gross saving ($I_Y > 0$, $I_r < 0$) ;

$S(Y, r)$ = savings ($S_Y > 0$, $S_r > 0$) ;

$\alpha$, $\beta$ = positive adjustment speeds .

The set of points $\{(Y_i, r_i) | I_i(Y_i, r_i) = S_i(Y_i, r_i)\}$ constitutes the $IS$-curve of the $i$th economy; the set of points $\{(Y_i, r_i) | L_i(Y_i, r_i) = M_i/p_i\}$ forms the $LM$-curve of the $i$th economy.

Equations (6.5.1) consist of a six-dimensional differential equation system that can also be written as a system of three independent two-dimensional systems which exhibit limit cycles (under appropriate conditions). As all of the three economies are oscillating, the overall motion of (6.5.1) constitutes a motion on a three-dimensional torus $T^3$, which is a geometric object in $R^6$.

Introducing international trade (export and import) into the isolated subsystems with functions $\mathrm{Ex}_i = \mathrm{Ex}_i(Y_j, Y_k)$, $i \neq j$, $k$, and $\mathrm{Im}_i = \mathrm{Im}_i(Y_i)$ yields

$$\frac{dY_i}{dt} = \alpha_i \left[ I_i(Y_i, r_i) - S_i(Y_i, r_i) + \mathrm{Ex}_i(Y_j, Y_k) - \mathrm{Im}_i(Y_i) \right] ,$$

$$\frac{dr_i}{dt} = \beta_i \left[ L_i(Y_i, r_i) - \frac{M_i^*}{p_i} \right] ,$$

$$i, j, k = 1, 2, 3, j, k, \neq i , \qquad (6.5.2)$$

where $M_i^*$ is the fixed money supply of the $i$th economy reflecting balance of payments equilibria.

The extended system (6.5.2) is made up of three coupled limit oscillators. As demonstrated by Newhouse, Ruelle and Takens (1978), a perturbation of a motion on a three-dimensional torus may result in a strange attractor.

Obviously, the existence of a strange attractor implies chaotic trajectories. It is not difficult to identify that the Newhouse-Ruelle-Takens theorem is satisfied for the system (6.5.2). We thus establish the existence of strange attractors in the international model.

**Proposition 6.5.1.** If all three autonomous economies are oscillating, the introduction of international trade may imply the existence of a strange attractor in the whole economy.

It was demonstrated by Lorenz that the existence of chaotic trajectories can be established in a reasonably specified model by simulation.

## 6.6 Chaos and Economic Forecasting

The great power of science lies in the ability to forecast the future. This ability is related to knowledge about the causal relationships among the elements under consideration. With regard to predictability, the arguments suggested by Laplace and

Poincaré are historically very important. And these points of view have significant influences on thinking about economic evolution.

The French mathematician Laplace proposed that the laws of nature imply strict determinism and complete predictability. It is only imperfections in observations that make the introduction of probabilistic theory necessary. According to Laplace "The present state of the system of nature is evidently a consequence of what it was in the preceding moment, and if we conceive of an intelligence which at a given instant comprehends all the relations of the entities of this universe, it could state the respective positions, motions, and general affects of these entities at any time in the past or future. ... The simplicity of the law by which the celestial bodies move, and the relations of their masses and distances, permit analysis to follow their motions up to a certain point; and in order to determine the state of the system of these great bodies in past or future centuries, it suffices for the mathematician that their position and their velocity be given by observation for any moment in time. Man owes that advantage to the power of the instrument he employs, and to the small number of relations that it embraces in its calculations. But ignorance of the different causes involved in the production of events, as well as their complexity, taken together with the imperfection of analysis, prevents our reaching the same certainty about the vast majority of phenomena. Thus there are things that are uncertain for us, things more or less probable, and we seek to compensate for the impossibility of knowing them by determining their different degrees of likelihood. So it is that we owe to the weakness of the human mind one of the most delicate and ingenious of mathematical theories, the science of chance or probability".

Poincaré argued that arbitrarily small uncertainties in the state of a system may be amplified in time and so predictions of the distant future are impossible. According to Poincaré "A very small cause which escapes our notice determines a considerable effect that we cannot fail to see, and then we say that the effect is due to chance. If we knew exactly the laws of nature and the situation of that same universe at the initial moment, we could predict exactly the situation of that same universe at a succeeding moment. But even if it were the case that the natural laws had no longer any secret for us, we could still only know the initial situation approximately. If that enabled us to predict the succeeding situation with the same approximation, that is all we require, and we should say that the phenomenon had been predicted, that it is governed by laws. But it is not always so: It may happen that small differences in the initial conditions produce very great ones in the final phenomena. A small error in the former will produce an enormous error in the latter. Prediction becomes impossible, and we have the fortuitous phenomenon."

We are following Poincaré's point of view on predictability. Although it is principally impossible to predict dynamic systems, this does not imply that we can say nothing about the future for any systems during any period. It may be concluded that our capacity to forecast is dependent on what phenomenon we are studying. On the basis of the laws of gravitation, for instance, eclipses can be predicted thousands of years in advance. However, weather forecasts are nowadays still impossible with certainty, though the movements of the atmosphere obey the laws of physics just as much as the movements of the planets. Our question is what makes the motion of one system much harder to anticipate than the motion of another system.

Like weather forecasting, the public has a very low opinion of economic forecasting. Both weather and economic forecasting try to predict the outcomes of very large systems, the components of which mutually interact in complex ways. Isolation and simplification methods, which play a central role in the development of sciences, are of limited use for the analysis of such systems. Moreover, as such large mutually interactional systems are often unstable, it becomes practically impossible to predict the long-term behavior of these systems.

To illustrate the difficulties of forecasting structural changes, consider an interesting question which has been recently readdressed by Dominguez, Fair and Shapiro (1988): Was the Depression forecastable? They have shown that neither contemporary forecasters nor modern time-series analysis could have forecast the large declines in output following the Crash. Their conclusion is based upon the Harvard and Yale forecasts, and modern time-series methods using the Harvard and Yale data and modern historical data.

The Harvard and Yale economic forecasting services were probably the two preeminent economic analysis and forecasting services available to businesses and members of the general public in the 1920s in the U.S. In the 1920s the Harvard Economic Service (HES) issued monthly reports on the current and expected future state of the economy. The HES used three indexes, representing speculation (the *A*-curve), business (the *B*-curve) and money (the *C*-curve), to help predict the future. The Harvard forecasts were then based on the relations determined to exist among the three curves during any given phase of the business cycle and on the magnitude of the movement from peak to trough of each curve. The resulting indexes are illustrated in Figs. 6.10 and 6.11 for, respectively, the years 1903–14 and 1919–31. All of these three indexes appear to lag the Crash, rather than anticipate it.

As explained by Dominguez, Fair and Shapiro (1988), the Harvard and Yale services neither predicted a downturn before the Crash nor became substantially more pessimistic about the economy following the Crash. To study the forecastability

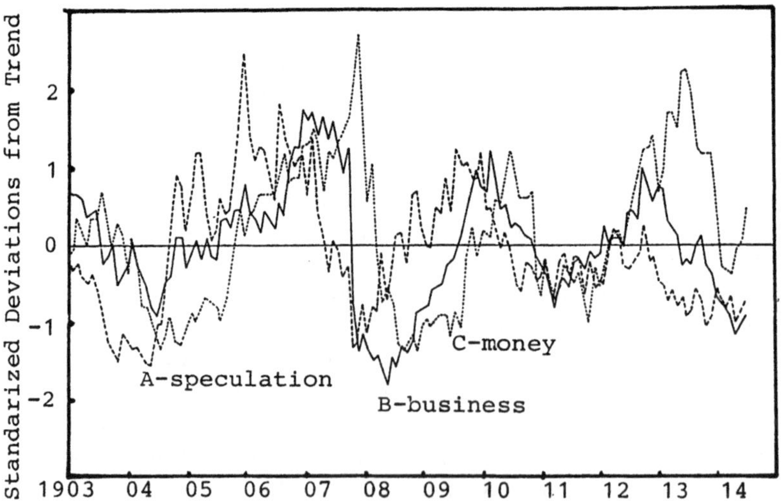


**Fig. 6.10.** HES Indexes, 1903–1904

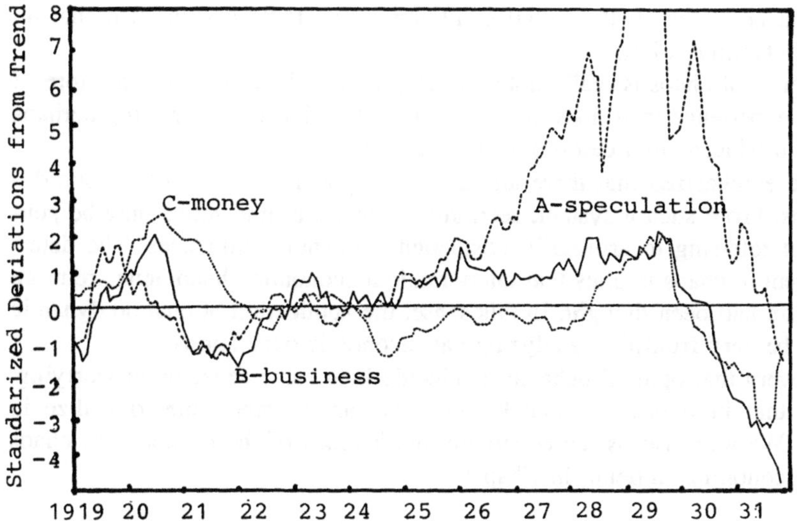

**Fig. 6.11.** HES Indexes, 1919–1931 (Source: Amer. Econ. Rev. 74, p. 597)

of the Depression, they estimated a number of vector autoregressive models using the Harvard, Fisher and modern historical data. The estimated results show that the large declines in output at the beginning of the Depression were essentially not predicted in either the months before the Crash or once the news of the Crash was taken into account. The Depression was unforecastable using the time-series methods. Therefore, the Harvard and Yale services, or an econometrician endowed with modern time-series methods and data, would be justified in appearing optimistic about the economy both on the eve of and in the months following the Crash.

As the leading explanations of the Depression are based on presumed structural stability and linearity, it is impossible to find the cause of such structural changes. Synergetic economics may proved a useful theory to explain the cause of the Depression.

## 6.7 Remarks

This chapter has been concerned with the existence of chaos (note that aperiodic solutions are not chaos). We discussed the concept of chaos and identified the existence of chaos in macro growth models, urban development, and regional and international economics.

It has been shown that economic chaos can arise even in quite simple differential equations. This striking discovery has altered our viewpoint about economic evolution. The discovery of chaos has created a new paradigm in economic modeling. The systems have the rather alarming property of giving rise to essentially unpredictable behavior. We have new fundamental limits on the ability to make economic predictions. To be sure, one can predict the future of such a system given its initial conditions, but any error in the initial conditions is so rapidly magnified that no practical predictability is left.

The existence of chaos is really not surprising to us. Chaos agrees with more of our everyday experience than does pure predictability – but it is surprising perhaps that it can be produced from deterministic equations.

It must be emphasized that there are also some positive aspects of chaotic dynamic systems. From such a system, seemingly stochastic time series may be generated without referring to arbitrarily postulated exogenous influences. The determinism inherent in chaos implies that many random economic phenomena are more predictable than had been thought. In this sense, the application of chaotic dynamic systems may be very fruitful in analyzing past economic oscillations.

The discovery that optimal behavior is chaotic is in some sense more worrying. This implies that individual rational behavior is almost impossible to realize in practical life. We will discuss the economic implications of the existence of chaos for different economies in detail in Chap. 9.

## Appendix: Some Criteria for Distinguishing Different Attractors

We refer to chaos as irregular motion stemming from deterministic equations. A difficulty rests in how to measure "irregular motion".

As shown from the previous chapters, different behavior, such as stable focus, limit cycles, and regular flucuations as subharmonic bifurcations, can be observed in deterministic systems. As these regular movements are also very complicated, it is therefore desirable to develop criteria to distinguish regular motion from chaos.

### A.1 The Lyapunov Exponents for Differential Equations

First, consider the following nonlinear equation

$$\frac{dx}{dt} = F(x) , \tag{6.A.1}$$

where $F$ is a nonlinear vector function of $x$. Let $x_0(t)$ be a solution of (6.A.1). Introducing $x(t) = x_0(t) + X(t)$, we obtain the linearized equations

$$\frac{dX(t)}{dt} = L\big(x_0(t)\big)X(t) , \tag{6.A.2}$$

where $L\big(x_0(t)\big) = \big(\partial F_i\big(x_0(t)\big)\partial x_j\big)_{n\times n}$. The Lyapunov exponents are defined by

$$z = \lim_{t\to\infty} \sup \left(\frac{1}{t}\ln |X(t)|\right) . \tag{6.A.3}$$

It can be shown that depending on different initial values of $X(t)$ at $t = t_0$, different Lyapunov exponents may exist, but not more than $n$ different ones. The following theorem can be found in, for instance, Haken 1983).

**Theorem.** At least one Lyapunov exponent vanishes if the trajectory $x(t)$ of an autonomous system remains in a bounded region for $t \to \infty$ and does not contain a fixed point.

It should be emphasized that the Lyapunov exponents are a special case of "generalized characteristic exponents". We know that if all the generalized characteristic exponents are negative, then the differential system is stable.

The Lyapunov exponents can be used to help us distinguish between different kinds of attractors. For instance, in one dimension, there are only stable fixed points, for which the Lyapunov exponents $z$ are negative. In two dimensions, the only two possible classes of attractors are stable fixed points and limit cycles. For a fixed stable point, the two Lyapunov exponents (which may coincide) are negative. For a limit cycle, $(z_1, z_2) = (-, 0)$. In three dimensions, we have some typical cases as follows

$$(z_1, z_2, z_3) = (-,-,-) \quad \text{for a stable fixed point} ,$$
$$(z_1, z_2, z_3) = (0,-,-) \quad \text{for a stable limit cycle} ,$$
$$(z_1, z_2, z_3) = (-,0,0) \quad \text{for a stable torus} ,$$
$$(z_1, z_2, z_3) = (+,0,-) \quad \text{for a strange attractor} .$$

Figure 6.12 shows the connection between the dimensions of simple attractors embodied in three-dimensional phase space and the signs of their Lyapunov exponents.

If one Lyapunov exponent is positive, chaos may appear. Since in a chaotic attractor at least one Lyapunov exponent is positive, neighboring trajectories de-

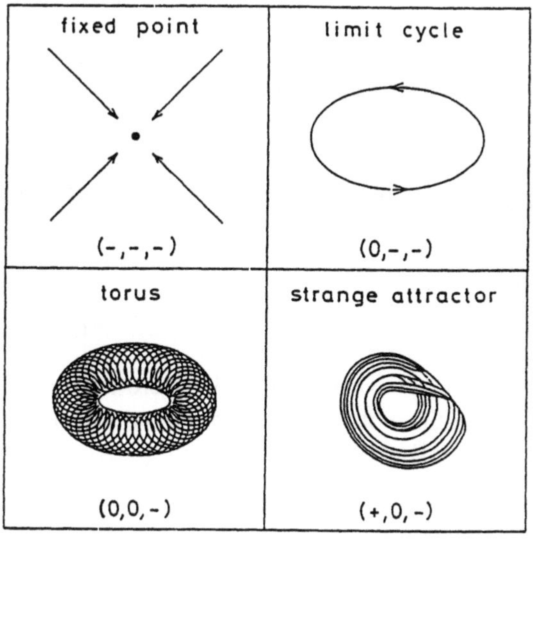


**Fig. 6.12.** Connection between the Lyapunov exponents and attractors

part very quickly from each other. For instance, $(z_1, z_2, z_3) = (+,0,0)$ may mean that we are dealing with an unstable torus. If an attractor possesses the exponents $(z_1, z_2, z_3) = (+,0,-)$, it is considered as a chaotic attractor. It should be emphasized that we still have very little knowledge about what Lyapunov exponents mean for attractors and how they can be determined.

## A.2 The Lyapunov Exponents for Discrete Maps

Consider a discrete map of the form

$$x_{n+1} = f(x_n) , \tag{6.A.4}$$

where $x_n$, $n = 1,2, \ldots$ are vectors in an $M$-dimensional space. We study the way in which we may formulate Lyapunov exponents for (6.A.4).

For a discrete map, the trajectory consists of the sequence of points $x_n$, $n = 0,1, \ldots$. We denote the trajectory whose neighborhood we wish to study by $x_n^0$. Let

$$x_n = x_n^0 + X_n , \tag{6.A.5}$$

where $x_n$ and $x_n^0$ satisfy (6.A.4) and $X_n$ are small perturbations. Substituting (6.A.5) into (6.A.4), we get the linearized system

$$X_{n+1} = L\left(x_n^0\right) X_n , \tag{6.A.6}$$

where $L(x_n^0) = (\partial f_k(x)/\partial x_j)$ at $x = x_n^0$. This equation may be solved by iteration. We have

$$X_n = L\left(x_{n-1}^0\right) L\left(x_{n-2}^0\right) \ldots L\left(x_0^0\right) X_0 .$$

The Lyapunov exponents are defined by

$$z = \lim_{n \to \infty} \sup \frac{1}{n} \ln |X_n| . \tag{6.A.7}$$

Depending on different directions of $X_0$, we may have different $z$.

For a one-dimensional map, we have

$$z = \lim_{n \to \infty} \sup \frac{1}{n} \sum_{m=0}^{n-1} \ln \left|f'\left(x_m^0\right)\right| . \tag{6.A.8}$$

For instance, consider the logistic map

$$x_{n+1} = a x_n (1 - x_n) . \tag{6.A.9}$$

Its Lyapunov exponent as a function of $a$ is shown in Fig. 6.13, where a positive value of the exponent corresponds to chaotic motion and negative values to the presence of regular (periodic) motion (see Haken 1983).

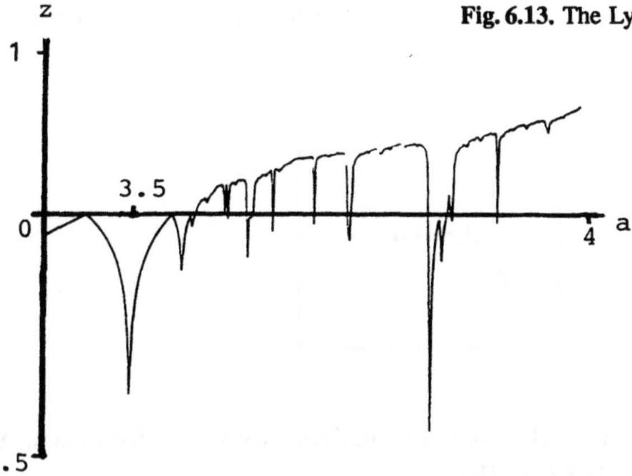


**Fig. 6.13.** The Lyapunov exponent for the logistic map

## A.3 The Signal, Power Spectrum, Autocorrelation Function and Poincaré Map

Now we mention some other possible criteria for chaotic motion. To distinugish between multiply periodic behavior (which can also look rather complicated) and chaos, it is often convenient to Fourier-transform the variable $x(t)$ :

$$x(w) = \lim_{T \to \infty} \int_0^T x(t) \exp(iwt)dt .$$

For multiply periodic motion, the power spectrum $P(w) = |x(w)|^2$ consists only of discrete lines of the corresponding frequencies, whereas chaotic motion (which is completely aperiodic) is indicated by broad noise in $P(w)$ that is mostly located at low frequencies.

To detect chaos, we may also define the change in the autocorrelation function

$$C(v) = \lim_{T \to \infty} \int_0^T c(t)c(t+v)dt$$

where

$$c(t) = x(t) - \lim_{T \to \infty} \int_0^T x(t)dt .$$

This function remains constant or oscillates for regular motion and decays rapidly if $x(t)$ becomes uncorrelated in the chaotic regime. It should be mentioned that $P(w)$ and $C(v)$ contain the same information.

The idea of the Poincaré map can be interpreted as follows. We may treat trajectories in an $n$-dimensional space and study the points where the trajectories cut through a hypersurface. This can be visualized in three dimensions as in Fig. 6.14a, in which the cross points are connected by a smooth curve (Fig. 6.14b). In this case, the Poincaré map belongs to the cross section of a two-dimensional plane with a trajectory in three dimensions.

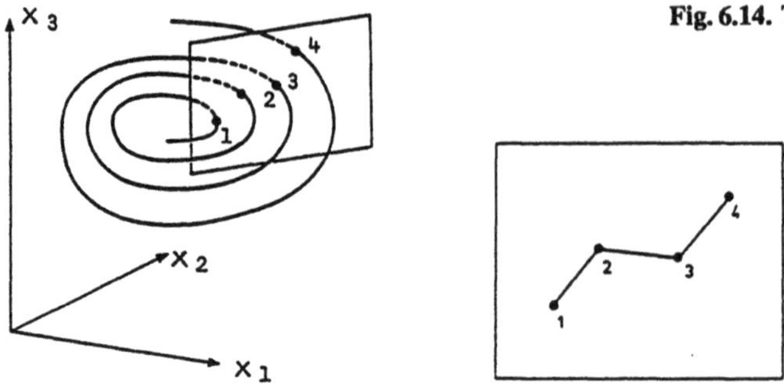


**Fig. 6.14.** The concept of the Poincaré map

The following table shows what the criterion numbers look like for chaos. The table is referred to in Schuster (1988, p. 10).

From this table, we see that for these four criteria of chaotic motion: (i) the time dependence of the signal "looks chaotic"; (ii) the power spectrum exhibits broadband noise; (iii) the autocorrelation function decays rapidly; and (iv) the Poincaré map shows space-filling points. The explanation of this table is given in detail in Schuster (1988).

**Table 6.1.** Detection of chaos in simple systems

| System | Equation of Motion | Indication | |
|---|---|---|---|
| Pendulum 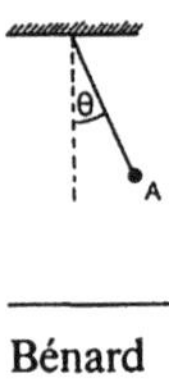  | $\ddot{\theta} + \gamma\dot{\theta} + g\sin\theta = A\cos\omega t$<br>$x = \theta,\ y = \dot{\theta},\ z = \omega t$<br>$\dot{x} = y$<br>$\dot{y} = -\gamma y - g\sin x + A\cos z$<br>$\dot{z} = \omega$ | 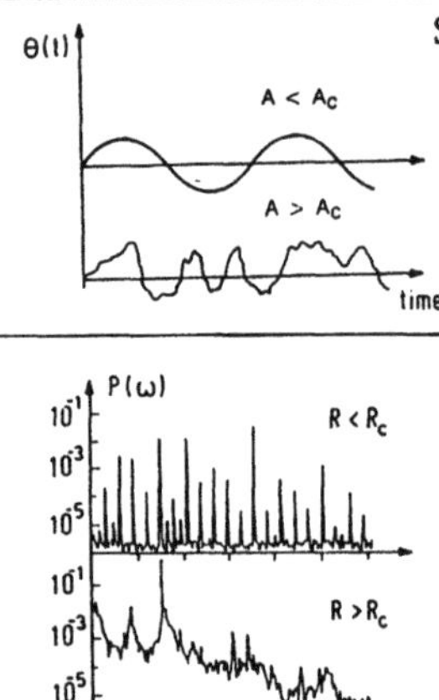  | Signal |
| Bénard Experiment 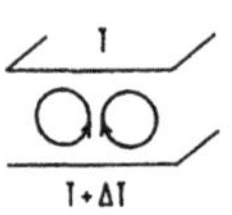  | $\dot{x} = -\sigma x + \sigma y$<br>$\dot{y} = rx - y - xz$<br>$\dot{z} = xy - bz$ | 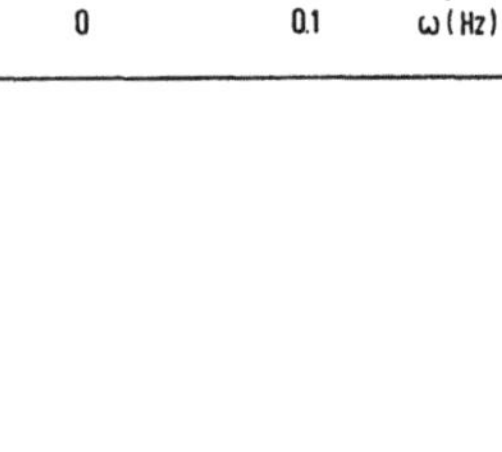  | Power Spectrum |

**Table 6.1.** (continued)

| | | |
|---|---|---|
| Belousov-Zhabotinsky Reaction<br>$Ce_2(SO_4)_3$<br>⋮<br>$Ce^{4+}$ | $\dot{\vec{x}} = \vec{F}(\vec{x}, \lambda)$<br>$\vec{x} = [c_1, c_2, \ldots c_d]$ | 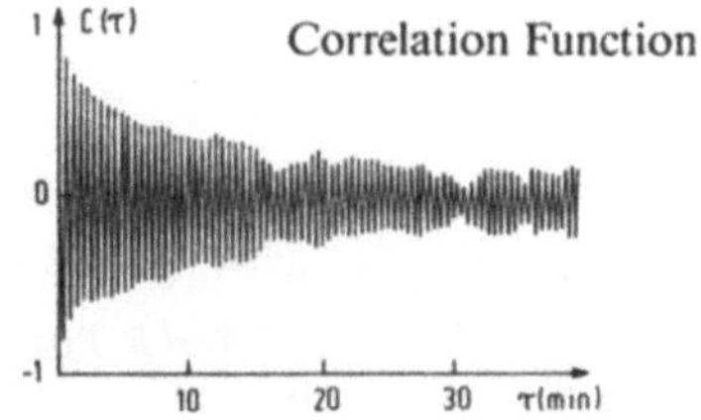 |
| Hénon-Heiles System | $H = \frac{1}{2} \sum_{i=1}^{2} (p_i^2 + q_i^2) +$<br>$+ q_1^2 q_2 - \frac{1}{3} q_2^3$<br>$\dot{\vec{p}} = -\frac{\partial H}{\partial \vec{q}}, \ \dot{\vec{q}} = \frac{\partial H}{\partial \vec{p}}$ | Poincaré Map<br>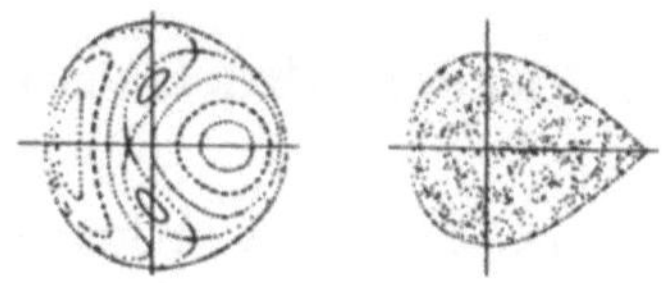 |

# 7. Stochastic Processes and Economic Evolution

... it is the aim of science to find satisfactory explanations of whatever strikes us as being in need of explanation.

*Karl R. Popper* (1972)

## 7.1 Random Processes and Economic Evolution

We have shown that random economic behavior may occur in relatively very simple differential equations. Regular and irregular oscillations may be created endogenously by relatively simple interactions among economic variables. There is order in chaos. However, as we have explained before, there is another way to explain economic fluctuations. Dynamic systems which are subjected to external noise can exhibit irregular behavior. For instance, consider a simple example of a periodically driven (pendulum) equation (see Schuster 1988)

$$\frac{d^2x}{d^2t} + \frac{r\,dx}{dt} + \sin x = A \cos(\omega t) \ ,$$

where $r$ is the damping constant, $A$ represents the amplitude and $\omega$ the frequency of a driving torque. This equation has been numerically integrated for different sets of parameters $(A, \omega, r)$, and Fig. 7.1 shows that the variation of the angle $x$ with time

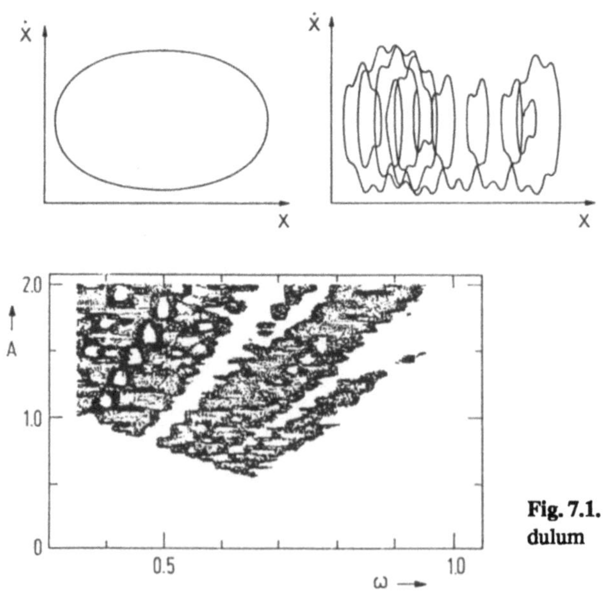


**Fig. 7.1.** Transition to chaos in a driven pendulum

simply looks chaotic if the amplitude reaches a certain value $A_c$. In Fig. 7.1, $r = 0.2$, $x(0) = 0$, $dx(0)/dt = 0$, and black points in Fig. 7.1c denote parameter values $(A, \omega)$ for which the motion is chaotic.

The introduction of noise into economics reflects the fact that some events are purely random, like lottery outcomes. It is considered that the economic system under consideration is often subjected to influences from the environment. Such influences are randomly distributed over time and space. Another point of view is that even if some events, such as crop yields, inventions and so on, may be potentially described by deterministic mechanisms, they may appear to be random from an economic point of view. If this view is the main reason for economists to consider a stochastic model as an appropriate approach, the model should be replaced by a deterministic one as soon as the mechanisms are discovered.

In this chapter, we are concerned with the "stochastic approach" to economic dynamics. We examine the effects of small fluctuations on economic evolution.

An important example of this approach in economics can be found in the work by Lucas (1975). The Lucas macroeconomic model is based upon a set of linear difference equations with stochastic terms. The economy is subdivided into "islands". Information flows between different islands are imperfect. This geographical characteristic has the effect that when agents see that prices are rising, they are not able to tell whether prices are rising only on the island on which they are located or over the economy. Hence purely nominal shocks of rising prices can induce people to invest, even though they make no difference in real profit opportunities. As capital invested during the nominal shock cannot be moved, real effects will be felt for some time after a disturbance. It can be shown that the model can generate pro-cyclical movements in both prices and the share of output devoted to investment. This gives a reasonable explanation of the auto-correlations observed in macroeconomic data.

In physics, we can find similar phenomena such as Brownian motion. The observation that, when suspended in water, small pollen grains are found to be in a very irregular state of motion, was first systematically investigated by Robert Brown in 1827. The "irregular" phenomenon took the name Brownian motion because of his fundamental pioneering work. The riddle of the motion was not satisfactorily explained until Einstein gave an explanation in 1905. There are two major in points in Einstein's solution to the problem: (i) The motion is caused by the exceedingly frequent impacts on the pollen grain of the incessantly moving molecules of liquid in which it is suspended; and (ii) The motion of these molecules is so complicated that its effect on the pollen grain can only be described probabilistically in terms of exceedingly frequent statistically independent impacts. A statistical theory is an effective way to explain fluctuations like this kind of motion.

The second approach to explain irregular fluctuations is developed by studying the behavior of deterministic systems. We have already given some examples of this approach. In general it is argued that deterministic equations, without being subjected to any fluctuations, are sufficient for describing economic dynamics for two reasons. First, fluctuations are of small intensity. Second, fluctuations occur on a time scale that is rapid compared to the macroscopic equation. In this chapter, we show that these points of view are valid only for some limited cases. Fluctuations of even average zero value can drive the system far away from equilibrium; small fluctuations can cause structural changes in dynamic systems. Thus fluctuations cannot be neglected in dynamic analysis. It should be emphasized that this important discovery is still related to the instability of dynamic systems which we have stressed in our approach.

Another question also arises. Between the approaches of instability and exogenous shocks, which one is the more desirable for explaining the observed irregularities in fluctuating data? This question cannot be principally answered by simulating economic models because, after all, both approaches involve extremely drastic simplifications of actual economic processes when actual data are input into the models. However, we have to provide some criteria for judging which approach is more acceptable. If we accept Friedman's argument (1953) that realism in economic models is not a desirable end in itself, then it should be incorporated in models to the extent that it tells us something useful about the economy. Nevertheless a good economic theory should indicate the essential mechanisms which cause economic phenomena. In this sense, it may not be so desirable to explain business fluctuations mainly through exogenous shocks as to understand them in terms of nonlinear interactions among economic variables.

It cannot, however, be denied that any economy is subject to random shocks. Such shocks may be due to our limited capacity to understand natural laws and other factors. For instance, we cannot accurately forecast weather, earthquakes and so on. One way to overcome these problems is to treat these factors as random factors in economic analysis.

It is argued that whether random shocks can have serious impacts is determined by characteristics of the system. In a stable system with the property of returning to its equilibrium quickly, the impact of an external shock will be small. As shown in this study, however, if the system is unstable, effects of random shocks, even if their average values are zero, are very complicated. The possibility of complexity may intuitively be understood from our previous discussions. We have shown that an unstable system may exhibit structural changes even when small changes in parameters occur. Hence, it can be intuitively agreed that if the system has the properties of "memory", a stochastic process even with zero averages may drive the system far away from the undisturbed equilibrium.

It is, therefore, essential to construct a theory of fluctuations around critical states to predict the behavior of systems. The remainder of this chapter studies the role of fluctuations in dissipative systems.

## 7.2 Stochastic Processes – An Introduction

This section defines some elementary concepts for the study of stochastic process. Before dealing with stochastic processes, we need some concepts of probability theory.

### 7.2.1 Some Concepts in Probability Theory

Let the set of events under consideration be denoted by $A^*$. If all the events can be indexed discretely with integers, then $A^*$ may be written as $A^* = \{x_1, x_2, \dots\}$, where $x_i$ is one of the events contained in $A^*$. Let $A$ denote any subset of $A^*$, and let $\dot{A}$ be the set of no events.

The probability of $A$, $P(A)$, is defined as a function of $A$ satisfying the following probability axioms:

i) $P(A) \geq 0$ for all $A$;
ii) $P(A^*) = 1$; and
iii) If $A_i$ $(i = 1, 2, \dots)$ is a countable (but possibly infinite) collection of nonoverlapping sets, then $P(\sum_i A) = \sum_i P(A_i)$.

The three axioms are sufficient for defining probability. From these we have:

iv) If $\bar{A}$ is the set of all events in $A^*$ which are not contained in $A$, then $P(\bar{A}) = 1 - P(A)$; and
v) $P(\dot{A}) = 0$.

Intuitively, the concept of probability means that if we choose an event from $A^*$ at random $N$ times, the relative frequency that the particular event chosen will belong to $A$ approaches $P(A)$ as the number $N$ of choices approaches infinity. The choices can be visualized as being done one after the other or at the same time.

The joint probability $P(A \cap B)$ is defined as

$$P(A \cap B) = P\{x \in A \text{ and } x \in B) \ ,$$

where $x$ is an event contained in both classes $A$ and $B$ under consideration. With the help of this concept, we can define the conditional probability $P(A|B)$ as

$$P(A|B) = P(A \cap B)/P(B) \ ,$$

which is the probability that all of the events in the set $A$ are contained in the set $B$. In probability theory, two sets of events $A$ and $B$ represent independent sets of events if the specification that a particular event is contained in $B$ has no influence on the probability of that event belonging to $A$, i.e., $P(A|B)$ is independent of $B$. That is, $A$ and $B$ are said to be independent if

$$P(A \cap B) = P(A)P(B) \ .$$

This can be similarly said to involve several sets of events.

The concept of a random variable is defined as follows. Suppose that we have an abstract probability space whose events are denoted by $x$, where $x$ may be continuous or discrete. The random variable $X(x)$ is defined as a function of $x$, which takes on certain values for each $x$. Random variables $X_1, X_2, \dots$ will be said to be independent if all values of the $X_i$ are assumed independently of those of the remaining $X_i$.

Consider the case when the basic events $x$ are continuous. The mean value of a random variable $X(x)$ is defined by

$$\langle X \rangle = \int_{X \in A^*} X(x)p(x)dx \ ,$$

where $p(x)$ is a probability function. Similarly, if the events $x$ are countable, then

$$\langle X \rangle = \sum_x P(x)X(x) \ .$$

For a single variable, the variance is defined as

$$\mathrm{var}\{X\} = \left\langle [X - \langle X \rangle]^2 \right\rangle \ .$$

The variance of a variable is a measure of the degree to which the values of $X$ deviate from the mean value $\langle X \rangle$. In the case of several variables, a similar concept is called the covariance matrix whose $i$–$j$ element is defined as

$$\langle X_i, X_j \rangle = \langle (X_i - \langle X_i \rangle) \left( X_j - \langle X_j \rangle \right) \rangle \ .$$

If the variables are independent, then the matrix is diagonal.

We consider some sequence of random variables $\{X_i\}$. The limit of the sequence as $n \to \infty$

$$X = \lim_{n \to \infty} X_n \ ,$$

"roughly" means that a random variable $X$ is approached by the sequence of $\{X_i\}$. Many definitions of limit can be suggested. For instance, we have almost certain limit, mean square limit, stochastic limit (or limit in probability), and limit in distribution (see, for example, Gardiner 1983).

As an application of the previous concepts, consider the law of large numbers. We measure the same quantity $N$ times, obtaining sample values of the random variable $X(n)$ $(n = 1, 2, \ldots)$. It is assumed that for every $n$, $X(n)$ has the same probability distribution. It should be noted that the $X(n)$ may be dependent. It can be proved that provided that $\langle X(n), X(m) \rangle$ vanishes sufficiently rapidly as $|n - m| \to \infty$, then we shall have

$$\lim_{N \to \infty} \bar{X}_N = \langle X \rangle \ ,$$

where

$$\bar{X}_N = \left( \frac{1}{N} \right) \sum_{n=1}^{N} X(n) \ .$$

It is clear that $\langle \bar{X}_N \rangle = \langle X \rangle$. We now calculate the variance of $\bar{X}_N$ and show that as $N \to \infty$ it vanishes under certain conditions. As $\mathrm{var}\{\bar{X}_N\}$ is equal to

$$\langle \bar{X}_N, \bar{X}_N \rangle = \langle \bar{X}_N, \bar{X}_N \rangle - \langle \bar{X}_N^2 \rangle = \left( \frac{1}{N^2} \right) \sum_{n,m=1}^{N} \langle X_n, X_m \rangle \ ,$$

provided that $\langle X_n, X_m \rangle$ falls off sufficiently rapidly as $|n-m|$ is large enough, we find

$$\lim_{N \to \infty} \left( \text{var}\left\{ \bar{X}_N \right\} \right) = 0 .$$

This implies that lim $\bar{X}_N$ is a deterministic variable equal to $\langle X \rangle$.

### 7.2.2 Stochastic Processes

By stochastic processes, we mean systems in which a certain time-dependent random variable $X(t)$ exists. In a loose sense, this means systems which evolve probabilistically in time. It is assumed that $x_1, x_2, \ldots$ of $X(t)$ at times $t_1, t_2, \ldots$ can be measured and that a set of joint probability densities exists

$$p(x_1, t_1, x_2, t_2, \ldots) .$$

Similarly to probability theory, we can define conditional probability densities. To illustrate how nonlinearity may affect the way of describing the motion of systems, consider the dynamics of a birth process $N(t)$, where $N$ is the number of individuals in a certain population. Let the probability that $N$ increases to $N$+1 in an infinitesimal time $\delta t$ be proportional to $N$ and $\delta t$, i.e.,

$$P\left\{ N \to N+1 \text{ in } (t, t+\delta t) \right\} = \alpha N \delta t ,$$

where $P$ denotes probability, and $\alpha$ is a constant. If we express the expected value of $N$ by $n = E(N)$, the following differential equation holds

$$\frac{dn}{dt} = \alpha n .$$

However, if a nonlinear term is introduced into the system

$$P\left\{ N \to N+1 \text{ in } (t, t+\delta t) \right\} = \alpha N \left( \frac{1-N}{\beta} \right) \delta t ,$$

where $\beta$ is a constant, then $n(= E(N))$ satisfies

$$\frac{dn}{dt} = \alpha n - \frac{\alpha E(N^2)}{\beta} .$$

Thus, unless $E(N^2) = \left( E(N) \right)^2$, we cannot write

$$\frac{dn}{dt} = \alpha n \left( 1 - \frac{n}{\beta} \right) .$$

Since the difference $E(N^2) = (E(N))^2$ is the variance of $N$, the approximation may be used when $N$ has small variance. Similar results are true for higher-order effects. If more than one variable is involved, we shall require covariances to be small.

The simplest kind of stochastic process is that of complete independence

$$p(x_1, t_1, x_2, t_2, \dots) = \Pi_i p(x_i, t_i) \ . \tag{7.2.1}$$

which means that the value of $X$ at time $t$ is completely independent of its values in the past (or future). Moreover, in (7.2.1) if $p(x_i, t_i)$ is independent of $t_i$, i.e., if the same probability law governs the process at all times, the process is called the Bernoulli trial. Different processes are named according to properties of their probability densities. For example, in the Markov process the knowledge about present situations determines the future. That is, the conditional probability of a Markov process is determined entirely by the knowledge of the most recent condition

$$\begin{aligned} &p\left(x_1, t_1, x_2, t_2, \dots \left| y_1, t_1^*, y_2, t_2^*, \dots\right.\right) \\ &\quad = p\left(x_1, t_1, x_2, t_2, \dots \left| y_1, t_1^*\right.\right) \ , \end{aligned}$$

in which $t_1 \geq t_2 \dots \geq t_1^* \geq t_2^* \dots$ .

As the following formulas are valid for all stochastic processes

$$\begin{aligned} p\left(x_1, t_1 | x_3, t_3\right) &= \int p\left(x_1, t_1, x_2, t_2 | x_3, t_3\right) dx_2 \\ &= \int p\left(x_1, t_1 | x_2, t_2, x_3, t_3\right) p\left(x_2, t_2 | x_3, t_3\right) dx_2 \ , \end{aligned}$$

if the Markov assumption is held for a process, then one has

$$\begin{aligned} &p\left(x_1, t_1 | x_3, t_3\right) \\ &\quad = \int p\left(x_1, t_1 | x_2, t_2\right) p\left(x_2, t_2 | x_3, t_3\right) dx_2 \ . \end{aligned} \tag{7.2.2}$$

Equation (7.2.2) is called the Chapman-Kolmogorov equation. We show that under appropriate conditions this equation can be written in the form of a differential equation. We require the following conditions for all $\varepsilon > 0$ :

i) $$\lim_{\delta t \to 0} \frac{p(x, t + \delta t)|z, t)}{\delta t} = W\left(x | z, t\right) \ , \tag{7.2.3}$$

uniformly in vectors $x$, $z$ and $t$ for $|x - z| < \varepsilon$;

ii) $$\begin{aligned} &\lim_{\delta t \to 0} \left(\frac{1}{\delta t}\right) \int_{|x-z|<\varepsilon} (x_i - z_i) \, p\left(x, t + \delta t | z, t\right) dx \\ &= A_i(z, t) + O(\varepsilon) \ ; \end{aligned} \tag{7.2.4}$$

iii) $$\begin{aligned} &\lim_{\delta t \to 0} \left(\frac{1}{\delta t}\right) \int_{|x-z|<\varepsilon} (x_i - z_i) \left(x_j - z_j\right) p\left(x, t + \delta t | z, t\right) dx \\ &= B_{ij}(z, t) + O(\varepsilon) \ ; \end{aligned} \tag{7.2.5}$$

the last two being uniform in $z$, $t$ and $\varepsilon$.

Under these conditions, it can be shown that all higher-order coefficients in the form of (7.2.4) and (7.2.5) must vanish. Under other appropriate conditions, the stochastic process under consideration can be expressed as follows (Gardiner 1983, Sect. 3.4)

$$\begin{aligned} \partial_t p\left(z, t | y, t'\right) = &- \sum_i \frac{\partial\left[A_i(z,t)p(z,t|y,t')\right]}{\partial z_i} \\ &+ \left(\frac{1}{2}\right) \sum_{i,j} \frac{\partial^2\left[B_{ij}(z,t)p(z,t|y,t')\right]}{\partial z_i \partial z_j} \\ &+ \int \left[W\left(z|x,t\right) p\left(x,t|y,t'\right)\right. \\ &\left. - W\left(x|z,t\right) p\left(z,t|y,t'\right)\right] dx \ . \end{aligned} \tag{7.2.6}$$

This equation is called the differential Chapman-Kolmogorov equation. If we appropriately specify $A(x,t)$, $B(x,t)$ and $W(x|y,t)$, a non-negative solution to the differential Chapman-Kolmogorov equation exists.

If we specify $A_i(z,t) = B_{ij}(z,t) = 0$, we have the Master equation:

$$\begin{aligned} \partial_t p\left(z,t|y,t'\right) = &\int \left[W\left(z|x,t\right) p\left(x,t|y,t'\right)\right. \\ &\left. - W\left(x|z,t\right) p\left(z,t|y,t'\right)\right] dx \ . \end{aligned} \tag{7.2.7}$$

In the next section, we derive this equation for birth-death processes and investigate properties of the equation.

If we assume that the quantity $W(z|x,t)$ is equal to zero, the differential Chapman-Kolmogorov equation is called the Fokker-Planck equation

$$\begin{aligned} \partial_t p\left(z, t | y, t'\right) = &- \sum_i \frac{\partial\left[A_i(z,t)p(z,t|y,t')\right]}{\partial z_i} \\ &+ \left(\frac{1}{2}\right) \sum_{i,j} \frac{\partial^2\left[B_{ij}(z,t)p(z,t|y,t')\right]}{\partial z_i \partial z_j} \ . \end{aligned} \tag{7.2.8}$$

and the corresponding process is known mathematically as a diffusion process. Applications of this equation to social systems are referred to in Weidlich and Haag (1983). The vector $A(z,t)$ is called the drift vector and the matrix $B(z,t)$ the diffusion matrix.

If only the first term in the differential Chapman-Kolmogorov equation is nonzero, then one has a special case of a Liouville equation:

$$\partial_t p\left(z, t | y, t'\right) = - \sum_i \frac{\partial\left[A_i(z,t)p(z,t|y,t')\right]}{\partial z_i} \ . \tag{7.2.9}$$

Some other forms of stochastic equations and different methods of solving these equations have been proposed. In what follows, we will provide some examples of applying stochastic methods to show how our emphasis on nonlinearity and instability can provide us with new insight into economic evolution processes.

## 7.3 Birth-Death Processes and the Master Equation

A wide variety of phenomena can be described by a particular class of process called birth-death processes (Gardiner 1983). The name stems from the modeling of human or animal populations in which individuals are born, or die. The prey-predator model mentioned in Chap. 3 is one of the most entertaining models. In this section, we will describe this process by the master equation on the basis of probability theory and dynamics. The description is rather brief because the analysis of the model is rather complicated and is well known in the literature.

It is assumed that the system consists of two kinds of animal, one of which preys on the other, and the other supplied with an inexhaustible food supply. Let $X$ symbolise the prey, $Y$ the predator, $A^*$ the food of the prey, $B^*$ the death of the predator. The process under consideration might be illustrated as

$$X + A^* \rightarrow 2X ,$$
$$X + Y \rightarrow 2Y ,$$
$$Y \rightarrow B^* , \tag{7.3.1}$$

in which the first equation symbolises the prey eating one unit of food, and reproducing immediately, the second symbolises a predator consuming one unit of the prey (which thereby dies – and this is the only way to die) and immediately reproducing, and the final equation symbolises the death of the predator by natural causes. In what follows, we use $x$ and $y$ to denote the number of $X$ and $Y$, respectively. If we assume that the first "reaction" symbolises a rate of production of $X$ proportional to the product of $x$ and the amount of food; the second equation the production of $Y$ and an equal rate of consumption of $X$ proportional to $xy$; and the last equation the death rate of $Y$, in which the rate of death of $Y$ is proportional to $y$, then the system can be described by the prey-predator equations

$$\frac{dx}{dt} = a_1 x - b_1 xy .$$

$$\frac{dy}{dt} = a_2 xy - b_2 y .$$

The properties of the system were studied in Chap. 3. We will discuss the cumulative effects of small random perturbations on such a system. It can be seen that small perturbations may cause the solution of the deterministic equations to wander about between trajectories until it finally meets one of the axes $x = 0$ or $y = 0$. If the system describes the population dynamics, this implies the extinction of prey or predator. Thus, the model cannot be considered adequate in describing long-term oscillatory behavior in prey-predator systems. There are limitations of the equations to realistic systems with oscillations.

If we want to include fluctuations appropriately in this approach, the simplest way is by means of a birth-death master equation (see, Nicolis and Prigogine 1977, Gardiner 1983). We assume a probability distribution, $P(x, y, t)$, for the number of individuals at a given time. We try to find a reasonable probabilistic law corresponding to the prey-predator equations.

Assume that in an infinitesimal time $\delta t$, the following transition probability laws hold

$$\begin{aligned} &\text{Prob}\,(x \to x+1\,;\, y \to y) = \alpha x\,\delta t\,, \\ &\text{Prob}\,(x \to x-1\,;\, y \to y+1) = bxy\,\delta t\,, \\ &\text{Prob}\,(x \to x\,;\, y \to y-1) = \beta y\,\delta t\,, \\ &\text{Prob}\,(x \to x\,;\, y \to y) = 1-(\alpha x + bxy + \beta y)\,\delta t\,. \end{aligned} \tag{7.3.2}$$

We thus can replace the dynamic rate laws by probability laws. We may write the probability at $t+\delta t$ as a sum of terms, each of which represents the probability of a previous state multiplied by the probability of a transition to the state. For simplicity, let $b = 1$. We have

$$\begin{aligned} &\frac{1}{\delta t}\left[P\,(x,y,t+\delta t) - P\,(x,y,t)\right] \\ &\quad = \alpha(x-1)P(x-1,y,t) + (x+1)(y-1)P(x+1,y-1,t) \\ &\qquad + \beta(y+1)P(x,y+1,t) - (\alpha x + \beta xy + \beta y)P(x,y,t)\,. \end{aligned} \tag{7.3.3}$$

which yields a stochastic differential equation when $\delta \to 0$. In (7.3.3) we assume that the probability of each of the events occurring can be sufficiently determined from the knowledge of $x$ and $y$. This is the Markov postulate. In economic application, this assumption must be used with great care since history cannot be taken into account by this hypothesis. In the population problem, the concept of heredity, i.e., that the behavior of progeny is related to that of parents, clearly contradicts this postulate. This assumption is valid to the extent that different individuals of the same species are similar.

These type of equations have found wide applications in science (for example, Gardiner 1983, Weidlich and Haag 1983). Besides birth-death processes, systems of molecules of various chemical compounds, electrons, biological systems, and political voting problems are examples of systems which can be modeled in this way. The particular choice of transition probabilities is made on various grounds determined by the degree to which details of the births and deaths involved are known.

Equation (7.3.3) has no simple solution. Its solutions determine both the gross deterministic motion and the fluctuations. Moreover, the fluctuations are typically of the same order of magnitude as the square roots of the numbers of individuals involved. However, as shown in Gardiner (1983) there is often a limit in which the solution of a master equation can be approximated asymptotically by a deterministic part, plus a fluctuating part.

To illustrate differences of the solution properties between the prey-predator equations and the stochastic equation, let us plot some simulation results in Fig. 7.2. (Figure 7.2a corresponds to Fig. 1.3a in Gardiner (1983), while Fig. 7.2b corresponds to Fig. 1.3c.) The solid line denotes the prey ($x$), the dashed line the predator ($y$).

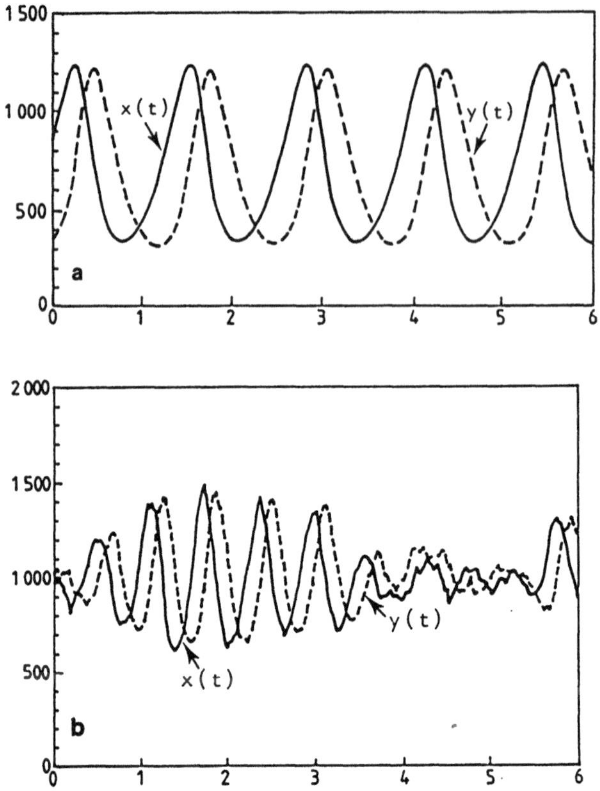


**Fig. 7.2a, b.** Time development in the system. **(a)** Deterministic equations, **(b)** stochastic equations

Now we will examine the way in which the equation (7.3.3) can be solved. The study of the equation is performed most conveniently in the generating function representation. The representation is defined as

$$F\left(s_x, s_y, t\right) = \sum_{x,y=0}^{\infty} s_x^x s_y^y P(x, y, t) . \tag{7.3.4}$$

Substituting (7.3.4) into (7.3.3) yields

$$\begin{aligned}\frac{\partial F}{\partial t} =& (s_2+1)(s_2-s_1)\frac{\partial^2 F}{\partial s_1 \partial s_2} \\ &+ a s_1 (s_1+1)\frac{\partial F}{\partial s_1} - b s_2 \frac{\partial F}{\partial s_2} ,\end{aligned} \tag{7.3.5}$$

which is termed the master equation. In (7.3.5), one has

$$s_1 = s_x - 1 , \quad s_2 = s_y - 1 . \tag{7.3.6}$$

Equation (7.3.5) is a partial differential equation with variable coefficients. In general, it is not easy to solve this equation. However, it can be approximately solved in some cases (Nicolis and Prigogine 1977). For instance, if we are mainly concerned with macroscopic dimensions of the problem, then it is reasonable to set

$$F = \exp\left[N f(s_1, s_2, t)\right] , \tag{7.3.7}$$

where the functions $f$ and $N$ are to be determined. As the local aspects of certain phenomena, for example, fluctuations in small "volumes", are to be neglected, the asymptotic solution may be obtained by letting $N$ be positively infinite. In this case, substituting (7.3.7) into (7.3.5) yields

$$\begin{aligned} \frac{\partial f}{\partial t^*} &= A s_1 (s_1 + 1) \frac{\partial f}{\partial s_1} - B s_2 \frac{\partial f}{\partial s_2} + (s_2 + 1)(s_2 - s_1) \\ &\quad \times \left[ \left(\frac{\partial f}{\partial s_1}\right) \frac{\partial f}{\partial s_2} + \left(\frac{1}{N}\right) \frac{\partial^2 f}{\partial s_1 \partial s_2} \right] , \end{aligned} \tag{7.3.8}$$

where $\alpha = AN$, $\beta = BN$, $t^* = tN$ and $A, B = \mathrm{O}(1)$. As $N$ is very large, the second derivative term multiplied by $1/N$ can be omitted. We seek solutions of (7.3.8) in the form of

$$\begin{aligned} f &= a_1(t)s_1 + a_2(t)s_2 + \tfrac{1}{2} b_{11}(t)s_1^2 + b_{12}(t)s_1 s_2 \\ &\quad + \tfrac{1}{2} b_{22}(t)s_2^2 + \dots , \end{aligned} \tag{7.3.9}$$

where the expansion coefficients are related to the moments of the probability distribution. For instance, $a_1 = E(x)/N$, $a_2 = E(y)/N$, and $b_{ij}$ are the variances of $x$ and $y$. Substituting (7.3.9) into (7.3.8), we obtain

$$\begin{aligned} \frac{da_1}{dt^*} &= A a_1 - a_1 a_2 - \frac{b_{12}}{N} , \\ \frac{da_2}{dt^*} &= -B a_2 + a_1 a_2 + \frac{b_{12}}{N} . \end{aligned} \tag{7.3.10}$$

If we neglect the terms containing $1/N$, (7.3.10) allows the steady solution

$$a_1 = B , \quad a_2 = A , \tag{7.3.11}$$

which is identical to the corresponding equilibrium in the prey-predator equations. However, for our model it is more important to know properties of the variances. We have

$$\frac{d}{dt^*} \begin{pmatrix} b_{11} \\ b_{12} \\ b_{22} \end{pmatrix} = \begin{pmatrix} 2AB \\ -AB \\ 2AB \end{pmatrix} + \begin{pmatrix} 0 & -2B & 0 \\ A & 0 & -B \\ 0 & 2A & 0 \end{pmatrix} \begin{pmatrix} b_{11} \\ b_{12} \\ b_{22} \end{pmatrix} . \tag{7.3.12}$$

It can be shown that these equations do not admit a time-independent solution. Assume that at $t = 0$ the system was described by a factorizable Poisson distribution for $x$ and $y$. This assumption implies that $b_{ij}(t = 0) = 0$. The solution of (7.3.12) satisfying the initial conditions is

$$\begin{aligned} A b_{11} + B b_{22} &= 2AB(A + B)t^* , \\ b_{12} &= \frac{A - B}{2} \left[ 1 - \cos\left( (4AB)^{1/2} t^* \right) \right] \\ &\quad - \frac{(4AB)^{1/2}}{2} \sin\left( (4AB)^{1/2} t^* \right) . \end{aligned} \tag{7.3.13}$$

We see that although the initial variances are zero and the system is macroscopically at a steady state, the variances $b_{ii}$ $(i = 1,2,)$ increase in time and deviate immediately from the initial values. It is impossible for the variances to reach a new steady state. Stochastically, the macroscopic equilibrium (7.3.11) is meaningless even in the limit of small fluctuations corresponding to the truncation performed on the moment equations by neglecting the $1/N$ terms. The system exhibits abnormal fluctuations that increase linearly in time with a periodic "background noise" whose frequency is twice the frequency of the macroscopic motion. Eventually, these fluctuations alter the order of magnitude of the $1/N$ terms, which may no longer be neglected in the moment equations. As a result, the average values are driven by the fluctuations to a time-dependent regime far from the steady state. This implies that the fluctuations play a decisive role by qualitatively altering the prediction of the macroscopic analysis.

The possibility of spontaneous deviations from the regime of the fluctuations provides a striking illustration of the breakdown of the laws of large numbers. As pointed out by Nicolis and Prigogine (1977), this entirely new situation is a consequence of the "coupling", as a result of which the transitions undergone by the stochastic variables are not statistically independent events, even in the limit of a large system.

There are some limitations in birth-death formalism. For example, the approach in which the transition probabilities are computed in terms of aggregated variables referring to the entire system implies that only very exceptional fluctuations are retained in the description. Treating the system as a whole may result in dismissing the importance of fluctuations associated with such properties as the size, the range over which they extend, and the correlation length over which two parts of the system can "feel" each other.

## 7.4 A Non-equilibrium Model of the Schumpeter Clock

It is one of the fundamental problems of economics to derive the properties of multi-component systems on the macroscopic level from their constituent components on the elementary microscopic level. One of the purposes of this approach is to explain which macro variables may be relevant under given circumstances for describing the dynamics of the system. For example, it is very popular to derive each macroeconomic theory on the basis of assumptions of rational behavior of households and firms. How the sum of parts and the whole are related to each other is an essential question in economics.

It is generally accepted in economics that even if the micro behavior of each firm (or household) is derived from uncertain mechanisms, the macro behavior of the system can be described by a few aggregated (mean value) variables, which make the analysis possible. Although some economic theories which take uncertainties into account have been proposed, most such studies are limited to static analysis.

Recently, Weidlich and Haag (1983) have suggested a very general quantitative approach to dynamic processes in social systems. This "statistical physics" approach is developed on the basis of the concepts of attitude space, socio-configuration and situation space. They try to describe dynamics of macroscopic variables using a phenomenological probabilistic description of the microscopic world. The phenomena dealt with by this approach belong to the field of socio-political psychology of individuals on a microscale and the consequential collective material, economic and abstract structures on a macroscale.

We now illustrate this approach and provide an example of applying it to economics.

Let the society under consideration consist of $N$ individuals and be subdivided into $P$ subgroups $P_k$ ($k = 1, \dots, P$) each with $N_k$ members: $N = N_1 + \dots N_k + \dots N_p$. As there are birth/death and immigration/emigration processes, $N_k$ is changeable. It is assumed that there is a set of $A$ different "aspects of life" related to such things as politics, religion, education, consumption and production, in which the individuals assume their roles. To each aspect $a$ ($a = 1, \dots, A$), there are $d_a$ possible attitudes ($i_a = 1, \dots, d_a$). The attitude space $G$ of $A$ dimensions consists of the $A$ different aspects. An individual's attitudes are denoted by the attitude vector $i = \{i_1, \dots, i_A\}$. The number of possible combinations of attitudes, $L$, is given by

$$L = \sum_{a=1}^{A} d_a \, .$$

Individuals have different attitudes. Let $n_{ki}$ be the number of members of $P_k$ having attitude $i$. The number of individuals with attitude $i$ is denoted by $N_i$. According to the definitions, we have

$$N_i = n_{1i} + \dots + n_{Pi} \, ,$$

$$N_k = n_{k1} + \dots + n_{kL} \, .$$

As $n_{ki}$ is changeable, we can define the socio-configuration at time $t$ by

$$n(t) = \{n_{ki}(t) \, , \, k = 1, \dots, P \, , \, i = 1, \dots L\} \, ,$$

which describes the microstate of the system. It consists of the $C = PL$ non-negative integer elements.

The attitudes and the socio-configuration are mainly related to the psychology and activities of the individual. In addition to these variables, to describe the development of society we have to take the material situation of the society into account. The material situation consists of quantitative measures such as the relevant variables in economics, the degree of efficiency in government and so on. Assume that there are quantitative measures $y_b$ ($b = 1, \dots, S$) which span an $S$-dimensional situation space $H$. The situation is described by the situation variable

$$y(t) = \{y_1(t), \dots, y_s(t)\} \, ,$$

belonging to $H$.

Thus the dynamics of the society are described by temporal changes of the socio-configuration and the situation vector. Obviously, the interactions between

$n(t)$ and $y(t)$ may lead to rather complicated behavior. Weidlich and Haag apply these concepts to explain processes of migration and/or birth-death in populations, opinion formation, and industrial evolution. In what follows, we present an industrial dynamic model proposed by Weidlich and Haag (1983, Chap. 5).

The model is limited to the "Schumpeter goods sector" which is largely identical with private and public industries. We are mainly concerned with the industrial investor (and innovator) and his strategic behavior under conditions of rivalry, whereby the causes and effects of macro-economic, demand-side "induced" investment are neglected. The model is to provide a partial theory for the non-equilibrium motion of an industrial system of nations or regions. It is defined within the framework of the "Schumpeter Clock" in the sense that its moving parts, driving mechanism and control devices are typically Schumpeterian. The Schumpeter Clock model stresses the existent, explicitly active pushing micro-economic forces and powerful supply-side checks and balances in explaining the short term non-equilibrium motions of an economy. The model is built on the basis of micro-economic differences, i. e., heterogeneity of products and production processes. These differences come into play at the subunit level (firms, markets, industries) of the economic system. The creation of such differences is the objective of strategic investments of entrepreneurs, which are classified according to their respective purposes as "expansionary" or "rationalizing". Alternative shifts from a predominantly expansionary investment portfolio to a predominantly rationalizing investment result in industrial fluctuations. During this cyclical process innovators and pioneering entrepreneurs in search of monopoly profits take the lead in the anticyclical redirection of investment strategies.

Relations of this special model to the general concepts just defined should be mentioned. The investors' configuration to be defined is a special case of a socio-configuration. The individuals considered here are the small groups of entrepreneurs in a position to make investment decisions. The economic decisions made are directly connected with material variables – the "investment structure index" to be defined.

First, consider strategic investment, and then the investors' "configuration". It is assumed that the investor's strategic choice set contains only two types of alternative: expansionary or rationalizing investment projects. Hence, the total volume of strategic investment $I(t)$ is

$$I(t) = E(t) + R(t) , \tag{7.4.1}$$

where $E$ and $R$ are the (non-negative) volumes of expansionary and rationalizing investment, respectively. If we denote by $E_0(t)$ and $R_0(t)$ the long term average values of $E(t)$ and $R(t)$, respectively, we may decompose $E$ and $R$ as

$$E(t) = E_0(t) + B(t) ,$$

$$R(t) = R_0(t) - B(t) ,$$

where $B(t)$ is called the oscillating shift, $-E_0 < B(t) < R_0$. An investment structure index $Z(t)$ is defined as

$$Z(t) = Z_0(t) + z(t) = \frac{E(t) - R(t)}{I} , \tag{7.4.2}$$

where $Z_0 = (E_0 - R_0)/I$, $z = 2B/I$, $-1 < Z(t) < 1$. The performance of the Schumpeter Clock will be demonstrated by observing the non-equilibrium motion of the investment structure index $Z(t)$ (or $z(t)$).

For convenience in explaining the inventor's configuration, it is assumed that each firm can only undertake one project and all projects (the total number of which is $2N$) have the same financial volume.

We consider a fictitious "neutral" investor who behaves according to the average long term investment trend. His investment project of $i = I/2N$ is composed of expansionary investment $e_0$ and rationalizing investment $r_0$ as follows

$$i = e_0 + r_0 \,, \tag{7.4.3}$$

where $e_0 = E_0/2N$, $r_0 = R_0/2N$. Actual investors, however, behave differently from the neutral investor. There are $E$-type ($R$-type) investors who favor the expansionary (rationalizing) investment against the average trend. For the $E$-type and $R$-type investors, the projects of financial volume $i$ can be written as

$$i = e_E + r_E = e_R + r_R \,, \tag{7.4.4}$$

where $e_E = e_0 + \beta$, $r_E = r_0 - \beta$, $e_R = e_0 - \beta$, $r_R = r_0 + \beta$, $\beta > 0$. For the $E$-type investors, $\beta$ is the surplus of expansionary investment in comparison with the neutral one, while for the $R$-type ones it is the surplus in rationalizing investment. For simplicity, we assume the same surplus for the two types of investors.

Let $n_E$ denote the number of $E$-type investors and $n_R$ the number of $R$-type investors, respectively. We have

$$n_E(t) + n_R(t) = 2N \,. \tag{7.4.5}$$

The investment structure is characterized by $[E(t), R(t)]$, and the investor's strategic investment activities by $\{n_E(t), n_R(t)\}$. We call $\{n_E(t), n_R(t)\}$ the investors' configuration and define the investors' configuration index as

$$x(t) = \frac{n_E - n_R}{n_E + n_R} = \frac{n}{N} \,, \tag{7.4.6}$$

where $n(t) = [n_E(t) - n_R(t)]/2$, $-1 \le x(t) \le 1$. The integer $n(t)$ may increase or decrease by one if the investors' configuration changes according to transitions

$$\{n_E, n_R\} \rightarrow \{n_E + 1, n_R - 1\}$$

or

$$\{n_E, n_R\} \rightarrow \{n_E - 1, n_R + 1\}$$

i. e., if an $R$-type investor becomes an $E$-type investor or vice versa. Multiple-unit motions ($n \rightarrow n - \sigma$, $\sigma$ is an integer) within the investor's configuration are also possible.

From the definitions, we see that the total expansionary and total rationalizing investments are given by

$$E = n_E e_E + n_R e_R \quad (= E_0 + 2\beta N x) ,$$

$$R = n_E r_E + n_R r_R \quad (= R_0 - 2\beta N x) , \tag{7.4.7}$$

which combined with (7.4.2), gives

$$Z = \frac{E_0 - R_0}{I} + \frac{2N\beta x}{I} ;$$

or

$$z = \frac{2N\beta x}{I} = gx , \quad g = \frac{4N\beta}{I} . \tag{7.4.8}$$

In (7.4.8) it is stated that the fluctuating part $z(t)$ of the investment structure index $Z(t)$ is proportional to the investors' configuration index $x(t)$. Thus oscillations of the investors' configuration will show up in oscillations of the investment structure index. In (7.4.8) relations between the investment structure and investors' configuration are defined.

In what follows, we derive equations of motion – for the investors' configuration and for the investors' propensities – of the two components of investment which formulate the changes of the industrial economy.

The micro-economic approach to changes in the investors' configuration $\{n_E, n_R\}$ incorporates the notion of individual transition probabilities for investors turning from an $R$-type to an $E$-type, and vice versa. A stochastic approach is used to describe behavior of the investors in risky uncertain environments.

The transition from one investors' configuration $\{n_E, n_R\}$ to another can be a single unit motion connected with a product innovation or a process innovation of one investor, or a multiple unit motion which is most often an imitation process. Such processes are uncertain, since risk and other factors affect the behavior. Let us define

$p_{\rightarrow}(n)$ = probability per unit time for turning from $R$-type to $E$-type investment ;

$p_{\leftarrow}(n)$ = probability per unit time for turning from $E$-type to $R$-type investment .

The individual transition probabilities yield the total probabilities for changes of the entire investors' configuration. The transition $\{n_E, n_R\} \rightarrow \{n_E+1 , n_R - 1\}$ takes place with the total transition probability

$$w_{\rightarrow}(n) = n_R p_{\rightarrow}(n) = (N - n) p_{\rightarrow}(n) . \tag{7.4.9}$$

Similarly, for the transition $\{n_E, n_R\} \rightarrow \{n_E - 1 , n_R + 1\}$, one has

$$w_{\leftarrow}(n) = n_E p_{\leftarrow}(n) = (N + n) p_{\leftarrow}(n) . \tag{7.4.10}$$

The probability that at time $t$ the investors' configuration $\{n_E, n_R\}$ is denoted as

$$p[n_E, n_R ; t] = p(n; t) , \quad \text{for} \quad -N \le n \le N . \tag{7.4.11}$$

As one of the configurations is always realized, we have

$$\sum_n p(n;t) = 1 \;, \tag{7.4.12}$$

for any time $t$. The master equation describes the motion of $p(n;t)$. The probability $p(n;t)$ of the configuration $n$ can increase by means of transitions from either one of the neighboring configurations $(n-1)$ or $(n+1)$ into configuration $n$, while $p(n;t)$ may simultaneously decrease by transitions from configuration $n$ into these neighboring configurations. From such balance considerations one immediately has the following master equation

$$\begin{aligned} dp(n;t) = & \left[w_{\rightarrow}(n-1)p(n-1;t) + w_{\leftarrow}(n+1)p(n+1;t)\right] \\ & - \left[w_{\rightarrow}(n)p(n;t) + w_{\leftarrow}(n)p(n;t)\right] \;, \end{aligned} \tag{7.4.13}$$

where the first term describes the probability flux per unit time into the configuration $n$ and the second term the flux from the configuration $n$. The master equation (7.4.13) represents $2N+1$ coupled differential equations for $p(n;t)$ and are generally difficult to solve. For simplicity, we assume that the $p(n;t)$ are sharply peaked and unit-modal around their mean values $\langle n \rangle_t$:

$$\langle n \rangle_t = \sum_n np(n;t) \;. \tag{7.4.14}$$

Taking the time derivative of (7.4.14) and inserting (7.4.13) into it yields

$$\frac{d\langle n \rangle_t}{dt} = \sum \left[w_{\rightarrow}(n) + w_{\leftarrow}(n)\right] p(n;t) \;,$$

which can approximately be written as

$$\frac{d\langle n \rangle_t}{dt} = w_{\rightarrow}\left(\langle n \rangle_t\right) + w_{\leftarrow}\left(\langle n \rangle_t\right) \;. \tag{7.4.15}$$

From (7.4.6) and (7.4.15) we can obtain the mean value equation for $x$

$$\frac{d\langle x \rangle_t}{dt} = K\left(\langle x \rangle_t\right) \;, \tag{7.4.16}$$

where the driving force $K\left(\langle x \rangle_t\right)$ of $\langle x \rangle_t$ is

$$K\left(\langle x \rangle_t\right) = \frac{1}{N}\left[w_{\rightarrow}\left(\langle n \rangle_t\right) + w_{\leftarrow}\left(\langle n \rangle_t\right)\right] \;.$$

We may rewrite (7.4.16) as

$$\frac{dx}{dt} = K(x) \;, \tag{7.4.17}$$

where $x$ represents $\langle x \rangle_t$. Weidlich and Haag specify (7.4.17) as

$$\frac{dx}{dt} = 2\delta\left[\sinh(a+kx) + x\cosh(a+kx)\right] = K(x\,;\,a,k) \;. \tag{7.4.18}$$

The driving force of $x$ is assumed to depend on the actual configuration $x(t)$ of all the investors in the system and the transition probabilities for change of this configuration. These in turn depend upon all investors' propensities parameterized by $a$ and $k$,

$$2\sinh(y) = \exp(y) - \exp(-y) ,$$

$$2\cosh(y) = \exp(y) + \exp(-y) ,$$

and $\delta$ is a scalar used in actual computations for time scaling purposes.

In (7.4.18), $a$ is the "alternator" representing the investor's switch in his preference for either $E$- or $R$-type investment under the given circumstances. In the specification of transition probabilities the alternator $a$ is chosen as a parameter obeying the following rule: positive $a$ implies that transitions to $E$-type investment are favoured; negative $a$ favors transitions to $R$-type investment. The alternator is assumed to be time-dependent and plays a role in creating cyclic (Schumpeter Clock) motion. The parameter $k$ is the "coordinator" representing the individual investor's interaction intensity under the given circumstances. In other words, $k$ describes the investor's inclination to conform to other investors' behaviour. The coordination effect will become visible as a synchronization and imitation of investments undertaken by various investors.

Now, we will describe the dynamics of the alternator $a(t)$ – the investors' strategic choice parameter in the driving force $K$. First, we observe that if the majority of investors tend to maximize profits at a given point in time by expanding (rationalizing) their business operations so that $x(t) > 0$ ($x(t) < 0$), then some innovators or pioneers (trend setters) will try to improve their market position by adopting a non-conformist strategy in an attempt to capture quasi-rent due to differentiation. When an upswing is well under way due to expansionary investments undertaken by a majority of investors, these trend setters tend to redirect their effort and to start pushing back the cost frontier by means of cost-reducing investment. In this case, the others are forced to imitate and also to undertake rationalizing investments in the expectation of further cost reduction. Similarly, when a downswing is well under way due to the contrast effects of rationalizing investments undertaken by a majority of investors, the trend setters start moving towards the quality section of the best practice frontier. This introduces better products and implements investment plans for expanding facilities and thereby forces others to imitate. Their expansionary and quality updating behavior thus creates the synchronization to be observed in the occurrence of the product life cycle. According to this discussion, we see that the equation of motion for the alternator $a(t)$, which measures in aggregate terms the differentiation acitvities of entrepreneurs in various fields of industrial investment, should generate switches under the circumstances stipulated above. As discussed by Weidlich and Haag (1983), a potentially suitable specification of the dynamic behavior may be given by

$$\begin{aligned} \frac{da(t)}{dt} &= -2\mu\left[a_0\sinh(\gamma x) + (a - a_1)\cosh(\gamma x)\right] \\ &= L(x\,;\,a, k) , \end{aligned} \tag{7.4.19}$$

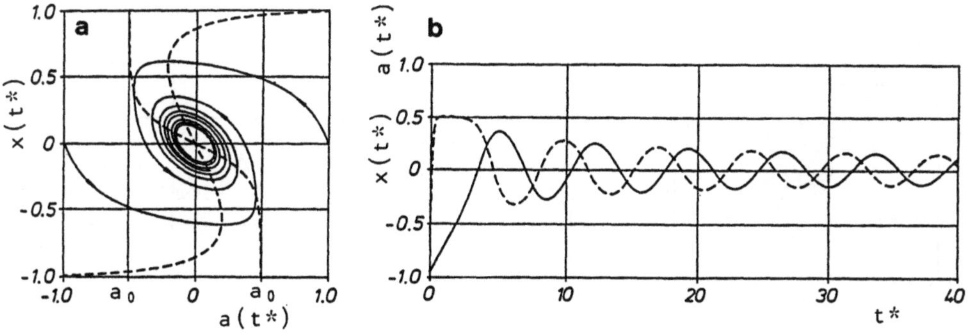


**Fig. 7.3a, b.** Marginally stable focus at the origin. (a) $a$-$x$ plane, (b) paths with time

where $L$ is the strategy reformation driving force, $\mu$ is the strategic flexibility parameter describing the flexibility of the investors in turning their strategies from expansionary to rationalizing, and vice versa, $\gamma$ is the trend reversal speed parameter, $a_1$ is a strategy bias parameter which is positive or negative according to whether the entire trend period is heavily biased towards expansion or towards rationalization, and $a_0$ is the strategic choice amplitude used as an operative scaling constant.

The whole dynamic system consists of (7.4.18) and (7.4.19). For simplicity, we introduce: $t^* = 2\delta t$, and $\varepsilon = \mu/\delta$. The system can be written as

$$\frac{dx}{dt^*} = \sinh(a + kx) - x\cosh(a + kx) = K^* ,$$

$$\frac{da}{dt^*} = -\varepsilon\left[\sinh(a + kx) - x\cosh(a + kx)\right] = L^* . \qquad (7.4.20)$$

The existence of equilibria can be easily guaranteed, although uniqueness cannot be demonstrated. In fact, there may be one, three or five equilibrium points which depends upon the values of the parameters. Furthermore, applying the Poincaré-Bendixon theorem, Weidlich and Haag identify the existence of limit cycles in the system.

We present some of these simulation results. Choose a parameter combination of $k = 1.5$, $a_0 = 0.5$, $a_1 = 0$, $\gamma = 4.0$, $\varepsilon = 0.5$. In this case, $k = 1.5$ is a threshold value for the transition to another type of solution. There occurs a critical reduction of the damping effect. Although the origin is still the only equilibrium and it is stable, the oscillatory relaxation takes very long time. The behavior is shown in Fig. 7.3 (see Fig. 5.5 in Weidlich and Haag).

If we take the following values of the parameters: $k = 1.6$, $a_0 = 0$, $a_1 = 0$, $\gamma = 4.0$ and $\varepsilon = 0.5$, then sustained cycles occur. The origin is an unstable focus of the motion of the economy. The behavior is illustrated in Fig. 7.4 (see Fig. 5.6 in Weidlich and Haag).

Weidlich and Haag also applied their model to the economy of the Federal Republic of Germany during the interval 1955–1980. Figure 7.5 shows the result of the empirical analysis which consists of a piecewise application of the model to a certain set of parameters for the three intervals 1955–1965, 1967–1971 and 1973–1980. The explanation in detail is given in Weidlich and Haag (1983).

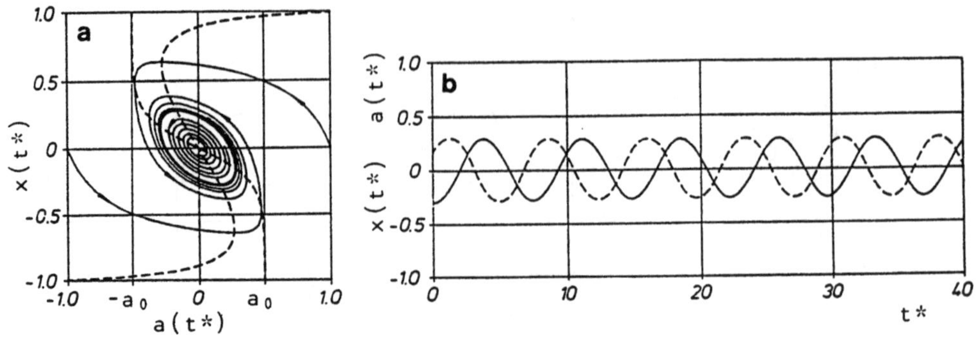


Fig. 7.4a, b. An unstable focus and one cycle. (a) $a$-$x$ plane, (b) paths with time

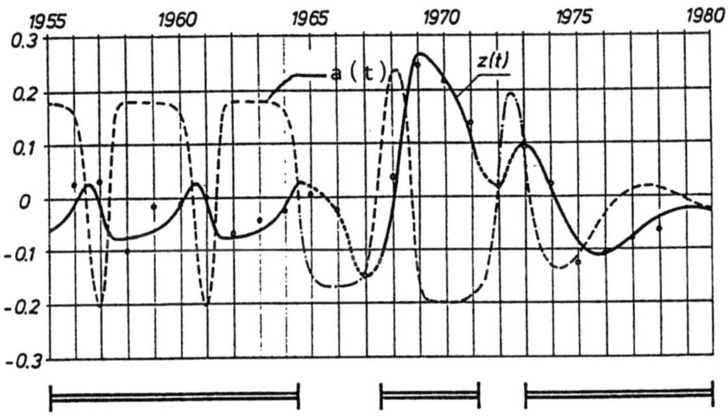


Fig. 7.5. A comparison between the simulation result and observed data

## 7.5 Effects of Noise on the Nonlinear Stochastic Systems Close to Critical Points

We have argued that an economic model which takes stochastic influences into account has to reflect on the degree to which these exogenous forces may potentially determine the results of the model. If the results from an economic model are crucially dependent on exogenous stochastic forces and are little influenced by interactions of economic variables, the model is of less interest. On the other hand, if the introduction of stochastic effects has hardly any effects on the qualitative results of the model, stochastic factors can be completely neglected in the analysis. However, as mentioned in Sect. 7.1, fluctuations may play a crucial role in economic development which is mainly "governed" by deterministic mechanisms. Effects of fluctuations on deterministic development cannot be neglected when the purely deterministic equations are located near critical points.

In the previous sections, we have derived macroscopic process equations with regard to microscopic processes. In discussing the master equation, we pointed out that these microscopic processes cannot be completely neglected as they give rise to fluctuating driving forces which drive the system far from equilibrium. Here, we directly define the dynamics of aggregated variables, considering such microscopic fluctuating forces as noise sources satisfying appropriate requirements.

The temporal evolution depends on causes which cannot be predicted with absolute precision. Usually, such causes are taken as fluctuating forces $F(t)$. The dynamic system (3.1.2) can thus be written as

$$\frac{dx}{dt} = f(x) + F(t) , \tag{7.5.1}$$

where $F(t)$ is given. This form of fluctuating forces acting on differential equations is termed additive noise. A randomly fluctuating environment may take on other forms. For instance, if the population growth rate appears to be fluctuating, then the population dynamics are given by

$$\frac{dP(t)}{dt} = \alpha(t)P(t) ,$$

where $P$ is the population and $\alpha(t)$ is the random growth rate. This type of fluctuation is called multiplicative noise. In this section, we are interested in the effects of additive noise upon the dynamics of the corresponding deterministic equations: $dx/dt = f(x)$ close to unstable points.

It is assumed that the function $F$ is comparatively small in the sense that it does not change the character of the transition appreciably. This means that instability is not induced by fluctuating parts but by the deterministic part $f(x)$.

A typical case for (7.5.1) are the Langevin equations for Brownian motion:

$$\frac{dx_i}{dt} = \frac{p_i}{m} ,$$

$$\frac{dp_i}{dt} = -rp_i + F_i , \quad i = 1,2,3 , \tag{7.5.2}$$

where $x_i$ and $p_i$ are the position and momentum of a small "Brownian" particle suspended in a gas. The force exerted on the Brownian particle is split into a "systematic" part $rp_i$, and a "random" fluctuating component $F_i$. If we neglect fluctuations, the Brownian particle performs a damped motion and finally comes to rest. The effects of fluctuations may lead to a ceaseless irregular motion of the particle (Fig. 7.6).

To further illustrate the effects of fluctuations, we introduce the so-called ensemble standpoint: an ensemble of macrosystems with equal composition and described by the same set of macrovariables $x_i$ is considered. Each member is denoted by $j$. We consider the case in which each member may be subjected to different microscopic fluctuations $F_i$. It may be expected that different paths $x_i^j(t)$ for the members of the ensemble will be observed, even if we assume that for all the initial conditions $x_i^j(0)$ (for $j = 1,2,\ldots$) the variables are the same. Let $\langle x_i(t)\rangle$ denote the ensemble mean value, i.e.

$$\langle x_i(t)\rangle = \left(\frac{1}{N}\right) \sum_{j=1}^{N} x_i^j(t) .$$

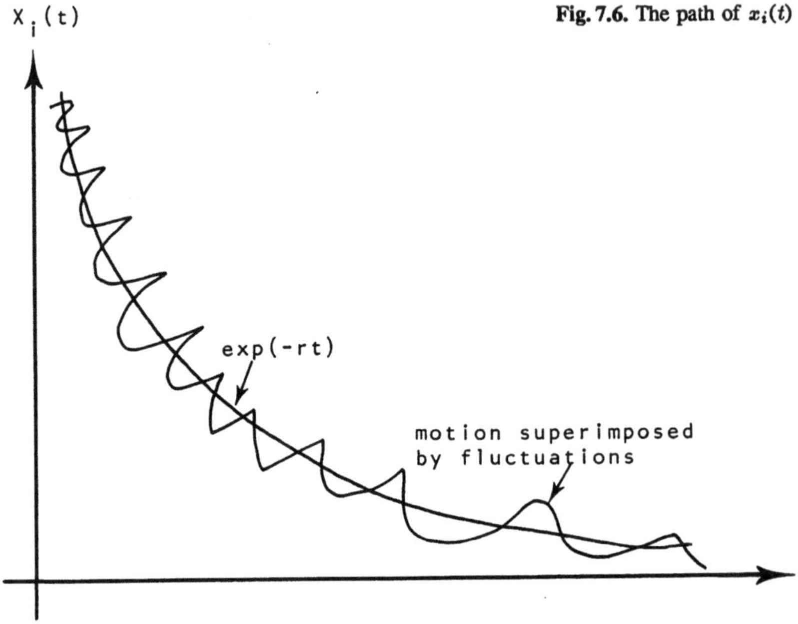


**Fig. 7.6.** The path of $x_i(t)$

Fluctuations may be specified in different forms. We assume that $\langle F_i(t)\rangle = 0$ for all $i$, although covariances between them are not identically equal to zero. We may have two structurally different cases:

(i) Suppose that the solutions of each $x_i^j(t)$ with the same initial value do not deviate essentially from their mean values $\langle x_i(t)\rangle$. In this case, we obtain the approximate equations for $\langle x_i(t)\rangle$ as

$$\frac{d\langle x_i(t)\rangle}{dt} = f\left[\langle x(t)\rangle\right] \ , \quad i = 1, 2, \dots .$$

This is a set of autonomous differential equations. Its solutions do not deviate essentially from the true paths $x_i^j(t)$. In this case, the macrovariables $x_i^j(t)$ approximately obey a closed self-contained subdynamics. Thus quickly fluctuating random forces exerted by microvariables only lead to small fluctuations of the macrovariables around the smoothly developing ensemble average (see Fig. 7.7).

(ii) Consider the corresponding deterministic equation

$$\frac{dx_i}{dt} = f_i(x) \ , \quad i = 1, 2, \dots . \tag{7.5.3}$$

A large number of examples in the preceding chapters show that bifurcations may occur in such systems. This means that solutions of (7.5.1) starting with slightly different initial values $x_i(0)$ in the vicinity of a critical state may have completely different paths $x_i(t)$. Thus, adding fluctuations to such systems may lead to completely different paths from those determined by the corresponding deterministic equations. In other words, infinitesimally small differences of certain "causes" may lead to very large differences of effects. In this case, the individual paths $x_i^j(t)$ can

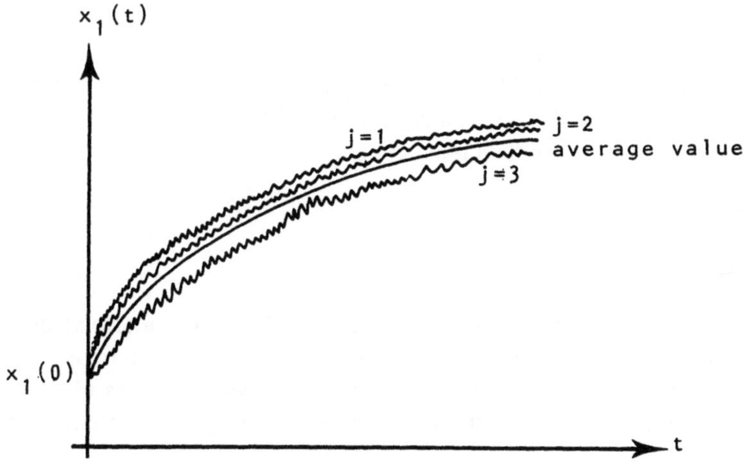


Fig. 7.7. Small fluctuations lead to small differences

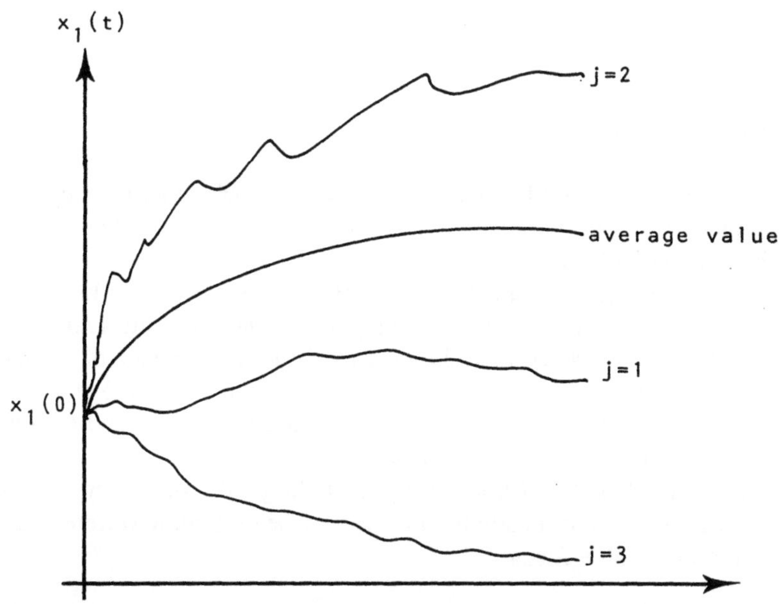


Fig. 7.8. Deviations from the average due to small fluctuations

deviate significantly from the mean value $\langle x_i(t) \rangle$. Therefore, the mean value dynamic equations are not valid to describe dynamic evolution of the system (see Fig. 7.8).

Consider simple predator-prey models in stochastic environments. The class of models under consideration is

$$\frac{dx_i}{dt} = x_i\left\{s_i(x_i) + \Sigma_j a_{ij} x_j\right\} + r_i(t) x_i \,,$$

$$i = 1, \ldots, n \,, \qquad (7.5.4)$$

where $x_i$ represents the $i$th species density, $a_{ij}$ the usual Volterra-Lotka inter-specific rate constants, and $s_i(x_i)$ includes all intraspecific interactions. It is assumed that if $i$ is a predator, then $s_i(x_i) = u_i$ (the Malthusian growth rate); while if $i$ is a prey, $s_i(x_i) = u_i + c_i x_i - d_i x_i^2$. In (7.5.4), $r_i(t)$ is a random variable affecting the average coefficients $u_i$ in the presence of unpredictable events such that

$$\langle r_i(t) \rangle = 0 ,$$

$$\langle r_i(t) r_i(t') \rangle = \sigma_{ij}^2 \delta_{ij}(t - t') , \tag{7.5.5}$$

where $\langle . \rangle$ means ensemble average. We assume that $r_i$ values are uncorrelated to each other and have a delta correlation time ($\delta$) with a constant dispersion ($\sigma$). Obviously, the Goodwin model and its generalized forms are special cases of these dynamics. We can thus interpret the results for these dynamics systems in terms of economics.

It is interesting to study the effects of a random environment with zero average upon the behavior of the corresponding deterministic system with $r_i(t) = 0$ holding identically. In what follows, we assume that the corresponding deterministic equations of (7.5.4)

$$\frac{dx_i}{dt} = x_i \left\{ s_i(x_i) + \Sigma_j a_{ij} x_j \right\} ,$$

have limit cycle solutions. It is well known that in these types of model limit cycles may occur. Applying the Hopf bifurcation theorem, we can provide the precise conditions for the existence of limit cycles.

Some very interesting results about the effects of fluctuations on the behavior of (7.5.4) have been obtained. For instance, applying an adaptation of the averaging method, Lin and Kahn (1977) proved the following result: in the presence of a limit cycle, (a) the radius of the limit cycle decreases as noise increases; (b) if noise dispersion is larger than the deterministic radius, no limit cycle exists; and (c) the dispersion of the angular variable increases linearly in time.

The conclusion (b) implies that if noise is relatively large, the stationary probability distribution of a small deterministic limit cycle may be difficult to differentiate from the distribution of a stable focus.

## 7.6 Effects of Random Environment on a Two-Dimensional Deterministic System Near Critical Points

We have just provided an example to show how the behavior of deterministic equations in the presence of a limit cycle can be affected in a noisy environment of zero average value. On the other hand, we identified the existence of economic oscillations in different systems in Chap. 5. However, as the results in the previous section hold for the predator-prey models, it is important to discuss the problem for the general case.

A two-dimensional system under consideration is generally described by

$$\frac{dx_i}{dt} = f_i(x_1, x_2, r) , \tag{7.6.1}$$

where $r$ is a parameter. In some cases, the system may have multiple limit cycles. This system is completely deterministic. Fluctuations are omitted. Mangel (1980) investigated the effects of fluctuations on systems with single and multiple limit cycles. Four types of periodic behavior are considered. They are: (1) a fixed, stable limit cycle, surrounding an unstable focus (Fig. 7.9a); (2) a fixed unstable limit cycle, surrounding a stable focus and enclosed by a stable limit cycle (Fig. 7.9b); (3) the Hopf bifurcation problems; and (4) a "dual" bifurcation in which an unstable cycle and stable focus coalesce (Fig. 7.9c). All of these behaviors are well observed in dynamic systems.

Mangel introduced noise into (7.6.1) in the following way:

$$\frac{dX_i}{dt} = f_i(X, r) + \left(\frac{\varepsilon}{\alpha^2}\right)^{1/2} F_i(X) Y_i \left(\frac{t}{\alpha^2}\right) ,$$
$$i = 1, 2 , \tag{7.6.2}$$

in which $X_i$ is the corresponding random variable with respect to $x_i$, and $Y$ is a stationary, zero mean process. The parameter $\varepsilon$, $0 < \varepsilon \ll 1$, characterizes the intensity of the fluctuations. If $\alpha$ is sufficiently small, $X(t)$ converges to a diffusion process.

When fluctuations are superposed upon the deterministic equations, a number of questions can be asked. Each question should be dependent on the nature of the deterministic equations. For instance, when the system always tends towards the stable limit cycle as in Fig. 7.9a, it is important to know how fluctuations may drive the system away from the limit cycle.

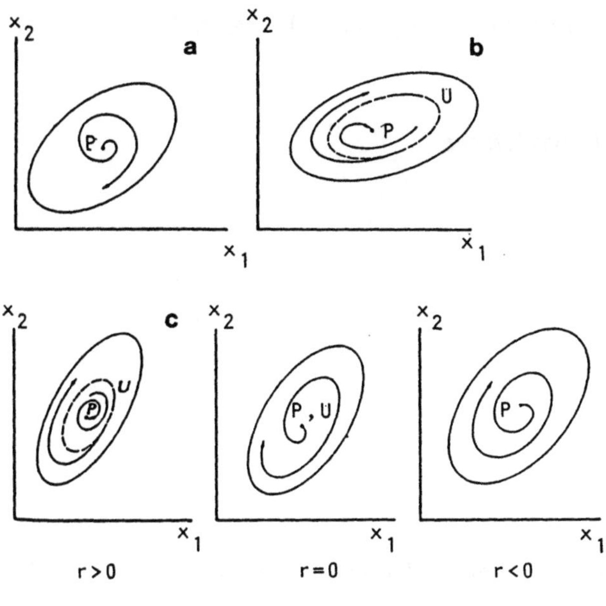


**Fig. 7.9.** (Line – stability, dashed line – instability)

To describe the behavior of the stochastic equations, let us introduce

$$\Theta(t,x)\delta x = \Pr\{x \le X(t) \le x + \delta x\} , \tag{7.6.3}$$

where $x$ is determined by (7.6.1), $X(t)$ by (7.6.2) and $\Theta(t,x)$ is the probability density for $X(t)$. If we let $t \to +\infty$, then $\Theta(t,x) \to \Theta(x)$, the equilibrium or stationary density, which gives the probability of eventually finding the process in the interval $(x, x+\delta x)$. Associated with $\Theta(t,x)$ is an initial density, $\Theta(0, X(0))$ which characterizes the distribution of the random variable $X(0) = x(0)$.

In the case of Hopf bifurcation, we are interested in the density for $X(t)$, which is denoted by $\Theta(t, x\,;\,r)$ with $r$ as the bifurcation parameter. Consider the dual Hopf bifurcation system (Fig. 7.9c). For small $r$, a phase point will leave a neighborhood of $P$ or $U$ and approach $L$ with a certain probability. The singularity at $P/U$ for $r=0$ will be evidenced by very slow deterministic repulsion from $P$. Let $L^*$ be a neighborhood of the stable limit cycle and let

$$\begin{aligned} T(x) = E\{t:\ & X(t) \in L^*,\ X(s) \notin L^*,\ s < t | X(0) \\ & = x, X(t)\,\text{reaches}\,L^*\} . \end{aligned} \tag{7.6.4}$$

Thus, $T(x)$ is the expected time to reach $L^*$, given that $X(0) = x$.

The initial phase point is crucial when considering an unstable limit cycle $U$ surrounded by a stable cycle as in Fig. 7.10.

A phase point initially in the vicinity of $U$ leaves any neighborhood of $U$ with probability 1. Even if $X(0)$ is on $U$, fluctuations will drive the trajectory away from $U$. As the problem of calculating the probability that the phase point reaches a neighborhood of the node $P$ rather than the stable limit cycle $L$ is too difficult, an alternative problem is posed as follows. Let $s$ measure distance normal to the unstable limit cycle, chosen so that $s > 0$ corresponds to an outward orientation. Let

$$S_1 = \{x : s(x) = s_1\}, \quad S_2 = \{x : s(x) = s_2\}, \quad s_1 > 0, \quad s_2 < 0 . \tag{7.6.5}$$

Consider the probability

$$\phi(t,x) = \Pr\{\text{by time } t,\ X(t)\,\text{has left the}$$

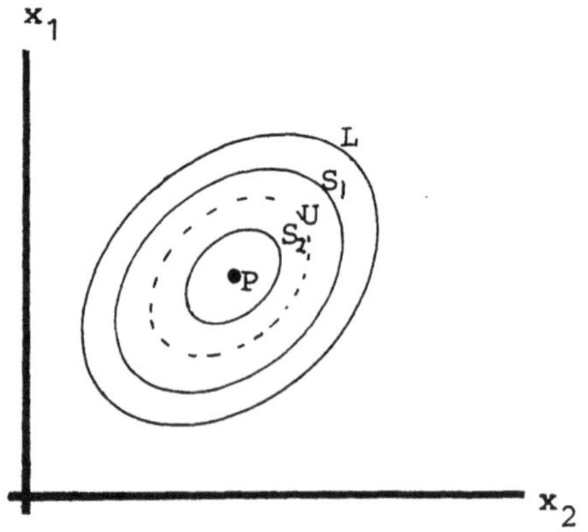


**Fig. 7.10.** The first exit problem

$$\text{annulus}\,(S_1, S_2)\,\text{through}\,S_1\,|X(0) = x\} \,. \tag{7.6.6}$$

The stationary version of $\phi(t, x)$ is $\phi(x)$, which is the probability that $X(t)$ first exits from $(S_1, S_2)$ through $S_2$.

It is important to calculate the above quantities from the system (7.6.2). Mangel obtains these quantities by the diffusion approximation method of Papanicolaou and Kohler (see Mangel 1980). In this approximation, $\Theta(t, x)$, $\phi(t, x)$ and $T(x)$ all satisfy deterministic partial differential equations. For instance, the equations determining the stationary versions of $\Theta$ and $\phi$ for the "canonical" problems are given by

$$0 = \frac{\varepsilon(A\Theta)_{xx}}{2} - (f\Theta)_x \,,$$

$$0 = \frac{\varepsilon A\phi_{xx}}{2} + f\phi_x \,, \tag{7.6.7}$$

subject to appropriate initial and boundary conditions. In (7.6.7), the functions $A(x)$ can be explicitly determined from $F$ and $Y$. We can thus analyze the behavior of the system by the traditional methods for partial differential equations.

The problems corresponding to stable and unstable limit cycles, and the Hopf bifurcation problems are analyzed in Mangel (1980). The functions $\Theta(t, x)$ and $\phi(t, x)$ are calculated for different cases by the use of formal asymptotic methods. It can be seen that such an analysis can be carried out similarly for the dynamic systems described in Chap. 5, where the models exhibit limit cycles.

## 7.7 Conclusions

In macroeconomics, usually only a few aggregated variables are chosen to describe economic evolution. It is explicitly or implicitly assumed that macroscopic description deals generally with average behavior and that probablistic elements or random fluctuations (with zero average) play no role. In economics this method has been emphasized after the Keynesian revolution. Surely, this point of view is held as long as stability is presumed in economic analysis. This is due to the properties of stable systems that small shifts of parameters (environment) can only result in small changes of the variables. However, if economic systems are unstable, we must be careful about the effects of random fluctuations. Small fluctuations may drive systems far away from old trajectories. This can be understood intuitively, as the examples in the previous chapters showed that nonlinear systems with instabilities are very sensitive to small changes in parameters. Often, structural changes (or phase transitions) are observed in the system in the presence of small shifts in parameters. From this one might expect that fluctuations, although measurable, should remain small compared to the macroscopic variables, if the system is stable, though this cannot be held at the points of phase transitions or "revolution". That is, in the presence of a critical point small fluctuations are amplified, attain a macroscopic level, and drive the system to a new phase. In the critical region around critical points, the system exhibits a marked coherent behavior, frequently accompanied by long-range fluctuations.

We have shown that there are macro-systems in which fluctuations and probablistic description play an essential role in affecting evolution paths. From these examples, we can demonstrate that in the neighborhood of a critical point any small fluctuations may play a significant role in economic development. They may drive the behavior of the system away from the average. This is the meaning of the concept of order through fluctuations (Nicolis and Prigogine 1977).

Haken (1983) once said: "fluctuations render the phenomena and problem of bifurcations (which are difficult enough) into the still more complex phenomena and correspondingly more difficult problems of nonequilibrium phase transitions". The examples provided in this chapter show the way in which complicated economic phenomena may occur under influences of small fluctuations.

To further illustrate the ideas given in this chapter, we finally look at the behavior of the logistic equation subject to external noise (see Schuster 1988)

$$x_{n+1} = r x_n (1 - x_n) + \Theta_n \,, \tag{7.7.1}$$

where $r$ is a parameter, $\Theta_n$ are fluctuations satisfying

$$\langle \Theta_n, \Theta_{n'} \rangle = \sigma^2 \delta(n, n') \,. \tag{7.7.2}$$

In (7.7.1), $\Theta_n$ are Gaussian-distributed and $\sigma$ measures the intensity of the white noise. Figure 7.11a shows the behavior of (7.7.1) without external noise ($\sigma = 0$) and Fig. 7.11b shows the behavior with noise ($\sigma = 10^{-3}$). In these figures, $z$ represents the corresponding Lyapunov exponents. It should be noted that although the noise washes out the fine structure of the logistic map and its Lyapunov exponent, there is still a sharp transition to chaos which is indicated by the change in the sign of $z$ in Fig. 7.11b.

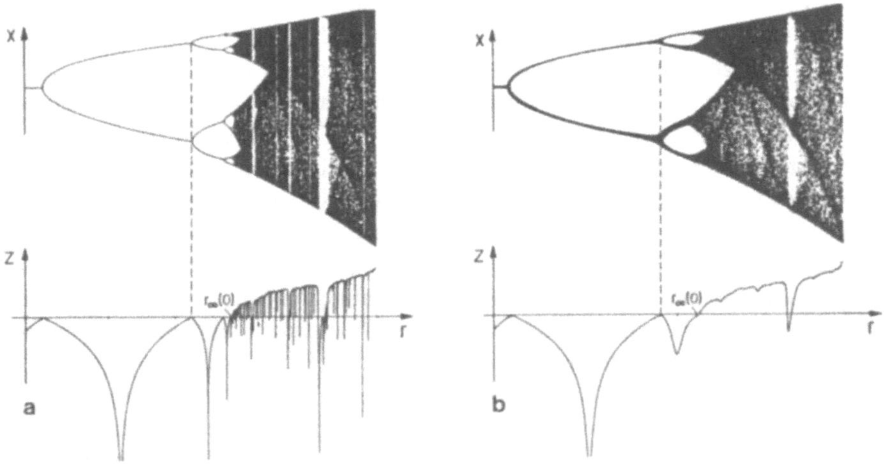


**Fig. 7.11.** Behavior of (7.7.1); **(a)** The logistic map and its Lyapunov exponent $z$, **(b)** the logistic map with external noise and its Lyapunov exponent

# 8. Urban Pattern Formation Process – Stability, Structural Changes and Chaos

> If, therefore, those cultivators of the physical science from whom the intelligent public deduce their conception of the physicist ... are led in pursuit of the arcane of science to the study of singularities and instabilities, rather than the continuities and stabilities of things, the promotion of natural knowledge may tend to remove that prejudice in favour of determinism which seems to arise from assuming that the physical science of the future is a mere magnified image of that of the past.
>
> *James Clerk Maxwell* (quoted after Schuster 1988)

An important aspect of evolutionary systems is related to space; any economic activities occur in a certain time and space. Interactions between different economic variables become strong over space as transportation and communication conditions are improved. It is important to understand the characteristics of such spatial interactions.

In the previous chapters we explicitly neglected the role of space in economic development. In a sense, this simplification is valid only if the spatial variables can effectively be represented by "aggregated" variables. This chapter is concerned with the spatiotemporal economic evolutionary process. In particular, we examine self-organized spatial evolution which exhibits structural changes and complicated dynamic behavior. We show that long-run urban patterns may be connected with regular or irregular time-dependent oscillations.

## 8.1 Continuous Spatial Economics and Description of Urban Pattern Formation

Urban problems have become very complicated as a result of technological progress and change in the behavior of human beings. The urban systems of our epoch are characterized by increasing spatial and temporal variation. Urban centralization has been observed in both developing and developed countries; decentralization has begun to appear in some developed countries. Rather than homogeneous urban patterns, we have observed heterogeneous ones. Metropolitan areas such as New York, Stockholm, Paris and Tokyo, provide examples of the complexity of modern urban forms.

Many urban models have been suggested to explain and forecast urban pattern formation in urban economics, regional science and geography. There are three main approaches in the recent literature. The first, called neoclassical urban economics, has been developed by urban economists. Since Alonso (1964) published his work, many urban models have been built within this framework. Like equilibrium theory, delicately reformed by Debreu, Arrow and others, the works in urban economics have deepened our insight into economic mechanisms of urban development. However, this approach is mainly limited to equilibrium analysis, and stability is still presumed in these models.

The second approach is mainly carried out by the researchers in regional science and geography (e.g., Wilson 1981). Time and space are treated significantly by this approach. However, as this represents space by discrete zones, it is invalid to explain internal structures of urban areas (see Beckmann and Puu 1985).

The third approach, called the interactional spatial dynamic approach, uses continuous space to study dynamic urban problems (e.g., Beckmann 1952, Beckmann and Puu 1985, Puu 1987, Andersson and Zhang 1988, Zhang 1988e, 1990). The collaboration between Beckmann and Puu provided a modern version of the von Thünen approach to classical location and spatial economics. In contrast to the recent development in regional economics where the spatial structure has been suppressed and replaced by mere matrices of abstract distance, Beckmann and Puu based their approach to spatial economics on the von Thünen tradition: activities are space consuming. The economic activities are described by their spatial densities. Evolution of internal urban structure is the main focus of this approach. Urban problems are thus often described by a set of partial differential equations subject to appropriate boundary and initial conditions. In this chapter we investigate effects of interactions among variables in urban pattern formation processes within the framework of this approach.

In order to explain this approach, we consider a transportation model defined in a continuous space. This model and its extensions are fully developed by Beckmann and Puu (1985). The model provides an example of the way in which space is treated by the interactional spatial dynamic approach.

It is assumed that the economic system is extended in a two-dimensional continuous plane. Let the study area be closed and be denoted by $A$. Suppose that the quantities produced and the quantities consumed of a commodity are given for every location. We are interested in whether the equilibrium prices, and the size and direction of the shipment can be determined as solutions to equilibrium conditions in a competitive but spatially extended market, and in the conditions under which the competitive economy is stable if the equilibrium exists.

Assume that supply and demand of a commodity are given in terms of area densities for each location. The difference between demand and supply densities at each location $(x_1, x_2)$ is $q = q(x_1, x_2)$, which is a given function of location. As the region is closed, the condition for an equilibrium of the spatial market is

$$\int\int q(x_1, x_2) dx_1 dx_2 = 0 \,. \tag{8.1.1}$$

We assume that $q(x_1, x_2)$is not identically equal to zero over space. (8.1.1) implies that if there are places of surplus with negative excess demand then there must also be places of deficit with positive excess demand. Accordingly, there exists movement of the commodity in the direction from points of excess supply to points of excess demand. We describe the commodity movement by a continuous flow field. The flow vector is denoted by $U = U(x_1, x_2) = (U_1(x_1, x_2), U_2(x_1, x_2))$. The relationship between the flow fields and the local excess demand is given by

$$-q(x_1, x_2) = \operatorname{div} U = \frac{\partial U_1}{\partial x_1} + \frac{\partial U_2}{\partial x_2}, \tag{8.1.2}$$

which gives a necessary condition for any spatial system in which the stock of the commodity is preserved. The equation can be derived as follows. The inflow into and outflow from a rectangular cell of side lengths $\delta x_1$ and $\delta x_2$ are broken into horizontal and vertical components as shown in Fig. 8.1.

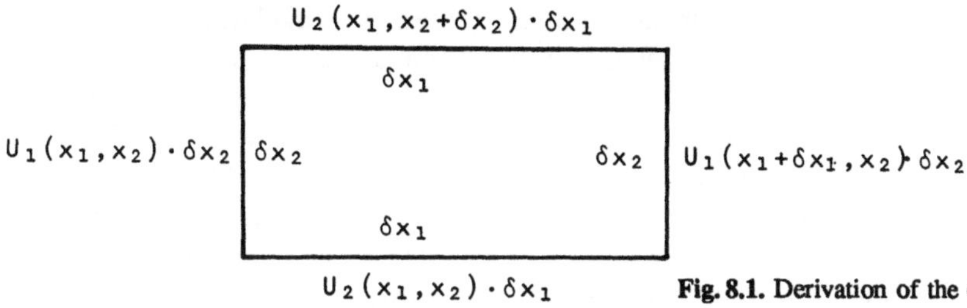


**Fig. 8.1.** Derivation of the divergence law

These components are given as: (i) horizontal inflow = $U_1(x_1, x_2)\delta x_2$, (ii) horizontal outflow = $U_1(x_1 + \delta x_2, x_2)\delta x_2$, (iii) vertical inflow = $U_2(x_1, x_2)\delta x_1$, and (iv) vertical outflow = $U_2(x_1, x_2 + \delta x_2)\delta x_1$. Thus the difference of inflow and outflow is

$$\begin{aligned} &\left[U_1(x_1 + \delta x_2, x_2) - U_1(x_1, x_2)\right]\delta x_2 \\ &\quad + \left[U_2(x_1, x_2 + \delta x_2) - U_2(x_1, x_2)\right]\delta x_1 \,. \end{aligned} \tag{8.1.3}$$

As $U_1(x_1+\delta x_2, x_2) - U_1(x_1, x_2) = (\partial U_1/\partial x_2)\delta x_2$ and $U_2(x_1, x_2+\delta x_2) - U_2(x_1, x_2) = (\partial U_2/\partial x_1)\delta x_1$ approximately hold and the net supply in the small area is approximately equal to $-q(x_1, x_2)\delta x_1 \delta x_2$, flow equilibrium means that $(\operatorname{div} U + q)\delta x_1 \delta x_2 = 0$, i.e., (8.1.2) holds. Equation (8.1.2) corresponds to the divergence law in hydrodynamics and thermodynamics, which describes the relationship between a fluid flow and its sources and sinks.

A closed system means that

$$U_n = 0\,, \quad \text{in } \partial A\,, \tag{8.1.4}$$

where $n$ denotes the direction normal to the boundary and pointing in an outward dirction, and $\partial A$ denotes the boundary of $A$.

It should be emphasized that (8.1.2) contains no economic meaning but represents a physical constraint. Let $k(x)$ denote the cost of transporting a unit of the commodity over unit distance at location $x = (x_1, x_2)$ and $p(x)$ the price of the commodity. Let $D_U$ denote the directional derivative in the direction $U$. The gain from trading a unit of the commodity between two adjacent locations separated by a distance $ds$ is then equal to $D_U p(x) ds$; the cost of transporting the same amount of the commodity is $k(x)ds$. Under perfect competition, trade takes place only when no traders suffer losses. This means that $|\operatorname{grad} p(x)| = k$, where $\operatorname{grad} p(x) = (\partial p/\partial x_1, \partial p/\partial x_2)$. As the direction of trade that achieves a gain equal to transportation cost is the gradient direction, we have

$$k\frac{U}{|U|} = \operatorname{grad} p\,, \quad \text{wherever } U \neq 0\,. \tag{8.1.5}$$

When $U$ vanishes, we should have $|\operatorname{grad} p| \leq k$. Equation (8.1.5) is the gradient law. The divergence law defines a relationship between quantitative variables; the gradient law defines a relationship between the monetary variables. Equations (8.1.2) and (8.1.5) represent the conditions for price equilibrium in the spatial extended market. It can be shown that the commodity flow $U/|U|$ is uniquely determined over the space. If $U$ does not vanish anywhere (except at singularities on a set of measure'zero), then $p$ is uniquely determined up to an additive constant.

We have derived the transportation model under the perfectly competitive mechanism. This model can also be derived from a planning model with the object of minimizing the total transportation cost (see Beckmann and Puu 1985).

As we are mainly interested in dynamics, we will not examine the equilibrium problems in detail. Now, we specify possible dynamics in the competitive market.

Let the price and flow vector not be initially located in equilibrium. As in a traditional equilibrium approach (Arrow and Hahn 1971), we specify a dynamic adjustment mechanism to finally move the system to the equilibrium when the system is subjected to perturbations. This means that the suggested dynamics is stable in the large scale.

Let $p$ and $U$ be a feasible urban pattern satisfying the boundary conditions $U_n = 0$, but not necessarily satisfying the equilibrium conditions – the divergence and gradient laws. Choosing a suitable time unit, we specify the following rules of adjustment

$$\frac{dU}{dt} = \operatorname{grad} p - k\frac{U}{|U|}\,, \tag{8.1.6}$$

$$\frac{dp}{dt} = -(q + \operatorname{div} U)\,. \tag{8.1.7}$$

In (8.1.6), $\operatorname{grad} p - kU/|U|$ indicates the direction of shipment that yields maximum profit or minimum loss. In (8.1.7), $q + \operatorname{div} U$ is the sum of local net demand and net exports. If this is positive, there is excess supply from this location and therefore prices should be lowered. Equilibrium of (8.1.6) and (8.1.7) is determined by the divergence law and the gradient law.

To prove the stability of (8.1.6) and (8.1.7), we define the cost function $K$ as

$$K(U,p) = \int\!\!\int_A k|U| + p(\operatorname{div} U + q)\,. \tag{8.1.8}$$

Using $d|U|/dt = (U \cdot dU/dt)/|U|$, $U_n = 0$ and the Gauss integral theorem

$$\begin{aligned} &\int\!\!\int_A \left(p \operatorname{div}\frac{dU}{dt} + \operatorname{grad} p\frac{dU}{dt}\right) dx_1 dx_2 \\ &= \int_A\!\!\int \operatorname{div} p\frac{dU}{dt} dx_1 dx_2 = \int_{2A} p\frac{dU}{dt} ds = 0\,, \end{aligned} \tag{8.1.9}$$

we can obtain the following equation

$$\frac{dK(U,t)}{dt} = -\int\int \left|\operatorname{grad} p - k\frac{U}{|U|}\right|^2 + [\operatorname{div} p + q]^2 \, dx_1 dx_2 \,. \tag{8.1.10}$$

This is strictly negative as long as any one of the equilibrium conditions is violated. Let $K_{\min}$ be the minimum value of the transportation cost integral. Define the function $K[U,p] - K_{\min}$. This function plays the role of a Lyapunov function. It is non-negative, vanishes only at the equilibrium and decreases monotonically with time. Thus the system is stable.

## 8.2 The Implications of Structural Stability in the Two-Dimensional Economy

In Chap. 2 we defined the concepts of stability and structural stability and discussed the economic implications of stability for an economic dynamic system. Samuelson's correspondence principle states that if stability is presumed for economic systems, meaningful results can often be derived. Stable dynamic systems usually exhibit uniqueness and steady states in the long term. The examples provided showed that if we relax the assumption of stability, the behavior of dynamic systems become very complicated. However, we have not examined the economic implications of the concept of structural stability. In this section, we show that from the assumption of structural stability we can derive meaningful economic results.

The following discussion is fundamentally based on the works by Puu (1981) and Beckmann and Puu (1985, Chap. 4). First, we will present a continuous model of the space economy. This particular choice is not important since the reasoning is applicable equally well to a wide variety of models.

It is assumed that there is one commodity in the economy which is produced over the urban area and three inputs – capital $K(x_1,x_2)$,, labor $L(x_1,x_2)$ and land $M(x_1,x_2)$. Suppose that technology is not dependent on location, although the combination of the inputs is varied spatially. Production at each location is described by the Cobb-Douglas production function

$$Q = K^\alpha L^\beta M^\gamma \,, \tag{8.2.1}$$

where $Q[= Q(x_1,x_2)]$ is output at location $(x_1,x_2)$, $\alpha$, $\beta$, $\gamma \geq 0$, $\alpha+\beta+\gamma = 1$. We can rewrite (8.2.1) in the form of

$$q = k^\alpha l^\beta \,, \tag{8.2.2}$$

where $q = Q/M$, $k = K/M$, $l = L/M$. According to neoclassical theory, the following conditions hold

$$\frac{rk}{\alpha} = \frac{wl}{\beta} = \frac{g}{\gamma} = pq \,, \tag{8.2.3}$$

where $r$ is interest, $w$ wages, $g$ land rent, and $p$ the product price. If $r$, $w$, $g$ and $p$ are given, then $k$ and $l$ are determined. We assume that interest $r$ is homogeneously given in the study period. This assumption holds if capital is mobile and the capital market is in a perfectly competitive equilibrium. As we will show below, product price and wages are dependent on location and land rent is determined residually from the local profitability of production.

To describe the movement of goods and labor, we introduce two vector fields

$$U = [U_1(x_1, x_2), U_2(x_1, x_2)] ,$$

$$V = [V_1(x_1, x_2), V_2(x_1, x_2)] , \tag{8.2.4}$$

where $U$ and $V$ represent the flows of goods and labor, respectively. Assume that the local demand of goods $q^*$ and the local supply of labor $l^*$ are given. Then the divergence laws for goods and labor are given respectively by

$$\operatorname{div} U = q - q^* ,$$

$$\operatorname{div} V = -(l - l^*) . \tag{8.2.5}$$

The gradient laws are defined as follows

$$f \frac{U}{|U|} = \operatorname{grad} p ,$$

$$f = \frac{V}{|V|} = \operatorname{grad} w , \tag{8.2.6}$$

where the scalar field $f(x_1, x_2)$ denotes the local cost of hauling goods or labor across the location. The cost function is made equal for both goods and labor by a suitable choice of the units. From (8.2.2) to (8.2.6) and certain given boundary conditions, we can determine the equilibrium structure of the spatial economy. It should be noted that by taking squares of (8.2.6) we can obtain partial differential equations for the product price and for the wage rate.

For simplicity, we assume that the flows satisfy

$$\frac{U}{|U|} + \frac{V}{|V|} = 0 , \tag{8.2.7}$$

which means that the two fields have opposite direction, labor flowing from residences to industry, and finished goods in the opposite direction. This assumption allows us to write $p(x)$ and $w(x)$ as

$$p = p' - y(x_1, x_2) ,$$

$$w = w' - y(x_1, x_2) , \tag{8.2.8}$$

where $y$ is a potential. From (8.2.6), we see that the flow lines are ultimately determined by the (given) local transportation cost function $f(x_1, x_2)$. Hence, we may assume that the resulting potential function $y(x_1, x_2)$ is known instead. As the flow lines are gradient to the potential, we can choose a suitable parametrization of the flow lines so that they are solutions to the differential equations

$$\frac{dx_1}{ds} = \frac{\partial y}{\partial x_1}\,, \quad \frac{dx_2}{ds} = \frac{\partial y}{\partial x_2}\,, \tag{8.2.9}$$

where $s$ is the "distance parameter". A structural stability means that if for another potential function $y^*(x)$, differences between the first and second partial derivatives of $y$ and $y^*$ with regard to $x_1$ and $x_2$, respectively, are sufficiently small, then the solution curves for the two potentials are structurally not different. The precise definition is given in Sect. 3.5.

Although it is very difficult to obtain general characteristics of structurally stable systems, there is fortunately a characterization theorem on structurally stable flows in the plane. It indicates that a stable flow has only a finite number of isolated singular points of very few specified categories, and is laminar everywhere else. There are also global results on how the few singularities may be connected. This makes it possible to draw a very precise picture of the structurally stable flow and the spatial organization of the economy corresponding to such a flow.

According to Peixoto's theorem (Peixoto 1977), the consequences of structural stability can be summarized in the following theorem.

**Theorem 8.2.1.** If the system (8.2.9) is structurally stable, then we have

i) regular flow everywhere, except at a finite number of isolated singular points; or equivalently, the flow is topologically equivalent to a set of parallel straight lines;
ii) singular points as nodes (sources and sinks) and ordinary saddles;
iii) no trajectory joining the saddle points. At each saddle point there are four incident trajectories, one pair directed inward, and one pair outward. No outward trajectory can thus be incident to another saddle, nor can it return after a loop to the same one in the inward direction.

Regularity means that only one trajectory passes through each point. We now consider the interpretation of these results in terms of the organization of the spatial economy.

A source is a point from which all trajectories in a surrounding basin of repulsion diverge. They form a set of radiating trajectories that are orthogonal to circular, concentric price contours. The economic organization is one of concentric rings of various activities, and the routes of transportation are radial (see Fig. 8.2). The picture is very similar to the von Thünen case or the case of new urban economics. In the case of a sink, all trajectories in a surrounding basin of attraction converge to the singular point, the price contours are again concentric closed curves, and the spatial organization is in terms of concentric rings. The directions of flow in the two cases are opposite. The sources may be considered to be productive centers, and the sinks consumptive centers.

In (i) of Theorem 8.2.1, except at the sinks and sources we have ordinary saddles. For each saddle there are two pairs of incident trajectories, one ingoing and the other outgoing. At a saddle, the surrounding space is separated into four sectors, each containing hyperbolic trajectories, attracted toward the singularity but missing it. The set of price contours, to which these hyperbolic trajectories are orthogonal, is itself a

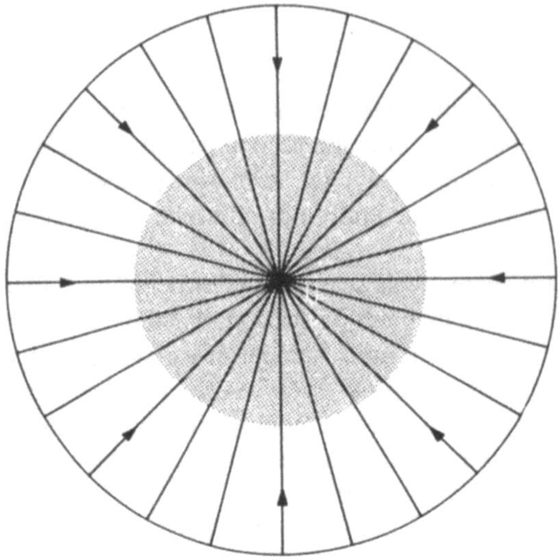

Fig. 8.2. Flow and spatial organization around a node singularity

set of hyperbolas. The various zones of economic activity are now contained between pairs of hyperbolas in opposite sectors, and the organization of space is sectoral around a saddle singularity. Since all the routes of transportation are thus deflected from the straight line, transportation must be very favorable in the neighborhood of a saddle singularity. The saddle points are "condensation nuclei" of empty space with particularly favorable conditions for transportation (see Fig. 8.3).

We now construct the global picture of flows and price contours on the principle that no trajectories join saddle points. As trajectories in only four directions at a saddle are actually incident, following any one of these will result in either crossing the boundary or ending up at a singularity that must be a node. We note that two

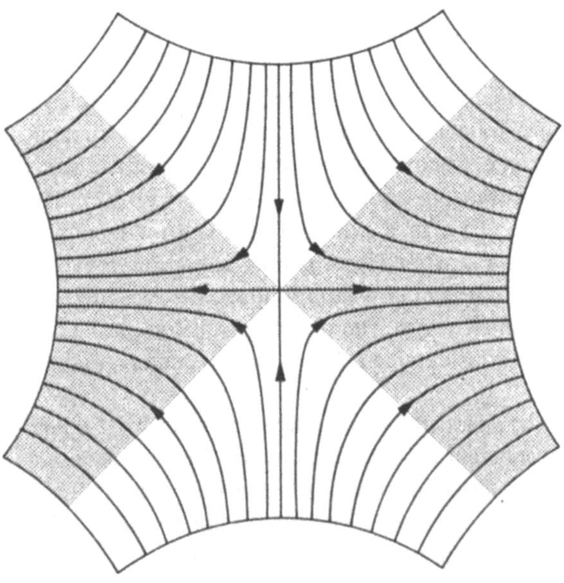

Fig. 8.3. Flow and spatial organization around a saddle singularity

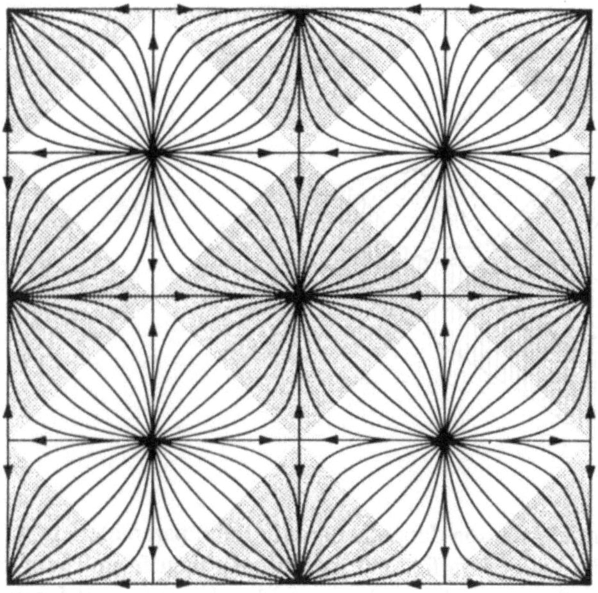

Fig. 8.4. The global pattern of structurally stable flows

of the nodes must be sources and two of them must be sinks. We can thus arrange global patterns on a square grid made up by trajectories incident to saddle points alone. The singular points are the intersection points in this grid.

Starting out from any saddle point, we can describe the whole urban pattern. As each saddle is surrounded by two stable and two unstable nodes, we can conclude that in the diagonal directions from the initial saddle there are again saddle points, since there are both ingoing and outgoing trajectories from them. The spatial organization is a chessboard pattern of square areas of industrial and residential character as shown in Fig. 8.4. Obviously, the spatial organizations in Figs. 8.2 and 8.3 are the local organizations around a node and a saddle point in the global picture.

The corresponding potential surface or "price landscape" is shown in Fig. 8.5. Labor is considered to flow downhill and goods uphill in the gradient direction. Choices of the combination of labor and capital inputs are dependent on location.

From the discussion above, we see that the mere assumption of structural stability makes it possible to give a precise description of the spatial structure of the economy in a topological sense. Similarly to the correspondence principle from which we obtain qualitative information, we have topological information on spatial structures of the economy from the assumption of structural stability. Such topological information may tell us more than might be imagined at first sight. For instance, if the system is structurally stable, then the Christaller-Lösch paradigm of spatial organization is invalid since their hexagonal pattern cannot be transformed into the structurally stable flow characterized below by any topological transformation.

The discussion in this section has been limited to the case when the system is structurally stable. It is natural for us to pose the question: what are the possible structures of the spatial economy if the assumption of structural stability is relaxed? Using catastrophe theory, Puu (1981) solved the problem.

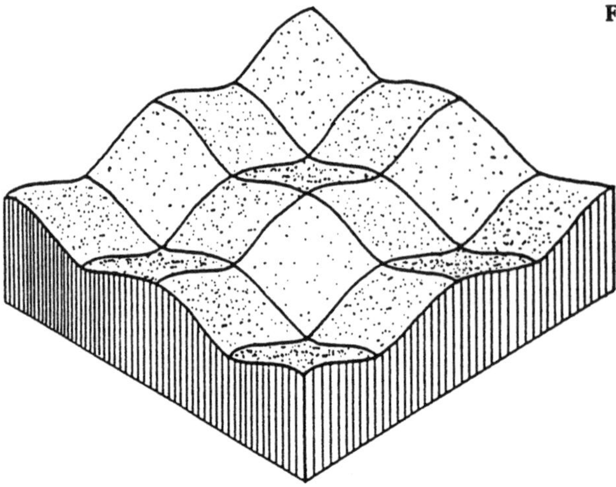

Fig. 8.5. The price landscape

If we assume that the potential function depends on only three parameters (e.g., road construction and repair, the load of traffic, and fuel prices), then from Thom's classification theorem we know that we only need to consider the canonical forms of the elliptic and hyperbolic umbilics

$$y = x_1^3 - 3x_1x_2^2 + \alpha\left(x_1^2 + x_2^2\right) - \beta x_1 - \gamma x_2\,, \tag{8.2.10}$$

$$y = x_1^3 + x_2^3 + \alpha x_1x_2 - \beta x_1 - \gamma x_2\,, \tag{8.2.11}$$

where $\alpha$, $\beta$ and $\gamma$ are parameters. By studying all the phenomena that can occur with the gradient fields to these $y$-functions for all possible value combinations of $\alpha$, $\beta$ and $\gamma$, we obtain a complete description of the possible structures. Figures 8.6 and 8.7 illustrate the canonical forms of the elliptic and hyperbolic umbilic catastrophes, respectively. At the top of each figure is the bifurcation manifold in parameter space. Below are the flow fields for different parameters. As long as the parameter combination moves in the $(\alpha, \beta, \gamma)$-space without crossing the bifurcation manifold, smooth shifts in the parameters only cause smooth changes in the structure. Near the bifurcation manifold, a sudden change of flow pattern will take place.

## 8.3 Economic Cycles in Puu's Spatial Multiplier-Accelerator Business Model

In the preceding section, we globally examined urban patterns by applying the concepts of structural stability and instability. The objective of this section is to study the unstable urban pattern formation process. We present a spatial dynamic economic model suggested by Puu (1986). The model is an extended form of the celebrated multiplier-accelerator model of business cycles, which was formulated by Samuelson (1939), and later elaborated by Hicks (1950) and others.

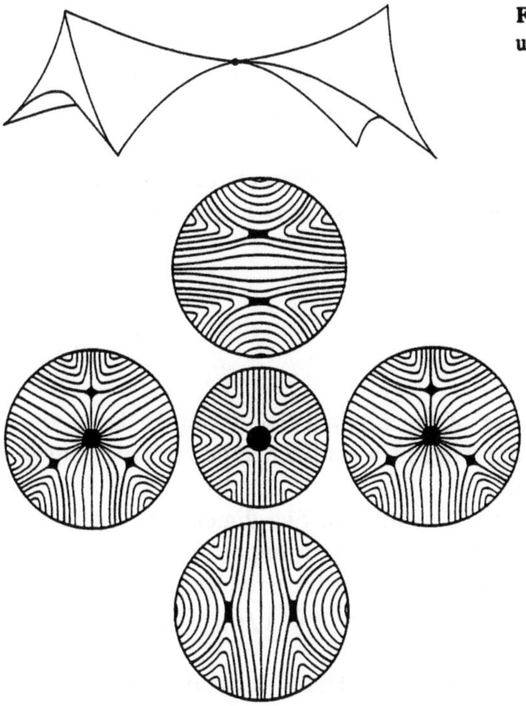

**Fig. 8.6.** The possible structures in the elliptic umbilic case

**Fig. 8.7.** The possible structures in the hyperbolic umbilic case

The essential elements of the model are saving (or consumption) and "induced" investments. Savings are assumed to be a given proportion $s$ of the income $Y$. Induced investments are proportional to the rate of change of income $dY/dt$, the proportionality factor (the accelerator) being denoted by $\Theta$. Equating savings to investments, $sY = \Theta dY/dt$, yields the Harrod model of balanced growth. The income and investment change rates may be specified, respectively, as: $dY/dt = I - sY$, and $dI/dt = \Theta dY/dt - I$. Obviously, this system is a special case of the models discussed in Sect. 6.6.

From these equations, we have

$$\frac{d^2Y}{d^2t} + (1 + s - \Theta)\frac{dY}{dt} + sY = 0\,. \tag{8.3.1}$$

The solutions of the model are like those of the original Samuelson-Hicks model formulated for discrete time, i.e., they are those of a simple damped or antidamped harmonic oscillator of one defined period.

We now generalize (8.3.1) in the setting of continuous two-dimensional space by introducing exports and imports. Denote the propensity to import by $m$. Applying Gauss' integral theorem, we can show that the proper measure of spatial income differences is the Laplacian of income $\nabla^2 Y = \partial^2 Y/\partial x^2 + \partial^2 Y/\partial y^2$ (Beckmann and Puu 1985).

To introduce nonlinearity in the model, we accept Hicks' reasoning of a "floor" for disinvestment where no capital is replaced and capital is depreciating at its natural rate, and a "ceiling" on investments where inputs other than capital become binding and their own rate of growth limits investments. Hicks (1950) introduced the constraints as linear inequalities. For simplicity of analysis, we introduce a continuous function, $\Theta \tanh dY/dt$, having the properties of a non-linear accelerator with a floor and a ceiling to replace $\Theta dY/dt$. Around zero this function is almost linear in the argument, but for large negative or positive $dY/dt$ it goes asymptotically to +1 or $-1$.

Under these assumptions, (8.3.1) can be rewritten in the form

$$\frac{d^2Y}{d^2t} + (1 + s)\frac{dY}{dt} - \Theta \tanh\frac{dY}{dt} + sY - m\nabla^2 Y = 0\,. \tag{8.3.2}$$

This is a nonlinear partial differential equation. One solution is $\nabla^2 Y = 0$. The urban pattern is homogeneous.

The qualitative behavior of this model can be inferred by sketching a phase diagram in $Y$-$dY/dt$ space. For large $Y$ or $dY/dt$, the system is damped. In the case $\Theta > (1 + s)$, there is a neighborhood of the origin in phase space where there is an antidamping zone; otherwise there is a damping zone. The combination of antidamping in the centre and damping in the periphery may result in the existence of a limit cycle. This is guaranteed by the numerical analysis (see Fig. 8.8) from Puu (1986). The Hopf bifurcation theorm can evidently be applied to identify conditions for the existence of cycles.

We can identify the existence of a stationary urban pattern by setting $d^2Y/d^2t = dY/dt = 0$. From these two special cases, we see that there may be a trade-off

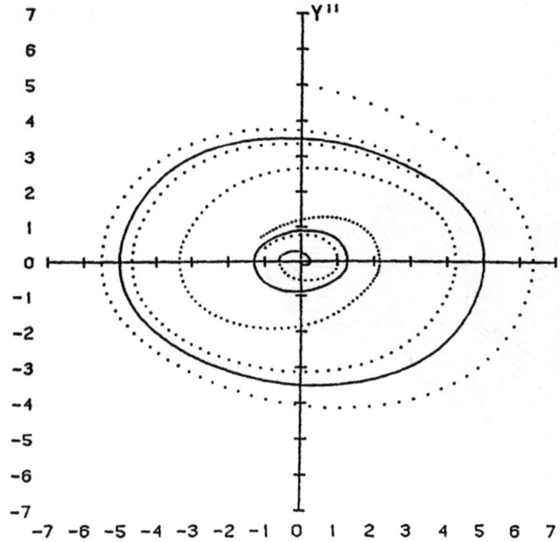


**Fig. 8.8.** Stable and unstable limit cycles

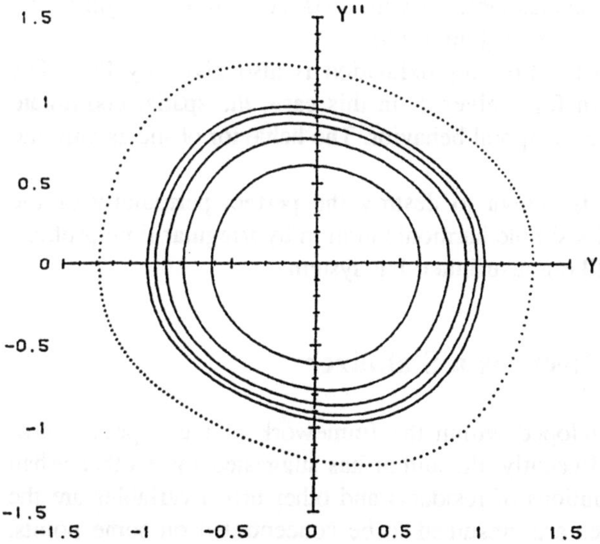


**Fig. 8.9.** Oscillations at different distances ($r = 1.5, 2, 2.5, 3, 3.5$)

between temporal change and spatial heterogeneity. As an example, let $m = s = 0.5$, $\Theta = 2$, $r = (x^2 + y^2)^{1/2}$. The system (8.3.1) is thus written as

$$Y'' + Y = 4 \tanh Y' - \left(3 + \frac{1}{r}\right) Y' , \tag{8.3.3}$$

where $Y'$ and $Y''$ denote the derivatives of $Y$ with respect to $r - t$. Figure 8.9 illustrates the oscillations for various values of $r$. For radii less than unity, the system is damped and there exist no limit cycles. For larger radii, limit cycles with finite and increasing amplitude emerge, the outermost one corresponding to an infinite radius.

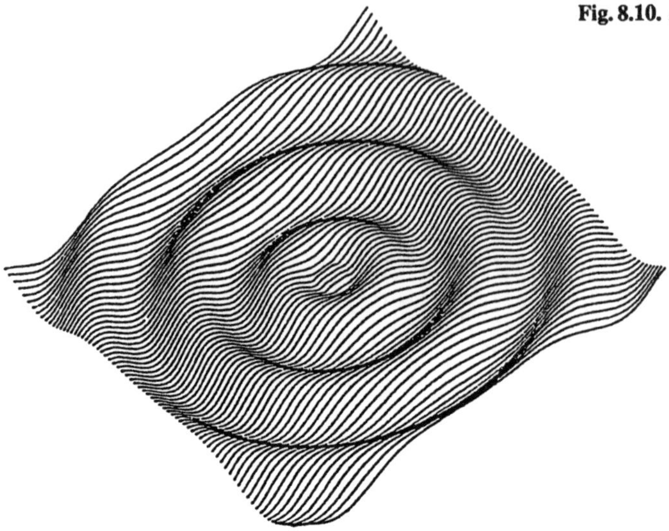

Fig. 8.10. An asymptotic spatial pattern

Although the amplitudes of the cycles increase with distance from the origin of the space, the periods seem to be the same (Puu 1986).

The asymptotic solution to the first approximation is also given by Puu. The solution defines a spatial pattern for a given $t$. In this case, the spatial coordinate acts as a bifurcation parameter in temporal behavior. The behavior of the asymptotic solution is shown in Fig. 8.10.

The introduction of space is shown to destroy the perfect periodicity of the original model, and to replace its simple harmonic motion by irregular time profiles. In the latter aspect space acts like a distributed lag system.

## 8.4 Spatial Diffusional Effects as a Stabilizer

The preceding models are developed within the framework of the approach proposed by Beckmann and Puu. Recently, the author has suggested some other urban models in which spatial distributions of residents and other urban variables are the main focus. Economic activities are presumed to be concentrated on some points. The remainder of this chapter will examine some urban models suggested by this approach.

We are interested in modeling behavior of households over space. It is well observed that distribution of people and their socioeconomic activities show a tendency towards aggregation and regionalization. A reason for regionalization may be the existence of "scale factors" for people and their activities. The results of these tendencies is that residents and their socioeconomic activities are not homogeneously distributed in time and space. This section reports the model recently developed by Zhang (1988c) to show spatial diffusion effects on urban pattern formation.

The urban system consists of three parts: the central business district (CBD), the suburban area, and the boundary of the urban area. The CBD is the area where main socioeconomic activities are carried out. There may exist socioeconomic activities in other areas of the urban system. For simplicity, we treat the CBD as a point. The boundary of the urban area is assumed to adjoin an area of agricultural land use. Between the CBD and the boundary lies the region called the suburban area, where people can build their houses and other socioeconomic activities can be carried out.

The urban pattern formation process is described by motion of the residential density and land rent over time and space. Define

$$x(r,t) = \text{the residential density at } (r,t) \, ,$$

$$y(r,t) = \text{the land rent at } (r,t) \, ,$$

where $r$ is the distance from the CBD to a point in the suburban area. When the variables $x(r,t)$ and $y(r,t)$ are independent of $r$, we say that the urban pattern is homogeneous, and when dependent on $r$, heterogeneous. According to Dendrinos and Mullally (1985), if the urban area is appropriately small, the dynamics can be described by the predator-prey model as discussed in Sect. 3.5

$$\frac{dx}{dt} = \alpha(y_1 - y)y \, ,$$

$$\frac{dy}{dt} = \beta(x - x_1)x \, , \tag{3.5.1}$$

where the parameters are defined in Sect. 3.5. We know that the solution is oscillatory and that the system is structurally unstable. Our question is whether the system can be stabilized by diffusional effects.

Under certain propositions about the "motion" of land rent and population, Zhang (1988c) added diffusional effects to the predator-prey urban model in the following way

$$x_t = \alpha(y_1 - y)y + \Theta_1^2 x_{rr} \, ,$$

$$y_t = \beta(x - x_1)x + \Theta_2^2 y_{rr} \, , \quad 0 \le r \le r_0 \, , \tag{8.4.1}$$

where $r_0$ is the distance from the CBD to the urban boundary, and $\Theta_1$ and $\Theta_2$ are the parameter of avoidance of crowding of the residents and the "rent-conductivity coefficient", respectively. The boundary conditions are

$$x_r(r_0,t) = x_r(0,t) = 0 \, ,$$

$$y_r(r_0,t) = y_r(0,t) = 0 \, . \tag{8.4.2}$$

(8.4.2) means that at the boundary there is no "flux" between the urban area and the agricultural area. Let the initial urban pattern be described by

$$x(r,0) = m(r) \, , \quad y(r,0) = n(r) \, . \tag{8.4.3}$$

The urban model consists of (8.4.1–8.4.3). It can be shown that there exist positive solutions to this system.

First, consider the case when $\Theta_1^2 = \Theta_2^2 (= \Theta)$. Equations (8.4.1) can be rewritten as

$$x_t = \alpha(y_1 - y)y + \Theta x_{rr} ,$$
$$y_t = \beta(x - x_1)x + \Theta y_{rr} . \qquad (8.4.4)$$

**Theorem 8.4.1.** If we assume $\Theta_1 = \Theta_2$, then as $t \rightarrow \infty$, the urban pattern becomes homogeneous, i.e., $x_r(r,t) = y_r(r,t) = 0$ as $t \rightarrow \infty$.

The proof can be found in Zhang (1988c). This result is proved by introducing

$$\phi(r,t) = \beta(x - x_1 \ln x) + \alpha(y - y_1 \ln y) ,$$

with which the problem can be rewritten as

$$\phi_t = \Theta\phi_{rr} - \Theta F ,$$

subject to appropriate boundary and initial conditions. The function $F$ is dependent on $x$ and $y$ as

$$F = \beta x_1 \left(\frac{x_r}{x}\right)^2 + \alpha y_1 \left(\frac{y_r}{y}\right)^2 .$$

Applying the maximum principle for the parabolic operators, we can show that $F \rightarrow 0$ as $t \rightarrow \infty$, which is equivalent to the urban pattern being homogeneous when $t$ is sufficiently large.

This result is very interesting if we compare it with the results obtained by Dendrinos and Mullally (1985). They did not consider the influence of spatial diffusion upon the urban pattern by assuming that the urban area is very small. Theorem 8.4.1 shows that their treatment of the urban system by aggregation is valid under the strict requirement of $\Theta_1 = \Theta_2$. In this case, if we are only interested in steady states, the introduction of diffusional effects does not affect the equilibrium pattern, though stability may be affected.

**Theorem 8.4.2.** If $\Theta_1$ and $\Theta_2$ are nonzero, then there is no periodic solution in the system (8.4.1).

Consider the function

$$W(t) = \int_0^{r_0} \left[\frac{1}{\alpha}\left(x - x_1 - x_1 \ln\frac{x}{x_1}\right) + \frac{1}{\beta}\left(y - y_1 - y_1 \ln\frac{y}{y_1}\right)\right] dr . \qquad (8.4.5)$$

It can be easily checked that $dW(t)/dt \leq 0$. That is, $W(t)$ is a monotonically non-increasing function of time $t$. Moreover, it is bounded below by zero, and thus Theorem 8.4.2 holds.

Consequently, the introduction of geographical factors destroyed the possibility of oscillations in the predator-prey urban model. This implies that the long-run urban pattern observed from the model will converge to a steady state.

To understand the characteristics of the urban system with diffusion effects, we have to study its stability. Local stability analysis shows that the time-independent homogeneous urban pattern is stable. Hence, the urban system is stabilized by the introduction of diffusional effects.

## 8.5 Separation and Coexistence of Residents

This section is still related to spatial and temporal distributions of residents. However, the residents are classified into different groups according to economic and individual characteristics. For instance, we may classify the residents by their skin color into white and black, or by their incomes, or education levels. In some areas of the United States and other countries, coexistence and separation of residents of different races have caused serious social problems and have been studied in different fields. For simplicity, we consider only two groups, denoted by group 1 and group 2, respectively. It is assumed that the two groups are interactional in the sense that their relationships affect the behavior of the residential distributions. Their relationship may be friendly, averse or "neutral". It is shown that separation and coexistence of the groups depend on these relationships. We denote the residential densities of the two groups by $X(r,t)$ and $Y(r,t)$, where $r$ is the distance from the CBD to a dwelling site. Here, the urban area is geographically similar to the one in the preceding section.

The remainder of this section is according to Zhang (1989c). The evolution of the two groups can be described as follows

$$X_t = \alpha X(a - bX - cY) - d_1 XY + \Theta_1 X_{rr} , \tag{8.5.1}$$

$$Y_t = \beta Y(a - bX - cY) - d_2 XY + \Theta_2 T_{rr} , \tag{8.5.2}$$

$$t > 0 , \quad r \in W ,$$

where $a$, $b$, $c$ and $d_i$ $(i = 1,2)$ are constant parameters, and $W$ is the domain of the suburban area.

In (8.5.1), $\Theta_1 X_{rr}$ measures the effects of geographical diffusion on residents. Geographical diffusional terms measure the human beings' propensity to live in less populated places. The parameter $\Theta_1$ may actually be dependent on variables and time and space. The term $\alpha X(a - bX - cY)$ describes the reaction of the residents to economic conditions. We interpret $a$ as a "physical" capacity of the urban area at location $r$. The parameter $a$ being constant means that the physical capacity is homogeneous over space. If we assume that $(bX + cY)$ measures how much area the two groups take up, then $(a - bX - cY)$ is the excess supply of physical capacity. If $(a - bX - cY)$ is greater than zero, then the dwelling site becomes more attractive to the residents. Obviously, the residents stop migrating when $a = bX + cY$ and $-d_1 XY$ and diffusional effects are neglected. In (8.6.1), $\alpha X$ is the adjustment speed of group 1 to its equilibrium value. If the residential density is higher, residents adjust slowly to equilibrium as the system is less well informed. The term $-d_1 XY$ is used to measure interaction between the two groups. Unlike the economic factors, this term illustrates social interactions. The coefficient $d_1$ is either positive, zero, or negative. If $d_1$ is positive, group 1 does not like to live with group 2. If $d_1 = 0$, there are no "racial" effects. If $d_1$ is negative, a high density of group 2 may attract residents of group 1. For instance, if we classify the residents according to their educational level, less educated people may tend to live with highly educated ones.

Equation (8.5.2) can be similarly interpreted.

It should be noted that this system can be extended in different ways. For instance, we may classify the residents into $N$ groups ($N > 2$). Similar to our basic model, the spatial and temporal distributions of the residents can be generally described by

$$X_{it} = X_i\left(\alpha_i - \Sigma_j \beta_j X_j\right) + \Theta_i \frac{\partial^2 X_i}{\partial^2 r} ,$$

subject to appropriate initial and boundary conditions. We can investigate different conditions for the coexistence and separation of different groups.

We may also take account of interactions among residents and other urban variables such as rent and housing quantity. The model in the preceding section is an example of this approach.

There may also exist some exogenous forces which affect the urban pattern. For instance, the government may accept special policies to guarantee the coexistence of the residents. Thus (8.5.3) may be generalized in the following form

$$X_t = \alpha X(a - bX - cY) - d_1 XY + E_1 + \Theta_1 X_{rr} ,$$

$$Y_t = \beta Y(a - bX - cY) - d_2 XY + E_2 + \Theta_2 Y_{rr} , \quad t > 0 , \quad r \in W ,$$

where $E_1(r,t)$ and $E_2(r,t)$ are exogenous "inputs". For instance, $E_1$ may be immigration of group 1 to the urban area. If $E_1$ is independent of distance, the amount of immigration is given by $E_1$ multiplied by the size of the urban area.

For simplicity, we confine ourselves to the simplest case (8.5.1–8.5.2). We rewrite the model as

$$X_t = X(a_1 - b_1 X - c_1 Y) + \Theta_1 X_{rr} ,$$

$$Y_t = Y(a_2 - b_2 X - c_2 Y) + \Theta_2 Y_{rr} , \tag{8.5.3}$$

where $a_1 = \alpha a$, $b_1 = \alpha b$, $c_1 = \alpha c + d_1$, $a_2 = \beta a$, $b_2 = \beta b + d_2$, $c_1 = \beta c$.

The initial and boundary conditions are respectively proposed to be

$$X(r,0) = F(r) , \quad Y(r,0) = H(r) , \quad r \in W ,$$

$$B_1(X) = p_1 \frac{\partial X}{\partial n} + q_1 X = 0 ,$$

$$B_2(Y) = p_2 \frac{\partial Y}{\partial n} + q_2 Y = 0 , \quad r \in \partial W , \tag{8.5.4}$$

where $n$ denotes the direction normal to the boundary and pointing in an outward direction, $\partial W$ is the boundary of the urban area, and $p_i$ and $q_i$ ($i = 1, 2$) are constants. The boundary conditions are determined by how the urban system interacts with the "outside world". If we consider $\partial X/\partial n$ and $\partial Y/\partial n$ as migration, the proposed boundary conditions say that migration is only dependent on residential density.

In Zhang (1989c), the case of $b = c = 0$ and $d_i > 0$, $i = 1,2$, is analyzed initially. This means that the amount of space taken up by residents does not affect migration, and interactions between the two groups are purely competitive. It was shown that the homogeneous urban pattern becomes heterogeneous after perturbations in initial conditions.

Here, we are interested in a general case when (8.5.3) is subjected to the Neumann boundary conditions, i.e., $p_1 = p_2 = 1$, $q_1 = q_2 = 0$, or

$$\frac{\partial X}{\partial n} = \frac{\partial Y}{\partial n} = 0\,, \quad \text{on } \partial W\,. \tag{8.5.5}$$

There is no migration between the urban area and the "outside world". Introduce

$$\mu = \max\left\{\frac{a_1}{c_1}, \frac{a_2}{c_2}\right\}\,,$$

$$\delta = \max\left\{\frac{a_1}{b_1}, \frac{a_2}{b_2}\right\}\,. \tag{8.5.6}$$

**Theorem 8.5.1.** Let $(X, Y)$ be solutions of (8.5.3) subject to (8.5.5), and let $0 < F(r) < \mu$ and $0 < H(r) < \delta$ be satisfied for all $r \in W$. Then we have

i) if $\dfrac{a_1}{a_2} > \max\left\{\dfrac{b_1}{b_2}, \dfrac{c_1}{c_2}\right\}$ then

$$\lim_{t\to\infty}\left[X(r,t)\,,\, Y(r,t)\right] \to \left(\frac{a_1}{b_1}, 0\right)\,; \tag{8.5.7}$$

ii) if $\dfrac{a_1}{a_2} < \min\left\{\dfrac{b_1}{b_2}, \dfrac{c_1}{c_2}\right\}$ then

$$\lim_{t\to\infty}\left[X(r,t)\,,\, Y(r,t)\right] \to \left(0, \frac{a_2}{b_2}\right)\,; \tag{8.5.8}$$

iii) if $\dfrac{c_1}{c_2} < \dfrac{a_1}{a_2} < \dfrac{b_1}{b_2}$ then

$$\lim_{t\to\infty}\left[X(r,t)\,,\, Y(r,t)\right] \to (m, n)\,; \tag{8.5.9}$$

where $m = (a_1c_2 - a_2c_1)/h$, $n = (a_2b_1 - a_1b_2)/h$, $h = (b_1c_2 - b_2c_1)$. It is assumed that $h$ is not equal to zero.

For simplicity of interpretation, let $\Theta_1 = \Theta_2 = 1$. $0 < F(r)$ says that the initial residential density of group 1 is not equal to zero anywhere in the urban area. If we note that $a_1/c_1 = a/(c + d_1)$, then $F(r) < a/(c + d_1)$ means that the initial density of group 1 is limited above by the capacity and the space units per dwelling site of group 2. As we have $a_2/c_2 = a/(c + d_2)$, we can simply explain $F(r) < a_2/c_2$.

The result (i) holds if $1 > \max\{b/(b+d_2), (c+d_1)/c\}$ is satisfied. This holds provided that $d_2 > 0$ and $d_1 < 0$. As $d_1$ ($d_2$) is a measure of the strength of action of group 2 (group 1) upon group 1 (group 2), we see that if group 1 likes to live with group 2, while group 2 is averse to group 1, group 2 will finally be driven out by group 1. We can see how the evolutionary process is carried out. The long-run equilibrium does not depend on the values of $d_1$ and $d_2$. This is natural since there is no classical struggle in the long run. As $X(r,\infty) = a/b$, we see that the capacity is used up by group 1. Otherwise, group 2 may locate their houses in the urban area. The case (ii) can be similarly interpreted.

Since the condition for (iii) may be rewritten in the form $(c+d_1)/c < 1 < b/(b+d_2)$, it is necessary to require that $d_1$ and $d_2$ are negative. There is no aversion among people. In practice, this requirement may be rather limited. Residents of different groups can live on the same dwellling site. It should be noted that (iii) is valid only when $h$ is positive. We will discuss the case when $h$ is negative below.

If the interactional terms $d_1XY$ and $d_2XY$ are regarded as affecting the efficiency of use of the urban area's capacity, the above results are intuitively acceptable. If $d_1$ is positive, then some urban area may not be effectively used. This can be understood if we let $\Theta_1 = 0$. In equilibrium, we have: $a = bX + cY + d_1Y$. Evidently, the term $d_1Y$ does not represent residential use.

**Theorem 8.5.2.** Let $(X, Y)$ be solutions of (8.5.3) subject to (8.5.5) and $b_1/b_2 < a_1/a_2 < c_1/c_2$. Then we have

i) if $\mu < F(r) < a_1/b_1$, $0 < H(r) < \delta$ then

$$\lim_{t\to\infty}[X(r,t), Y(r,t)] \to \left(\frac{a_1}{b_1}, 0\right) ; \tag{8.5.10}$$

ii) if $0 < F(r) < \mu$, $\delta < H(r) < \frac{a_2}{c_2}$ then

$$\lim_{t\to\infty}[X(r,t), Y(r,t)] \to \left(0, \frac{a_2}{b_2}\right) . \tag{8.5.11}$$

As $b_1/b_2 < a_1/a_2 < c_1/c_2$ is identical to $b/(b+d_2) < 1 < (c+d_1)/c$ and $h < 0$, it is necessary to have $d_1 > 0$ and $d_2 > 0$. Thus groups 1 and 2 are purely competitive. In this case, there can exist only one group in the long term. It can be further verified that the system has two stable and two unstable constant stationary solutions. The long-run equilibrium is dependent on initial states. For the case (i), as the density of group 1 is initially very high and that of group 2 is low, group 2 should finally be driven out by group 1. We can explain (ii) similarly. Here, it should be noted that by "driven out" we mean that residents disappear in some biological sense because no migration is permitted.

We have "isolated" the urban system by assuming the Neumann boundary conditions. It is interesting to see what will happen to the system if we "open" it. Consider the following boundary condition

$$p\frac{\partial X}{\partial n} + q = 0 ,$$

$$p\frac{\partial Y}{\partial n} + qY = 0\,, \quad t > 0\,, \quad \text{on } \partial W\,. \tag{8.5.12}$$

Let us define an eigenvalue problem as follows

$$\nabla^2 S + kS = 0\,, \quad r \in W\,,$$

$$B(s) = 0 \text{ on the boundary}\,, \tag{8.5.13}$$

where $k$ is an eigenvalue and $S$ an eigenvector associated with the problem. This is a partial differential equation of one dimension and the existence of solutions is well known. We denote the smallest eigenvalue by $k_0$ and the eigenvector of $k_0$ by $\phi(r)$. We normalize $\max \phi(r) = 1$. Moreover, it can be verified that $k_0$ and $\phi(r)$ are positive.

**Theorem 8.5.3.** Let the boundary conditions be (8.5.12). Then the trivial solution (0,0) of (8.5.3) is globally asymptotically stable (with respect to non-negative perturbations) if $a_1 < k_0\Theta_1$, $a_2 < k_0\Theta_2$, and is unstable if either $a_1 < k_0\Theta_1$ or $a_2 < k_0\Theta_2$.

If the capacity of the urban area is too small, neither group can exist in the urban area. The unstable condition implies that any small perturbation of the trivial solution will result in a new urban pattern. Surely, we are only interested in the case when the trivial solution is unstable. An equilibrium of (8.5.1) is defined as the solution of

$$X(a_1 - b_1 X - c_1 Y) + \Theta_1 X_{rr} = 0\,,$$

$$Y(a_2 - b_2 X - c_2 Y) + \Theta_2 Y_{rr} = 0\,. \tag{8.5.14}$$

**Theorem 8.5.4.** Let the system (8.5.3) be subjected to the boundary condition (8.5.12). Then

i) If $a_1 > k_0\Theta_1$, $a_2 < k_0\Theta_2$, (8.5.14) has a unique positive solution $(X^*(r), 0)$; and for any initial function $(F, H)$ with $F(r) \geq \varepsilon\phi(r)$, $0 \leq H(r) < k_0\Theta_1\phi(r)/c_1$ where $\varepsilon$ may be arbitrarily small, the solution of (8.5.3) satisfies

$$\lim_{t\to\infty} X(r,t) = X^*\,, \quad \lim_{t\to\infty} Y(r,t) = 0\,; \tag{8.5.15}$$

ii) If $a_1 < k_0\Theta_1$, $a_2 \geq k_0\Theta_2$, then there exists a positive solution $(0, Y^*(r))$ to (8.5.13); and for any initial function $(F, H)$ with $0 < F(r) < (a_2 - k_0\Theta_2)\phi(r)/b_2$, $H(r) \geq \varepsilon\phi(r)$, where $\varepsilon$ may be arbitrarily small, the solution of (8.5.3) satisfies

$$\lim_{t\to\infty} X(r,t) = 0\,, \quad \lim_{t\to\infty} Y(r,t) = Y^*\,. \tag{8.5.16}$$

As $a_1 > k_0\Theta_1$, condition $a_2 < k_0\Theta_2$ in (i) can be rewritten as $\Theta_1/k_1 < a/k_0 < \Theta_2/k_2$; these requirements imply that capacity is bounded above by the characteristics of group 1, and below by group 2. In this case, group 2 is driven out by group 1 in the long run.

Finally, it can be shown that if $a_1 > k_0\Theta_1$, $a_2 > k_0\Theta_2$, then for appropriate initial conditions the two groups can coexist in the system. The precise conditions can be found in Zhang (1989c).

## 8.6 Long-Term Traveling-Wave Urban Pattern

In the previous sections we analyzed different urban problems. It has been shown that the number of urban pattern forms created from structurally stable systems is rather limited. Instability increases the variety of urban patterns. To illustrate the way in which instability may increase the complexity of an urban system, we consider an urban model which exhibits traveling-wave-like behavior near unstable points. The model is suggested by Zhang (1989e).

The geographical characteristics of the urban system are similar to those in Sect. 8.4. It is assumed that the urban pattern is described by two variables:

$$n(r,t) = \text{the residential density at } (r,t) ,$$

$$q(r,t) = \text{the housing quality at } (r,t)$$

where $r$ is the distance from the CBD to a dwelling location.

It should be emphasized that the model suggested in Sect. 8.4 deals with different aspects of urban pattern formation process from the model to be developed.

It is assumed that the number of residents does not change during the study period. We neglect demographic changes and migration between the urban system and the "outside world". According to Zhang (1989e), the urban pattern is described by the following dynamics

$$n_t(r,t) = \alpha\left[f(q) - n\right] + \Theta n_{rr} , \tag{8.6.1}$$

$$q_t(r,t) = -\delta q + H\big(I(n,q)\big) , \quad r \in G , \tag{8.6.2}$$

where $G$ denotes the urban area, $\alpha$ represents an adjustment parameter, and $\Theta$ and $\delta$ represent the diffusional parameter of population and the rate of deterioration, respectively. The system is subjected to certain initial and boundary conditions. There is no diffusional effect on change in housing quality.

To interpret (8.6.1), we omit the diffusional term. We thus have: $dn/dt = \alpha[f(q) - n]$. The function $f(q)$ is considered to be determined by the rational behavior of households. It defines an equilibrium value of the density for the given housing quality.

Equation (8.6.2) describes how the housing quality changes over time. The term $-\delta q$ describes the effects of deterioration. It is assumed that houses are maintained by an agent who decides the maintenance expenditures, and the cost is dependent on the current income of the agent from housing units. Let the income be denoted by $I$. The income at a location is related to the residential density and housing quality: $I = I(n,q)$ where $I_n$ is uncertain and $I_q$ is positive. For a fixed level of $q$, $I_n$ is, in general, uncertain since whether an increase in the residential density makes the income increase or decrease depends on actual situations. $I_q$ is positive because an improvement in the housing quality should increase the income of the agent for a fixed level of residential density. The maintenance is assumed to be positively related to income, i.e., $dH/dI > 0$. For simplicity, we specify $H(I)$ as

$$H(I) = \frac{\mu n q^2}{1 + \sigma n} , \tag{8.6.3}$$

where $\mu$ and $\sigma$ are positive coefficients. If we interpret $q^2/(1 + \sigma n)$ as rent per dwelling unit, $nq^2/(1 + \sigma n)$ is the total income of the agent at the location. The parameter $\mu$ can be interpreted as the investment ratio on the maintenance from the income.

To analyse the behavior of the system, we make the following transformation

$$at \to t , \quad q = \frac{\mu Q}{\alpha \sigma} , \quad n = \frac{N}{\sigma} , \quad \gamma = \frac{\Theta}{\alpha} ,$$

$$v = \frac{\delta}{\alpha} , \quad g(Q) = \sigma f \frac{\mu Q}{\alpha \sigma} . \tag{8.6.4}$$

Under (8.6.4), the system can be rewritten as

$$N_t(r,t) = g(Q) - N + \gamma N_{rr} ,$$

$$Q_t(r,t) = -vQ + \frac{NQ^2}{1 + N} , \tag{8.6.5}$$

subject to the corresponding initial and boundary conditions. We assume the existence of a long-term equilibrium of (8.6.5). Here, we are interested in the existence of periodic traveling-wave-like solutions around the long-term static equilibrium. We will apply bifurcation theory to solve the problem.

The periodic traveling-wave-like solutions are very typical of partial differential equations and are well observed in physics, chemistry and biology. The behavior can be generally illustrated as in Fig. 8.11. If we interpret such behavior as an evolutionary process of the urban pattern over space, we see that in the initial stage the densest area, which is defined as the location with the largest difference between the actual and equilibrium density, is found near the CBD. As time passes, the densest area moves away from the CBD. It is then possible to find the densest area near the middle of the suburban area.

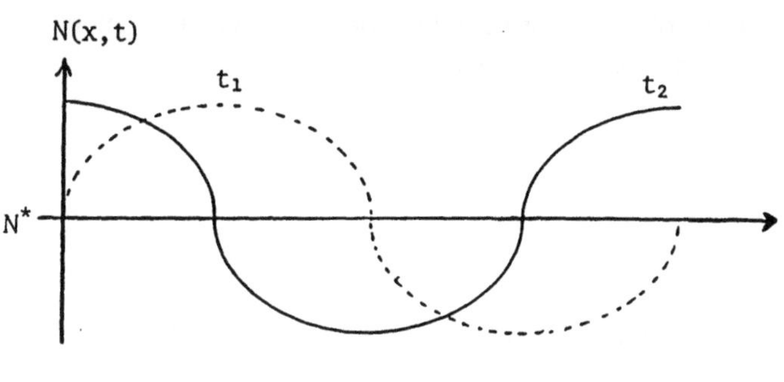


**Fig. 8.11.** Traveling-wave-like density – an illustration

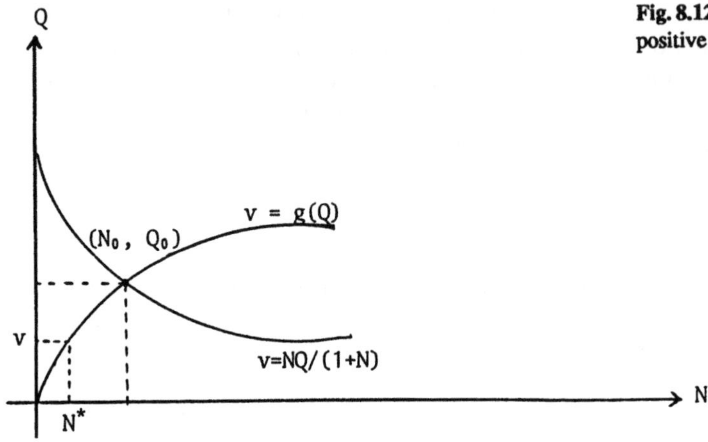


**Fig. 8.12.** The existence of a unique positive equilibrium

A traveling-wave-like solution of (8.6.5) can be written in the form

$$N(r,t) = N(r - \varepsilon t) \; ,$$

$$Q(r,t) = Q(r - \varepsilon t) \; , \tag{8.6.6}$$

where $\varepsilon$ is a positive parameter to be determined. A periodic traveling-wave-like urban pattern is defined as the solution which is periodic with respect to $r - \varepsilon t$. Introducing $W(r - \varepsilon t) = N'(r - \varepsilon t)$ and substituting (8.6.6) into (8.6.5), we obtain

$$N' = W \; ,$$

$$\gamma W' = N - \varepsilon W - g(Q) \; ,$$

$$\varepsilon Q' = vQ - \frac{NQ^2}{1+N} \; , \tag{8.6.7}$$

where the prime represents a derivative with respect to $(r - \varepsilon t)$. Thus our problem is reduced to proving the existence of limit cycles in (8.6.7). The Hopf bifurcation theorem is applied.

As shown in Zhang (1989e), an equilibrium of (8.6.7), denoted by $(N_0, W_0, Q_0)$ where $W_0 = 0$, is determined as in Fig. 8.12.

The proof of the existence of limit cycles can be guaranteed using the bifurcation method discussed in Chap. 5. The following theorem is proved in Zhang (1989e).

**Theorem 8.6.1.** There exists a set of meaningful values of the parameters such that there is a periodic traveling-wave-like urban pattern given by

$$N(r,t) = N(r - \varepsilon t) = N\left(r - \varepsilon t + \frac{2\pi}{D}\right) \; ,$$

$$Q(r,t) = Q(r - \varepsilon t) = Q\left(r - \varepsilon t + \frac{2\pi}{D}\right) \; ,$$

where $D$ is determined by the parameters and $\varepsilon$ is sufficiently small.

It should be noted that we choose $v$ as the bifurcation parameter. As $v = \delta/\alpha$, $v$ may be shifted either by changes in deterioration factors or by changes in the adjustment speed in the residential density equation. As discussed in Zhang (1989e), the theorem implies that when the system is subjected to small perturbations in the bifurcation parameter, a new urban pattern is formed. A static long-term urban pattern is bifurcating to a time-dependent traveling-wave-like pattern.

As $\varepsilon$ is sufficiently small, it may take a long time for the system to complete a cyclical movement.

## 8.7 Instabilities and Urban Pattern Formation

This chapter has investigated the behavior of different urban models developed within the framework of the interactional spatial dynamic approach. As in the previous chapters, we are interested in unstable urban evolution. The examples show that instability is a source of the complexity of urban evolution. Urban structure and structural changes are emphasized in this chapter. Well-ordered time-dependent heterogeneous urban patterns may be bifurcating from a homogeneous time-independent urban pattern due to small changes in the external environment.

Through the examples in this chapter, we see that diffusional factors have a significant role in the urban pattern formation process. An unstable urban system may be stabilized by introducing the space dimension. Surely, we can also find an opposite example. In fact, effects of urban size and diffusional factors upon urban evolution are still little known. It seems that small urban areas tend to be homogeneous, while large ones tend to be heterogeneous. As the urban extension is usually very slow, we may treat urban size as a bifurcation parameter. As this parameter increases over critical points, a more complicated urban pattern may result. This intuitive suggestion proposes another way to explain the variation and complexity of urban evolutionary processes. However, it seems necessary to develop more sophisticated urban models and analytical methods before we can explain such processes well.

## Appendix: Structural Changes in Two Pattern Formation Models

In this appendix, we provide two examples of pattern formation models. We show how different analytical methods can be applied to deal with the behavior of partial differential equations. The model for morphogenesis is an important example in synergetics for explaining the concept of order out of chaos (see Haken 1977). The Brusselator is well known as an example for illustrating the characteristics of self-organization in dissipative systems.

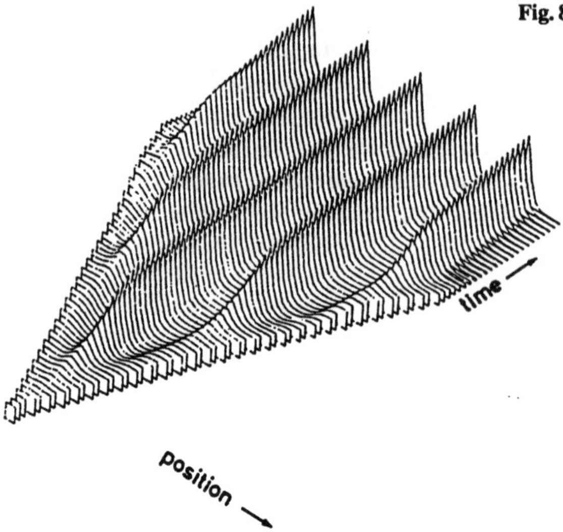


Fig. 8.13. Dynamics of activator concentration

## A.1 A Model for Morphogenesis

In the field of morphogenesis, an important model is related to the hydra – an animal a few millimeters in length, consisting of about 100 000 cells of about 15 different types. It is assumed that there are at least two types of chemicals: an activator and an inhibitor, which cause biological processes of morphogenesis. Denote the concentration of activator by $a$ and that of the inhibitor by $h$. The model is formulated as follows

$$\frac{\partial a}{\partial t} = r + \frac{k' a^2}{h} - sa + D_a \frac{\partial^2 a}{\partial x^2} ,$$

$$\frac{\partial h}{\partial t} = ca^2 - vh + D_h \frac{\partial^2 h}{\partial x^2} , \tag{8.A.1}$$

where $r$, $k'$, $s$, $c$, $v$, $D_a$ and $D_h$ are constant parameters, and $x$ represents a distance parameter. Figure 8.13 and Fig. 8.14 show some simulation results of the model (see Haken 1977). As we are not interested in the meanings of the parameter values, we just illustrate some possible patterns.

Introducing

$$\nabla = \frac{\partial^2}{\partial x^2} + \frac{\partial^2}{\partial y^2} , \quad t^* = vt , \quad A = \frac{k' a}{c} ,$$

$$H = \frac{vch}{k'^2} , \quad R = \frac{rc}{vk'} , \quad S = \frac{s}{v} ,$$

$$D = \frac{D_h}{D_a} , \quad X = \left(\frac{v}{D_a}\right)^{1/2} x , \quad Y = \left(\frac{v}{D_a}\right)^{1/2} y ,$$

we can rewrite (8.A.1) as

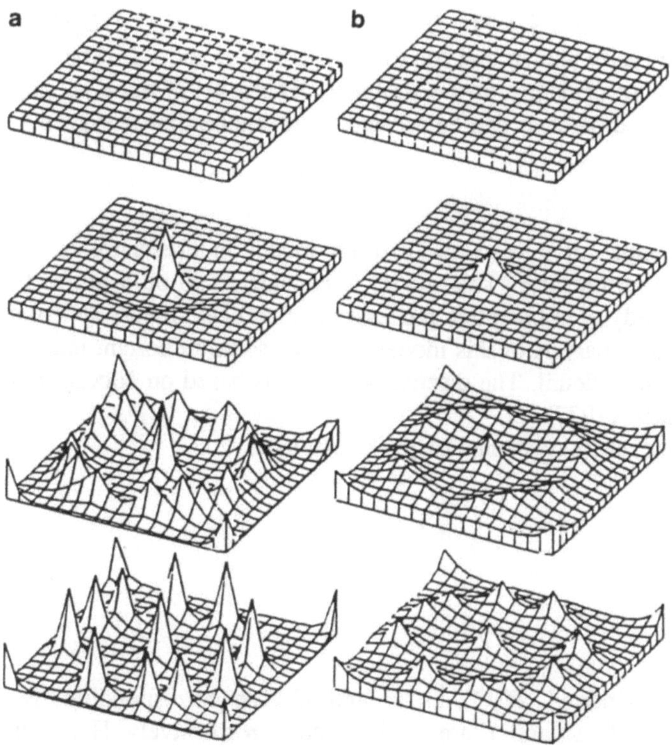


Fig. 8.14a, b. Behavior of the morphogenetic model in two dimensions. (a) Activator concentration, (b) Inhibitor concentration

$$\frac{dA}{dt^*} = R + \frac{A^2}{H} - SA + \nabla A ,$$

$$\frac{dH}{dt^*} = A^2 - H + D\nabla H . \tag{8.A.2}$$

The stationary homogeneous solution reads: $A_0 = (R+1)/S$, $H_0 = A_0^2$. Introducing $q_1 = A - A_0$ and $q_2 = H - H_0$, we have (8.A.2) in the form

$$q = K(\nabla)q + g(q) , \tag{8.A.3}$$

where $q = (q_1, q_2)^{\mathrm{T}}$ and

$$K(\nabla) = \begin{pmatrix} S(2/(1+R)-1)+\nabla & -S^2/(1+R)^2 \\ 2(R+1)/S & -1+D\nabla \end{pmatrix} .$$

For the linear analysis, dropping $g(q)$ and letting $q = \alpha \exp(\mathrm{i}\beta\phi + \Theta t)$ where $\alpha$, $\beta$ and $\phi$ $[= (X,Y)^{\mathrm{T}}]$ are vectors, we have the following eigenvalue equation

$$\Theta_{1,2} = \frac{m(\beta)}{2} \pm \left[\frac{m^2(\beta)}{2} - n(\beta)\right]^{1/2} ,$$

where $m(\beta) = -(D+1)\beta^2 + 2S/(R+1) - S - 1$, $n(\beta) = (\beta^2 + S)(1 + D\beta^2) - 2SD\beta^2/(R+1)$. It can be easily shown that for some values of $\beta$ and $R$, instability occurs. Applying the slaving principle to the problem, Haken (1977) analytically showed that very complicated patterns can be created from this nonlinear system.

## A.2 The Brusselator

Another well known example of a reaction-diffusion model in chemistry is the trimolecular model of the Brusselator. This model is considered to be a perfectly acceptable model for the study of cooperative processes in chemical kinetics (Nicolis and Prigogine 1977). As the analysis of this model involves some important mathematical skills, we explain it in detail. The discussion below is based on Nicolis and Prigogine (1977) and Haken (1977).

Consider the following reaction scheme of the Brusselator

$$A \to X$$

$$B + X \to Y + D ,$$

$$2X + Y \to 3X ,$$

$$X \to E$$

between molecules of the kinds $A$, $X$, $Y$, $B$, $D$ and $E$. The corresponding "reduced" concentrations to $A$, $B$, $X$ and $Y$ are $\alpha$, $\beta$, $x$ and $y$, respectively. Here, the concentrations $\alpha$ and $\beta$ are treated as fixed quantities, whereas $x$ and $y$ are treated as variables. According to Nicolis and Prigogine (1977), the "reduced" variables $x$ and $y$ in one dimension $r$ satisfy

$$\frac{\partial x}{\partial t} = \alpha - (\beta + 1)x + x^2 y + \Theta_1 \frac{\partial^2 x}{\partial r^2} ,$$

$$\frac{\partial y}{\partial t} = \beta x - x^2 y + \Theta_2 \frac{\partial^2 y}{\partial r^2} , \tag{8.A.4}$$

where $\Theta_1$ and $\Theta_2$ are diffusion constants, and $0 \leq r \leq R$. The boundary is denoted by $D$. We have two types of boundary conditions

$$x(t, r) = \alpha , \quad y(t, r) = \frac{\beta}{\alpha} , \quad \text{on the boundary} ,$$

$$\text{(Dirichlet conditions)} \tag{8.A.5}$$

or

$$\frac{\partial x}{\partial n} = \frac{\partial y}{\partial n} = 0 . \tag{8.A.6}$$

Here, we are only concerned with the Dirichlet conditions. The similar analysis for no-flux conditions is given in Nicolis and Prigogine (1977).

A unique homogeneous steady state solution is $x_0 = \alpha$, $y_0 = \beta/\alpha$. Introducing small perturbations $[q = (q_1, q_2)^{\mathrm{T}}]$ as

$$x = x_0 + q_1 \ , \quad y = y_0 + q_2 \ , \tag{8.A.7}$$

one finds the following form of linearized equations

$$\frac{\partial q}{\partial t} = Lq \ , \tag{8.A.8}$$

subject to $q_1 = q_2 = 0$ on the boundary. The linearized operator is given by

$$L = \begin{pmatrix} \beta - 1 + \Theta_1 \partial^2/\partial r^2 & \alpha^2 \\ -\beta & -\alpha^2 + \Theta_2 \partial^2/\partial r^2 \end{pmatrix} .$$

To analyze the asymptotic behavior of the solutions of (8.A.8), it is sufficient to find the eigenvalues $w_i$ and the eigenvectors $u_j$ [$= (u_{1j}, u_{2j})^T$] of $L$

$$Lu_j = w_j u_j \ , \tag{8.A.9}$$

subject to $u_j = 0$ ($j = 0, 1, \dots$) on the boundary. The solution of (8.A.9) is

$$u_i = (\gamma_1, \gamma_2)^T \cos \frac{i\pi r}{R} \ , \quad j = 0, 1, \dots \ .$$

The solution vector $q$ can be expressed as $q = \Sigma_i a_i \exp(w_i t) u_i$. Thus the reference state $(\alpha, \beta/\alpha)$ is asymptotically stable, if for all $i$, the eigenvalues $w_i$ obey $\mathrm{Re}\,(w_i) < 0$. If for some $i$, $\mathrm{Re}\,(w_i) > 0$, then the solution is unstable. At $\mathrm{Re}\,(w_i) = 0$ there is a bifurcation phenomenon provided that the eigenvalue is of odd multiplicity. It is not difficult to identify the following characteristic equation

$$w_j^2 + (v_j - p_j) + \alpha^2\beta - v_j p_j = 0 \ ,$$

with $v_j = \beta - 1 - j^2\pi^2\Theta_1/R^2$, $p_j = \alpha^2 + j^2\pi^2\Theta_2/R^2$. Solving this equation yields

$$w_{j(1'2)} = \frac{v_j - p_j}{2} \pm \left[\frac{(v_j + p_j)^2}{4} - \alpha^2\beta\right]^{1/2} .$$

Stability conditions can be easily found. We are interested in the behavior when the system is unstable. We discuss the case when $w_j$ is real. We have one positive root provided $v_j p_j - \alpha^2\beta > 0$ or

$$\beta > 1 + \alpha^2 \frac{\Theta_1}{\Theta_2} + \frac{\alpha^2 R^2}{\Theta_2 j^2 \pi^2} + \frac{j^2\pi^2\Theta_1}{R^2} = f(j^2) \ . \tag{8.A.10}$$

Figure 8.15 shows $\beta$ as a function of $j$ along the critical curve $\beta_j = f(j^2)$. As $\beta$ increases, the first instability occurs for $\beta = \beta_c$ corresponding to an integer $j_c$ closest to the minimum $(u^*, \beta^*)$ where $u^{*2} = \alpha R^2/\pi^2(\Theta_1\Theta_2)^{1/2}$, $\beta^* = [1 + \alpha(\Theta_1\Theta_2)^{1/2}]^2$. From the figure we see that the first bifurcation point $\beta_c$ lies in the neighborhood of minimum $\beta^*$ of the marginal stability curve (in general, $\beta_c$ is not equal to $\beta^*$).

We will find the conditions under which the eigenvalue $w_j$ is doubly degenerate. As (8.A.10) is a function of $j^2$, what we need is that the equation is written in the form of $(j^2 - j_1^2)(j^2 - j_2^2)$, where $j_1$ and $j_2$ are two positive integers. In particular,

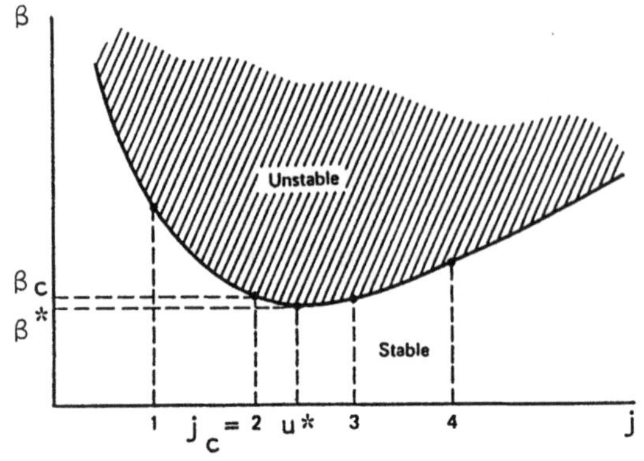


**Fig. 8.15.** Linear stability diagram with bifurcation of stationary solutions

if we take $j_1 = j_c \leq u^*$, then we necessarily have $j_2 = j_c + 1$. It can be identified that this condition is satisfied if

$$\frac{\alpha R^2}{\pi^2(\Theta_1\Theta_2)^{1/2}} = j_c\,(j_c+1)\ . \tag{8.A.11}$$

To get explicit forms of the steady-state solutions bifurcating beyond the critical value $\beta_c$, we decompose the operator $L$ as

$$L = L_c + (L - L_c) = L_c + \begin{pmatrix} \beta-\beta_c & 0 \\ -(\beta-\beta_c) & 0 \end{pmatrix} ,$$

where $L_c$ is the operator evaluated at the critical point of the first bifurcation. It can be shown that $L_c$ is a parabolic operator which admits a single null eigenvector $u_\sigma$ ($\sigma = j_c$) and otherwise has eigenvalues with negative real parts.

Setting $\partial u/\partial t = 0$, we can write (8.A.4) in the form

$$L_c q = \left(-h(q), h(q)\right)^{\mathrm{T}} , \tag{8.A.12}$$

where $h(q) = (\beta - \beta_c)q_1 + 2\alpha q_1 q_2 + \beta q_1^2/\alpha + q_1^2 q_2$. To find the solution, let

$$\begin{aligned} q &= nq_0 + n^2 q^1 + \dots\ , \\ \beta_* &= \beta - \beta_c = n\beta_1 + n^2\beta_2 + \dots\ , \end{aligned} \tag{8.A.13}$$

where $n$ is a small expansion parameter, $q^j = (q_1^j, q_2^j)^{\mathrm{T}}$. Substituting (8.A.13) into (8.A.12) and then identifying equal powers of $n$, we have

$$L_c q^k = (-a_k, a_k)^{\mathrm{T}}\ ,\quad k = 0, 1, \dots\ , \tag{8.A.14}$$

subject to $q^k = 0$ ($k = 0, 1, \dots$) on the boundary. The first few coefficients $a_k$ are

$$a_0 = 0\ ,$$

$$a_1 = \beta_1 q_1^0 + \frac{\beta_c q_1^{02}}{\alpha} + 2\alpha q_1^0 q_2^0\ ,$$

$$a_2 = \beta_2 q_1^0 + \left(\beta_1 + \frac{2\beta_c q_1^0}{\alpha} + 2\alpha q_2^0\right) q_1^1$$
$$+ 2\alpha q_1^0 q_2^1 + \left(\frac{\beta_1}{\alpha} + q_2^0\right) \left(q_1^0\right)^2 . \tag{8.A.15}$$

According to the Fredholm alternative (see Iooss and Joseph 1980) we know that the vector $q^k$ is a solution of (8.A.14) provided that the right-hand side of (8.A.14) is orthogonal to the null eigenvector of the adjoint operator $L_c^*$ where

$$L_c^* = \begin{pmatrix} \Theta_1 \partial^2/\partial r^2 + \beta_c - 1 & -\beta_c \\ \alpha^2 & \Theta_2 \partial^2/\partial r^2 - a^2 \end{pmatrix} .$$

We can solve (8.A.14). It can be shown (Nicolis and Prigogine 1977) that the Fredholm alternative is satisfied for the problem if

$$\int_0^R dr \alpha_k \left(q^{k-m}\right) \cos \frac{j_c \pi r}{R} = 0 , \tag{8.A.16}$$

where $0 < m \leq k$; $k = 1, \ldots$. From (8.A.16) we see that the solution is dependent on whether $j_c$ is even or odd.

These conditions (8.A.16) with (8.A.15) determine the coefficients $\beta_k$. From the second relation (8.A.13), $n$ is determined as a function of $b - b_c$. We thus can solve $q$ as a function of $\beta - \beta_c$. In what follows, we assume that we have calculated the first few terms $\beta_k$.

With these results, the following theorem can be proved (see Nicolis and Prigogine 1977).

**Theorem 8.A.1.**

i) Let $j_c$ be even. In the neighborhood of the critical point $\beta_c$, the new bifurcating solutions are asymptotically stable in the supercritical region $\beta > \beta_c$ ($\beta_2 > 0$). However, when $\beta_2 < 0$ the subcritical branches are unstable.
ii) Let $j_c$ be odd. Then in the neighborhood of $\beta_c$, the bifurcation solution is defined for $\beta$ on both sides of $\beta_c$. The new bifurcating solution is stable on the supercritical branch where $\beta < \beta_c$ and unstable on the subcritical branch where $\beta > \beta_c$.

Figure 8.16 shows case (i) in the theorem, whereas Fig. 8.17 shows case (ii).

It should be noted that the most important property of the dissipative structure of the bifurcations outlined above is their symmetry-breaking character. When a certain

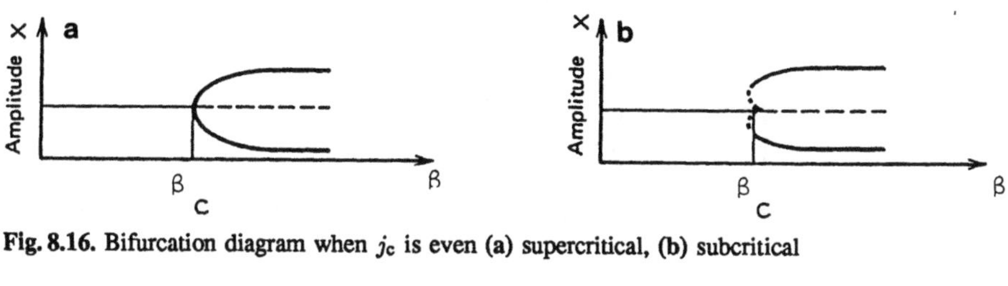


**Fig. 8.16.** Bifurcation diagram when $j_c$ is even (a) supercritical, (b) subcritical

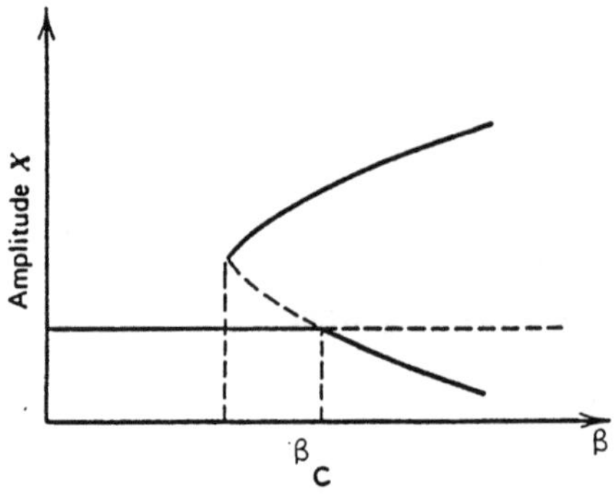


Fig. 8.17. Bifurcation diagram when $j_c$ is odd

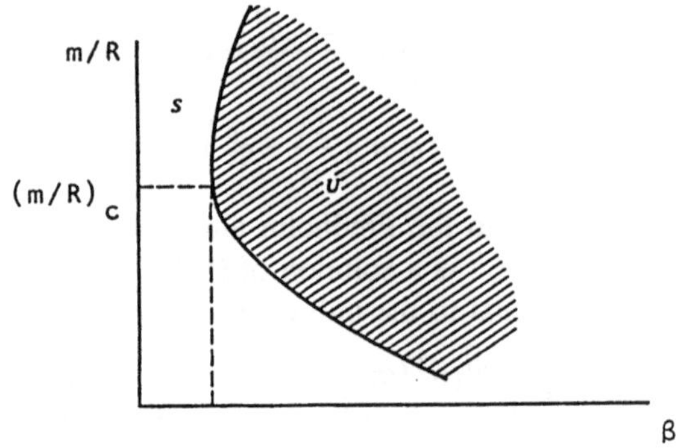


Fig. 8.18. The length as a bifurcation parameter

critical value $\beta_c$ of $\beta$ is crossed, the most symmetric solution ceases to be stable, and the system evolves to a regime with a less spatial symmetry. In the previous example when $j_c$ is even, the system has equal a priori probability to evolve to two different solutions beyond the transition, depending on the initial conditions.

Another interesting aspect of bifurcation analysis is to treat the length of the system $R$ as a bifurcation parameter. In problems involving gradual changes of shape or size of the system, it is of interest to consider the influence of the length on the formation of dissipative structures. This can be similarly analyzed as before. Introducing $r^* = r/R$, we have (8.A.4) as

$$\frac{\partial x}{\partial t} = \alpha - (\beta + 1)x + x^2 y + \frac{\Theta_1}{R^2}\frac{\partial^2 x}{\partial r^{*2}} ,$$

$$\frac{\partial y}{\partial t} = \beta x - x^2 y + \frac{\Theta_2}{R^2}\frac{\partial^2 y}{\partial r^{*2}} , \quad 0 \le r^* \le 1 .$$

Thus changing $R$ may be viewed as changing the diffusion coefficients in our problem. Similarly to Fig. 8.15 we can get a linear stability diagram as shown in Fig. 8.18 where $m$ is taken to be one of the integers compatible with the boundary conditions.

As shown in Nicolis and Prigogine (1977), we can also observe multiple bifurcations in this model. Figure 8.19 illustrates a typical case.

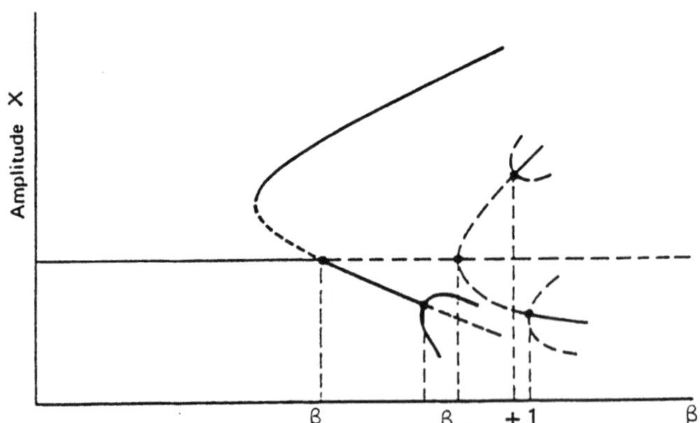


Fig. 8.19. Possible multiple bifurcations

# 9. The Haken Slaving Principle and Time Scale in Economic Analysis

> ... we are in fact raising a problem of understanding, and this always turns out to be a problem about a problem; that is to say a higher level problem.
>
> *Karl R. Popper* (1972)

The main objective of this book is to investigate the behavior of nonlinear unstable dynamic economic systems. In particular, we are interested in the behavior of dynamic systems near critical points. In the preceding examples, we showed that dramatic changes and chaotic economic behavior occur when linear stability is lost due to small shifts of parameters. However, analysis of nonlinear phenomena often involves sophisticated methods, especially when the problem is defined in many dimensions. It is rather desirable to develop some methods to reduce a high-dimensional problem to a lower one. The Haken slaving principle may be used to eliminate some of the variables when the system is close to points where linear stability is lost, so that the macroscopic behavior of the system is governed by very few degrees of freedom only. We can also find similar ideas in the center manifold theorem, although the slaving principle is more general. In this chapter, we are also interested in the role of adjustment speeds of economic variables and the time scale of the study in economic analysis.

## 9.1 The Haken Slaving Principle

This section will explain the Haken slaving principle, which is one of the important analytical methods in synergetics. First, let us consider a simple example provided by Haken (1977). The dynamic system consists of two equations

$$\frac{dx}{dt} = -r_1 x - \alpha xy \, , \tag{9.1.1}$$

$$\frac{dy}{dt} = -r_2 y + \beta x^2 \, , \tag{9.1.2}$$

where $r_2 > 0$. Obviously, (9.1.2) is damped in the absence of the system (9.1.1). We require $r_2 \gg r_1$. In this case, we may solve (9.1.2) approximately by putting $dy/dt = 0$ which results in

$$y = \frac{\beta x^2}{r_2} \, . \tag{9.1.3}$$

Since from (9.1.3) one can see that (9.1.2) follows (9.1.1), the system (9.1.2) is said to be slaved by the system (9.1.1). Substituting (9.1.3) into (9.1.1), we obtain

$$\frac{dx}{dt} = -r_1 x - \frac{\alpha\beta x^3}{r_2} . \tag{9.1.4}$$

It is easily seen that two completely different kinds of solutions of $dx/dt = 0$ occur depending on whether $r_1 > 0$ or $r_1 < 0$.

In a certain sense, $x$ describes the degree of "order" in a complex system. Haken refers to $x$ as the "order parameter". The technique of the adiabatic elimination of fast relaxing variables just illustrated is called the slaving principle (adiabatic approximation) by Haken. The validity of the procedure is proven in Haken (1983).

To generalize the preceding process, let us consider

$$\frac{dx_i}{dt} = -r_i x_i + g_i(x_1, \ldots, x_n) ,$$
$$i = 1, \ldots, n , \tag{9.1.5}$$

The indices are arranged in such a way that there are two distinct groups in which $i = 1, \ldots, m$ refer to modes with small damping which can even become unstable modes (i.e., $r_i$ being non-positive) and another group with $s = m+1, \ldots, n$ referring to stable modes. The functions $g_i$ are nonlinear with regard to the variables.

It should be noted that the procedure to be described can be generally applied to nonlinear systems

$$\frac{dx_i}{dt} = h_i(x) , \quad i = 1, \ldots, n ,$$

because the system can always be locally written in the "standard" form (9.1.5) after appropriate transformations.

In (9.1.5), if we assume $r_i$ ($i = 1, \ldots, m$) are very small, and $r_s$ ($> 0$, $s = m+1, \ldots, n$) are finite, then we can invoke the "adiabatic" approximation principle putting $dx_s/dt = 0$. In this case, we have to check that $|x_s|$ is much smaller than $|x_i|$. We can thus from the last $n-m$ equations of (9.1.5) determine $x_s$ ($s = m+1, \ldots, n$) as a function of $x_i$. The behavior of the system is approximated by

$$\frac{dx_i}{dt} = -r_i x_i + g_i^*(x_1, \ldots, x_m) ,$$
$$i = 1, \ldots, m . \tag{9.1.6}$$

The solutions of these equations then determine whether nonzero action of the subsystems is possible or not. Similarly we say that the modes $x_i$ ($i = 1, \ldots, m$) play the role of the order parameters which slave all the other modes.

As in many applications where only one or few modes become unstable, we can greatly reduce the dimension of the problem. Since all the damped modes follow the order parameters, the behavior of the whole system is determined by the behavior of very few order parameters. Thus even very complex systems may be dealt with very simply. Furthermore, as bifurcations occur in the order parameter equations, the complexity of bifurcation analysis is much reduced.

In practice, we often deal with a hierarchical structure, in which the relaxation constants can be grouped so that

$$r^{(1)} \gg r^{(2)} \gg r^{(3)} \dots .$$

In this case we can apply the procedure first to the variables connected with $r^{(1)}$, leaving us with the other variables. Then we can apply this method to the variables connected with $r^{(2)}$ and so on.

From the discussion above, we see that the slaving principle allows us to reduce the number of degrees of freedom considerably. Moreover, an interesting example of relations between chaos and the slaving principle can be found in Haken (1977, Sect. 12.4). Taking the Lorenz equations as an example, Haken shows that chaotic motion may occur when the slaving principle fails and the formerly stable mode can no longer be slaved but is destabilized. The principle can also be applied to the stochastic system and discrete equation. In the appendix to this chapter we provide an example to show how the principle can be applied to stochastic equations.

From an economic point of view, this principle implies that we can find a few (aggregated or transformed) variables which govern the dynamic behavior of the whole economic system near critical points. That is, other economic variables are slaved by these variables. Thus, the main application of the principle to economic analysis is the justification of the reduction of problems posed in spaces of high or infinite dimension to one or two dimensions. Analyses of these low-dimensional problems are often involved with bifurcation at simple eigenvalues. Thus, the bifurcation methods used in the preceding chapters are applicable to high dimensional problems.

It should be noted that we have some other mathematical methods to reduce the dimension of problems. For instance, the method of Lyapunov–Schmidt is used to decompose the space of solutions and equations into a finite-dimensional and an infinite-dimensional part. The infinite part can be solved and the resulting finite dimensional problem has all the information about bifurcation (see Chow and Hale 1982). Iooss and Joseph (1980) apply the implicit function theorem to justify the direct, sequential computation of power series solutions in an amplitude, using the Fredholm alternative, as an economical way to determine qualitative properties of the bifurcating solutions and to compute them. We have applied the bifurcation method of Iooss and Joseph in Chap. 5. We can also use the center manifold theorem (see below) to reduce the problems to finite dimensions. This method uses the fact that solutions are attracted to the center manifold, which is finite-dimensional. It should be emphasized that the choice of a method often depends on the characteristics of the problem at hand.

## 9.2 The Center Manifold Theorem

The Haken slaving principle provides an effective method to reduce the dimension of a dynamic system. Here, we will introduce the center manifold theorem for reducing dimensions of a system. This theorem is very useful for a number of applied problems.

As we have shown in Chap. 3, the linear part of $dx/dt = f(x)$ ($f(0) = 0$) determines the qualitative behavior of the system trajectories locally, provided that all the real parts of the eigenvalues at $x = 0$ are not equal to zero. In other words, if the linear part has no eigenvalues lying on the imaginary axis, then in a neighborhood of the origin the system behaves as if it were a linear system. This result suggests that any essentially nonlinear behavior of the system (multiple equilibria, limit cycles, hysteresis effects) are strongly connected with the eigenvalues lying on the imaginary axis. We can justify this conjecture, at least insofar as the stability properties of the system are concerned.

Consider the system of differential equations

$$\frac{dx}{dt} = Ax + f(x,y) ,$$

$$\frac{dy}{dt} = By + g(x,y) , \tag{9.2.1}$$

where $x \in R^n$, $y \in R^m$, and $A$ and $B$ are constant square matrices. We assume always that the eigenvalues of $A$ and $B$ have real parts which are zero and negative, respectively, and $f$ and $g$ are functions of class $C^k$, $k > 1$, which vanish together with their first derivatives at the origin 0.

**Definition 9.2.1.** (*Invariant manifold*). A set $S \in R^{n+m}$ is an invariant manifold for (9.2.1) if for any solution $(x(t), y(t))$, $(x(0), y(0))$ implies that for some $T > 0$, $(x(t), y(t)) \in S$ for all $t \in [0, T]$.

**Definition 9.2.2.** ( *Center manifold*). An invariant manifold $S = \{(x, h(x)) \| |x| < \delta\}$ for (3.10.1) is a center manifold if $h(0) = 0$, $Dh(0) = 0$, where $D$ is a derivative operator.

**Theorem 9.2.1.** There exists a center manifold $y = h(x)$, $|x| < \delta$ for (9.2.1) of class $C^k$. The flow on this manifold is governed by the $n$-dimensional system

$$\frac{du}{dt} = Au + f(u, h(u)) . \tag{9.2.2}$$

**Theorem 9.2.2.**

i) Suppose that the zero solution of (9.2.2) is stable. Then if $(x(t), y(t))$ is a solution of (9.2.1) with $(x(0), y(0))$ sufficiently small, there is a solution $u(t)$ of (9.2.2) such that for $t > 0$

$$x(t) = u(t) + r_1(t) ,$$

$$y(t) = h\big(u(t)\big) + r_2(t) ,$$

with $r_1(t)$ and $r_2(t) \to 0$ exponentially fast as $t \to \infty$. In particular, if the zero solution of (9.2.2) is asymptotically stable, then the zero solution of (9.2.1) is also asymptotically stable.

ii) Suppose that the zero solution of (9.2.2) is unstable. Then the zero solution of (9.2.1) is unstable.

If we substitute $y(t) = h(x(t))$ into the second equation of (9.2.1) and then eliminate $dx/dt$ using the first equation, we obtain

$$Dh(x)\left[Ax + f\left(x, h(x)\right)\right] = Bh(x) + g\left(x, h(x(t))\right) . \qquad (9.2.3)$$

This equation, together with $h(0) = 0$ and $Dh(0)$, forms the system to be solved for the center manifold. While we may not be able to find $h$ exactly from these equations, we may approximate it. Let us set

$$N(j)(x) = Dj(x)\left[Ax + f\left(x, j(x)\right)\right] - Bj(x) - g\left(x, j(x)\right) , \qquad (9.2.4)$$

for functions $j : R^n \to R^m$ of class $C^1$ on a neighborhood of $0 \in R^n$. By (9.2.3), $N(h) = 0$.

**Theorem 9.2.3.** If $j(0) = 0$, $Dj(0) = 0$, and

$$N(j)(x) = 0\left(|x|^\gamma\right) , \quad x \to 0$$

for some $\gamma > 1$, then

$$|h(x) - j(x)| \to 0 , \quad x \to 0 .$$

To illustrate how the theorems can be applied, let us consider the system of differential equations

$$\frac{dx}{dt} = Ax + f(x, y, r) ,$$

$$\frac{dy}{dt} = By + g(x, y, r) ,$$

$$\frac{dr}{dt} = 0 , \qquad (9.2.5)$$

where $A$, $B$, $f$ and $g$ have the same properties as in (9.2.1). $(x, y) = 0$ is a critical point for small $r$. It should be mentioned that our discussion may be applied to the general case: $dz/dt = f(z, r)$, where $z$ are $(m + n)$-dimensional variables and $r$ are $q$-dimensional parameters. We may require that $z = 0$ is a critical point for small $r$ and such that $D_z f(0, 0)$ has $n$ eigenvalues whose real parts are zero and $m$ eigenvalues whose real parts are negative. We can always transform this general case to (9.2.5).

For small perturbations in $r$ (from zero), such a system is "near critical". It has been recognized that an essential step in analyzing asymptotic behavior of small solutions of near-critical systems is a reduction of the system in $n + m$ dimensions to a system of $n$ dimensions by eliminating a part of the solution, presumed to decay exponentially with time, contributed by the $m$ eigenvalues whose real parts are negative. Perturbation techniques such as amplitude expansions based upon multiple time scales provide one method of obtaining reductions of this kind. Now, we show that the center manifold theorem provides a rigorous and unifying method for reducing "near critical" systems.

By Theorem 9.2.1, (9.2.5) has a center manifold $y = h(x,r)$, $|x| < r_1$, $|r| < r_2$, of class $C^k$. By Theorem 9.2.2 the behavior of small solutions of (9.2.5) is governed by the reduced system

$$\frac{du}{dt} = Au + f\big(u, h(u,r), r\big) ,$$

$$\frac{dr}{dt} = 0 . \tag{9.2.6}$$

Finally, by (9.2.4) the system governing $h$ is

$$D_x h(x,r)\left[Ax + f\big(x, h(x,r), r\big)\right] = Bh(x,r) + g\big(x, h(x,r), r\big) ,$$

$$h(0,0) = 0 , \quad D_x h(0,0) = 0 , \quad D_r h(0,0) = 0 ,$$

and we may use Theorem 9.2.3 to approximate its solution.

When attempting to solve (9.2.6) we may drop the equation $dr/dt = 0$ and regard $r$ simply as a parameter once again. However, when applying Theorem 9.2.2 we must examine the stability of the origin of (9.2.6) because $dr/dt = 0$ can have an important effect upon stability for this equation. For instance, the solution $u = 0$ of $du/dt = r^2u - u^3$ is unstable for each $r$ not equal to zero while the solution $(u,r) = (0,0)$ of $du/dt = r^2u - u^3$, $dr/dt = 0$, is in fact stable. Examples of how to apply the center manifold theorem can be found in Carr (1981), Carr and Muncaster (1983), and Casti (1985).

## 9.3 Singular Perturbations

A dynamic system contains certain parameters, both those appearing explicitly in the equations, and implicit ones. It may be that some parameters of a system are small. In this case, it is possible to neglect the terms in the equation which involve the small parameter and obtain a good approximation to the solution of the problem. A number of examples have shown that this sometimes does hold. Even when we cannot neglect a term in a small parameter, it may be possible to obtain an approximate solution to the problem by using the fact that the parameter is small. This section is concerned with some (asymptotic) methods for constructing such approximations.

The applications of asymptotic analysis are related to perturbations, or small changes, in mathematical problems. These may be the addition of an extra term into the equations or a change in one of the parameters of the problem. Let $\varepsilon$ be a measure of the size of the perturbation. According to whether the solution of the perturbed problem is close to that of the unperturbed one or not, one can define regular or singular perturbation.

**Definition 9.3.1.** Let $x(\varepsilon)$ satisfy the perturbed problem $P(\varepsilon)$ (where $P(\varepsilon)$ will normally consist of a system of differential equations in a spatio-temporal domain $Q$ and some initial and boundary conditions to be satisfied by $x$). Then the perturbation is regular if $x$ is a continuous function $\varepsilon$ at $\varepsilon = 0$, i.e., if $\|u(\varepsilon) - u(0)\| \to 0$ as $\varepsilon \to 0$, where $u(\varepsilon)$ is the solution and $\| . \|$ is an appropriate norm, and singular otherwise.

Note that a perturbation may be regular with respect to one norm but singular with respect to another. Solutions of regular perturbation problems may often be obtained as power series in $\varepsilon$. It is possible to look for an asymptotic expansion of $x(\varepsilon)$ as $\varepsilon \to 0$ in terms of the asymptotic sequence $\{\varepsilon^n\}$.

This section is only concerned with singular perturbations. Such problems may occur when one of the variables has the capacity to change much more quickly than another. In this aspect, perturbation analysis is very significant since adjustment speeds of economic variables are often very different.

To illustrate the problem, let us consider a one-sector economic growth model

$$\varepsilon\frac{dp}{dt} = f(p, k) ,$$

$$\frac{dk}{dt} = g(p, k) , \tag{9.3.1}$$

where $p$ is price, $k$ capital per capita, $f$ and $g$ are O(1) quantities, i.e., they are neither small nor large for general values of $p$ and $k$, and $\varepsilon$ is a small parameter. The adjustment of $p$ is faster than that of $k$. This is assumed in neoclassical growth theory in which wages and prices are always determined as soon as capital and (full) employment level are given.

The singular nature of such a problem is easily seen. We know from Picard's theorem that the system has a solution in the neighborhood of $t = 0$ as long as $f$ and $g$ satisfy the Lipshitz conditions in $p$ and $k$. However, the unperturbed ($\varepsilon = 0$) system clearly does not possess a solution unless the initial condition satisfies $f(p_0, k_0) = 0$. Furthermore, it can be shown that under appropriate conditions the fast variable may be neglected except in the short initial period.

It should be noted that some perturbation methods in applied mathematics have been proposed (e.g., Kevorkian and Cole 1981, Britton 1986). Here, we provide an example to show the way in which average methods can be applied to the van der Pol equation. The example is cited from Britton (1986).

Consider the van der Pol equation

$$\frac{d^2x}{d^2t} - \varepsilon\left(1 - x^2\right)\frac{dx}{dt} + x = 0 . \tag{9.3.2}$$

where $\varepsilon$ is a very small parameter, and $x$ is a scalar variable. This model is often used as an example in singular perturbations since it is one of the simplest second-order differential equations to exhibit the behavior – oscillations and specifically, limit cycles – in which we are interested.

When $\varepsilon = 0$ this has solution

$$x = a \exp(\mathrm{i}t) + \bar{a} \exp(-\mathrm{i}t) , \tag{9.3.3}$$

so that

$$\frac{dx}{dt} = \mathrm{i}a \exp(\mathrm{i}t) - \mathrm{i}\bar{a} \exp(-\mathrm{i}t) ,$$

where $a$ is a complex constant. The effect of the nonlinear term in the van der Pol equation is to make $a$ vary with time. We define a time-varying function $a(t)$ in terms of the function $x(t)$ by the equations

$$x = a(t) \exp(\mathrm{i}t) + \bar{a}(t) \exp(-\mathrm{i}t) , \tag{9.3.4}$$

$$\frac{dx}{dt} = \mathrm{i}a(t) \exp(\mathrm{i}t) - \mathrm{i}\bar{a}(t) \exp(-\mathrm{i}t) . \tag{9.3.5}$$

where $x$ satisfies the van der Pol equation and initial conditions. We require that the right-hand side of (9.3.5) be the differential of the right-hand side of (9.3.4). This is the same as to require

$$\exp(\mathrm{i}t)\frac{da}{dt} + \exp(-\mathrm{i}t)\frac{d\bar{a}}{dt} = 0 . \tag{9.3.6}$$

Differentiating (9.3.5) and substituting (9.3.6) into the equation yields

$$\frac{d^2x}{d^2t} = \left(\mathrm{i}\frac{da}{dt} - a\right) \exp(\mathrm{i}t) - \left(\mathrm{i}\frac{d\bar{a}}{dt} + \bar{a}\right) \exp(-\mathrm{i}t) .$$

The requirement that $x$ satisfies (9.3.2) leads to

$$\begin{aligned}\frac{da}{dt} =& \varepsilon \left[1 - a^2 \exp(2\mathrm{i}t) - 2a\bar{a} - \bar{a}^2 \exp(-2\mathrm{i}t)\right] \\ & \times \left[a - \bar{a} \exp(-2\mathrm{i}t)\right] /2 ,\end{aligned} \tag{9.3.7}$$

where we use (9.3.6). This is an exact differential equation for the function $a$ defined in (9.3.3). This function shows that $a$ is a slowly varying function of $t$, i.e., $da/dt$ is $O(\varepsilon)$. Hence, $a$ and its derivatives are constant to the first order over the period $\pi$ of the periodic functions. Taking an average of (9.3.7) over this period, we have

$$\frac{1}{\pi}\int_0^\pi \frac{da}{dt}dt = \frac{\varepsilon}{2\pi}\int_0^\pi \{(1 - |a|^2)$$
$$\left[a - \bar{a} \exp(-2\mathrm{i}t)\right] - a^3 \exp(2\mathrm{i}t) + \bar{a}^3 \exp(-4\mathrm{i}t)\} dt .$$

Using the fact that $a$ and $da/dt$ are approximately constant over the period yields

$$\frac{da}{dt} = \frac{1}{2}\varepsilon a \left(1 - |a|^2\right) , \tag{9.3.8}$$

which holds approximately. This result is correct up to terms of $O(\varepsilon)$. Defining $y = |a|^2$, we obtain

$$\frac{dy}{dt} = \varepsilon y(1 - y) . \tag{9.3.9}$$

Evidently, $y = 1$ is a limit cycle solution. If the initial condition $y(0)$ is not equal to 1, then we can solve $y$ in the form

$$y = \frac{y(0)\exp(t)}{1 - y(0)\left[1 - \exp(t)\right]} .$$

It can be seen that for $y_0 > 0$, $y \rightarrow 1$ as $t \rightarrow \infty$. In other words, the limit cycle is stable.

From this example it can be seen that the average method may be very effective for analysis of nonlinear phenomena. This method can also be applied to the oscillatory reaction scheme

$$\frac{dx_1}{dt} = -x_2 + f(\varepsilon, x) ,$$

$$\frac{dx_2}{dt} = x_1 + g(\varepsilon, x) ,$$

in which Hopf bifurcations may occur under certain conditions. The van der Pol equation is a special case of this equation. Moreover, it is not difficult to see that some economic models defined in Chap. 5 can be rewritten in this form. We thus have another method to analyze regular economic oscillations.

Another class of perturbation methods useful in dealing with oscillatory ordinary differential equations and associated reaction-diffusion equations is the two-timing method. It is similar to the average method in that the amplitude and phase of the variable are considered to be slowly varying functions of time, but the approach is quite different.

To explain this method, consider the simple dynamics

$$\frac{dx}{dt} = y ,$$

$$\frac{dy}{dt} = -2\varepsilon y - x , \tag{9.3.10}$$

where $\varepsilon$ is very small. For $\varepsilon = 0$, there are oscillatory solutions. Referring to the Hopf bifurcation theorem, we know that for small $\varepsilon$ it is possible for the system to have closed orbits. The problem is how to investigate its behavior using perturbation methods, rather than bifurcation theory.

First, we rewrite (9.3.10) as

$$\frac{d^2x}{d^2t} + 2\varepsilon\frac{dx}{dt} + x = 0 . \tag{9.3.11}$$

A "regular" perturbation expansion of the equation of the form

$$x(t;\varepsilon) = x_0(t) + \varepsilon x_1(t) + \ldots ,$$

gives

$$x(t;\varepsilon) = (1 - \varepsilon t)a\exp(it) + \bar{a}\exp(-it) + \ldots ,$$

which looks as if it is unbounded as $t \rightarrow \infty$. However, the exact solution is given by

$$x(t;\varepsilon) = \exp(-\varepsilon t)\left\{a\exp\left[ig(t)\right] + \bar{a}\exp\left[-ig(t)\right]\right\} ,$$

where $g(t, \varepsilon) = t(1-\varepsilon^2)^{1/2}$. The exact solution agrees with the regular expansion on the assumption that $\varepsilon t$ is very small, which means that the time scale of the study should not be very long. Note that $x$ is periodic over a long time scale $g(t, \varepsilon)$ but is modulated over a short time scale $\varepsilon t$.

This leads to the idea that the dependence on fast and slow time scales should be considered simultaneously. Define

$$T_1 = Gt = \left(\varepsilon + \varepsilon^2 G_2 + \varepsilon^3 G_3 + \ldots\right) t\,,$$

$$T_0 = wt = \left(1 + \varepsilon^2 w_2 + \varepsilon^3 w_3 + \ldots\right) t\,.$$

In the above example $T_1$ is introduced to deal with $\varepsilon t$, and $T_0$ to deal with $G(t, \varepsilon)$. We then put $x(t; \varepsilon) = x(T_0, T_1\,;\, \varepsilon)$ and look for an asymptotic expansion in the form

$$u(T_0, T_1\,;\, \varepsilon) = x_0(T_0, T_1) + \varepsilon x_1(T_0, T_1) + \ldots\,.$$

This is known as the method of double time scales or the two-timing method. In Britton (1986), this method is applied to the van der Pol equation. The same result as with the average method is obtained.

## 9.4 Fast Variable Versus Slow Variable in Economic Analysis

One of the serious debates in economic dynamics is related to the adjustment speeds of economic variables, which means the time for the variable to adjust to an equilibrium. The usefulness of this concept is in fact related to the time scale of the study.

In different theories of economics, adjustment speeds of the same economic variable may be very different. We often find that what economists are arguing about is which variable is fast-adjusting in economic analysis. For instance, in Keynesian economic theory the wage rate is fixed, while in the neoclassical (see Zhang 1990b) model the wage rate is assumed to be a fast-adjusting variable. In the neoclassical approach, as soon as output and capital are given, wages are determined as a result of competition and fast adjustment. In a Keynesian economy it is assumed that wages adjust to an equilibrium at a very slow speed.

The adjustment speed of a variable is determined by many factors. For instance, labor unions have been assumed to have a significant role in determining how wages are adjusted in economic systems. The institutional systems of a country may also affect the speed. For instance, if we study the economic development during the period of the Cultural Revolution in China, it is reasonable to assume that all prices and wages were fixed. However, if we study the current economic evolution in China, the assumption is unacceptable because we are faced with inflation in the reforming process and wages are not fixed at all.

The adjustment speeds of economic variables are closely related to economic mechanisms dominating the country. A structural change in the economy (e.g., from "capitalism to socialism", or from "socialism to capitalism") is always associated with changes in the adjustment speeds of economic variables. From a "pure" economic point of view, all the economic systems in the world are mixed in the sense that no country is purely planned or perfectly competitive, although "mixture degrees" are different among countries (it may be very important to study the interactions between mixture degree and economic development). Prices in a perfectly competitive economy adjust faster than prices in a planned economy. Explicit awareness of such differences is very important for understanding the differences between various economic mechanisms.

In general, we may describe economic dynamics with distinct adjustment speeds as follows

$$\frac{dx_i}{dt} = s^j f_i(x) \,, \tag{9.4.1}$$

$$i = 1, \ldots, n \,, \quad j = 0, 1, \ldots, m, n \geq m \,,$$

where $x_i$ is a vector of economic variables such as monetary, quantitative, or technological aspects. The functions $f_i$ are assumed to be continuously differentiable. The parameter $s$ is a measure of the adjustment speeds of the economic variables. For simplicity, the parameter is given a sufficiently small value. The economic variables can thus be classified as different levels (groups) according to their adjustment speeds. As discussed in the following section, such a classification is very important in economic analysis because some variables can be reasonably omitted. It is not difficult to see that applying the Haken slaving principle or the center manifold theorem may reduce problems to low dimensions.

Economists have treated adjustment speeds of economic variables differently. For instance, Walrasian dynamics can generally be described by

$$\frac{dk}{dt} = s^{-1} f(k, p, z) \,,$$

$$\frac{dp}{dt} = g(p, k, z) \,,$$

$$\frac{dz}{dt} = s\, h(p, k, z) \,, \tag{9.4.2}$$

where $k$ represent quantitative variables such as capital and labor employment, $p$ monetary terms such as wages and prices, $z$ technological terms, and $s$ is a very small parameter. Here, $x$, $p$ and $z$ are vectors. As we are only interested in how the adjustment speeds of economic variables are treated in economic theories, we will not specify these functions.

In the Walrasian dynamic economy, quantities are considered as fast variables in comparison to their prices. Exchange takes place as soon as the price subsystem arrives at equilibrium. Technological changes are rarely considered as their change rates are assumed to be very small.

Setting $s = 0$ in (9.4.2) yields

$$f(k, p, z) = 0 \,,$$

$$\frac{dp}{dt} = g(p, k, z) \, ,$$

$$\frac{dz}{dt} = 0 \, , \tag{9.4.3}$$

From (9.4.3), we see that $z_0$ (= $z(0)$) is constant during the study period. Technology is invariant and can be treated as a constant parameter in economic analysis. Applying the implicit function theorem, we may find the conditions for $f(k, p) = 0$ to have a solution $k = f^*(p)$. Substituting this solution into the price dynamics we have

$$\frac{dp}{dt} = g\left(p, f^*(p)\right) = g^*(p) \, . \tag{9.4.4}$$

Thus the whole dynamic system is governed by the motion of monetary variables.

Obviously, our reduction is not very accurate because we already know that such reduction may be invalid when the system is unstable. We have provided some mathematical methods to treat the system. However, for a stable system under appropriate conditions the reduction qualifies as valid.

We can also characterize other economic theories by the adjustment speed of economic variables. It must be emphasized that the following classification is rather rough. However, it provides another way to analyze economic theories.

Marshallian dynamics can be generally described by

$$\frac{dk}{dt} = f(k, p, z) \, ,$$

$$\frac{dp}{dt} = s^{-1} \, g(p, k, z) \, ,$$

$$\frac{dz}{dt} = s \, h(p, k, z) \, . \tag{9.4.5}$$

In this system, the dynamics are governed by the motion of quantitative variables, whereas technology is kept fixed and monetary variables become functions of quantitative variables.

In Schumpeterian dynamics, interactions among variables may be described as

$$\frac{dk}{dt} = f(k, p, z) \, ,$$

$$\frac{dp}{dt} = g(p, k, z) \, ,$$

$$\frac{dz}{dt} = s \, h(p, k, z) \, . \tag{9.4.6}$$

In this system, quantitative and monetary variables adjust fast in comparison to technological change. However, technology is changeable in the study period. Innovation may cause structural change in the system. Obviously, this implies that the system is possibly unstable. Even if $s$ is very small, technological change can drive the system rather far away from the trajectory without technological change.

The Keynesian system, at least in the short run, can be described by

$$\frac{dk}{dt} = f(k, p, z) ,$$

$$\frac{dp}{dt} = sg(p, k, z) ,$$

$$\frac{dz}{dt} = s\,h(p, k, z) . \tag{9.4.7}$$

The monetary approach, e.g., the Tobin model, can be described as

$$\frac{dk}{dt} = f(k, p, z) ,$$

$$\frac{dp}{dt} = g(p, k, z) ,$$

$$\frac{dz}{dt} = s^i\,h(p, k, z) , \tag{9.4.8}$$

where $i$ is either 0 or 1 according to different authors (see Zhang 1990b).

Similarly the standard neoclassical growth model is given by

$$\frac{dk}{dt} = f(k, p, z) ,$$

$$\frac{dp}{dt} = s^{-1}g(p, k, z) ,$$

$$\frac{dz}{dt} = s^i\,h(p, k, z) , \tag{9.4.9}$$

where $i$ is either 0 or 1 according to different authors (see Zhang 1990b).

When discussing the adjustment speeds of economic variables, I remember that once I asked a mathematical economist (I have to say a very famous one) about what he thought about speeds in Marshallian and Walrasian dynamics (his papers are related to these topics). He gave me a definite answer: this is not a scientific question, you have to ask China's Chairman. Young as I was, the answer from this famous man certainly shocked me very much. It now seems to me that what the term "scientific" means depends on the knowledge structure of the people who are talking about it.

## 9.5 The Time Scale in Economic Analysis

There are relations between the time scale and the adjustment speeds of economic variables. For example, it has been argued that if the study period is not long enough, it may be reasonable to assume a fixed wage, while if the study period is a long term, the neoclassical approach may be appropriate. To describe relations among the time scale and the adjustment speeds, let us discuss the following system

$$\dot{x} = f(x, y) ,$$

$$\dot{y} = s\, g(x, y) , \tag{9.5.1}$$

in which $x$ is a vector of fast variables, $y$ a vector of slow variables, $f$ and $g$ are appropriate continuous functions, and $s$ is a measure of adjustment speed. It is assumed that $s$ is sufficiently small. We have just examined the characteristics of adjustment speeds of variables in different economic systems.

If the study is limited to the short term, $y$ may be treated as a constant. If the long-run behavior is concerned, let us make a transformation of time: $t^* = st$ under which the system can be rewritten as

$$s\dot{x} = f(x, y) ,$$

$$\dot{y} = g(x, y) , \tag{9.5.2}$$

in which the dot is with regard to $t^*$. If the functions $f$ and $g$ satisfy appropriate conditions (see, e.g., Gu, Nefedov and O'Malley 1989, O'Malley 1988), then $f(x, y) = 0$ holds almost everywhere. It is possible for us to obtain $x = f^*(y)$. Substituting $x$ into the second equation, the dynamics only consist of:

$$\dot{y} = g\left(f^*(y, y)\right) = g(y) . \tag{9.5.3}$$

Hence, the variables $x$ do not appear in the dynamic system. As soon as the values of $y$ are determined, those of $x$ are given by $f^*(y)$.

From this example, we see that if the study period is very short, the slow variables may be treated as constants in economic analysis; if the period is very long, the fast variables can be treated as functions of the slow variables, and thus in dynamic analysis the fast variables will not explicitly appear in the dynamic equations. It should be emphasized that the functional form $f(x, y)$ in the dynamics of the fast variables affects the reduced dynamics of the slow variables. This is due to the fact that in the long-run analysis, the fast variables are "governed" by the slow variables. That is, as soon as the slow variables are given, the fast variables accordingly adjust to the new equilibrium very fast.

As discussed in Zhang (1989), in economic development theory the time scale plays a central role in any study of economic growth and development. Approaches to long-run and short-run economic evolutionary processes may be completely different according to the time span under consideration. For a one-year period, if the system is stable, it may be sufficient to look at the dynamics of prices, wages, consumption and so on. However, in a long-term study, technologies and institutional systems become endogenous variables. The duration of the study period influences the choice of exogenous parameters and endogenous variables in a dynamic system. For instance, although it is reasonable to treat technologies as parameters for the short-run analysis, if we study economic problems in the long term, there should exist interactions between technology and economic variables. It may be said that in this case we are no longer faced with a problem of rationality, but with a process of learning. In fact, if we want to understand real economic evolution in the long run, economics seems to be narrow. Not only is technology changeable, but also institutions.

As an example of how the time scale may affect economic analysis, we consider interactions which occur between economic development and institutional systems. We should consider the case of economic development in China.

In China, one of the major current issues is the way in which the country will be opened to "Western" technologies and culture. We use "opening", denoted by $y$, to describe this policy. Although there are uncertainties in determining the speed of "opening", one may assume the existence of interactions between economic development and the policy. To illustrate this, we suggest the following dynamics:

$$\dot{x} = f(x, y) ,$$

$$\dot{y} = s(x, y, t)\,[g(x, y, t) - y] , \tag{9.5.4}$$

where $x$ is a vector of the economic variables, and $f$ and $g$ are continuous with respect to the variables. The other countries are exogenous to our system. Of course, this assumption depends upon the strength of China's role in the world. The first equation means that the economic variables are dependent upon the opening of the country. The Chinese have benefited economically and technologically since the country was "partially" opened, though the country is now experiencing new problems in its development. The second equation describes the behavior of the government. The function $g$ is the "fittest opening level" of the country, which depends upon $x$, $y$ and $t$. The functional form $g$ cannot be determined uniquely from an economic point of view. Very complicated factors can influence its form. The variable $s(x, y, t)$ is an adjustment spped. If $s$ is zero, the opening policy will not be changed. This case can be observed during the period of the Cultural Revolution. If $s$ is very large, the government adjusts to the optimal situation very rapidly.

If the study period is very short and the system is stable, then it may be possible to disregard the opening policy in the economic analysis without affecting the qualitative results. However, if we want to analyse the evolutionary processes of economic development in the long term, we have to consider this factor.

The introduction of "opening" may change the characteristics of the pure economic system where $y$ is fixed. For instance, the stability may seriously be affected. If we have a large $s$, the system may become destabilized, while if we close the country, it is possible to have a (local) stable society although the people suffer from poverty.

It is possible to specify the forms of $f$ and $g$ to a reasonable extent in order to obtain certain insights into the dynamic systems. However, as we are not concerned with the institutional systems, further analysis is omitted.

The time scale for economic modeling is a complex matter which requires exhaustive philosophical discussion in order to be understood. That which is wrong in the short term may be right in the long term, and vice versa. The relations between the long-term and short-term horizons are as important as the relations between aggregated and disaggregated variables, and the relations of the whole and the parts.

Finally, it should be emphasized that whether or not the reduction from (9.5.2) to (9.5.3) is valid is dependent on the characteristics of the system. For instance, if the system is unstable, the discussion may become very complicated since the behavior is very sensitive to small shifts of parameters.

## 9.6 Another Problem – Understanding a Dynamic Man

> **Interactions among a few elementary factors (such as love of power, lust for gold) plus changeable environment make human life chaotic. However, there is order in chaos. This is still rationalistic.**

Before writing this section, one question had confused me for a long time. On the one hand, why is human life so complicated to understand? Why it is possible that the same subject – love – can be written about a million times and new stories on it are expected to be created without limit? On the other hand, there are only a few basic words (variables) which are used to describe the characteristics of human beings. Carefully analyzing, we find that the complexities of behavior result from interactions of these basic elements plus various environments. However, the question is how such a wide variety of human behavior can result from interactions of these elements. Why is human behavior so complicated even though we well know that the behavior is controlled by a few elements? Is there any deterministic mechanism to explain human behavior? Is there any order in chaos? Synergetic economics may provide some hints on these problems.

We now discuss the dynamic behavior of a man in a changeable environment – a problem which may be classified as psychology. However, understanding of the problem is necessary even for economic analysis. It should be noted that we do not intend to analyze the problem precisely. Only some ideas are provided to illustrate another potential application of the concepts introduced in this book.

Assume that the man (with a given heredity) is characterized by a few elementary "variables" such as attitudes towards money (or material conditions), (sexual) love, friendship, and the work ethic. These variables are very slow in comparison to "behavioral" variables such as consumption and choice of leisure time. Let these slow variables be denoted by vector $x$. For an economist it is not necessary to measure the slow variables $x$ since they are treated as constant parameters in economic analysis. However, philosophers (and maybe artists) are usually concerned with changes of such slow variables.

Economists choose consumption set, time distribution, wage, locational choice of dwelling site and housing qualities and so on, which may be rapidly changed, as variables in their analysis. We denote these fast variables by vector $y$. It should be noted that in most cases such fast variables are measurable. That is, they can be scientifically analyzed. Of course, among these variables, some, such as time distribution, may be slowly changed (or even fixed). The classification of the adjustment speeds depends on the characteristics of the individual and environment.

We assume that the dynamic behavior of the man can generally be described by

$$\frac{dx}{dt} = s\, f(x, y, t) \,, \tag{9.6.1}$$

$$\frac{dy}{dt} = g(x, y, t) \,, \tag{9.6.2}$$

where $s$ is a very small parameter and $f$ and $g$ are appropriate continuous functions.

The functions $f$ and $g$ are explicitly dependent on time because the environment is changeable. We may derive this dynamic system upon the basis of different mechanisms. For instance, under the assumption of rationality, the behavior may be described by an optimal control problem, whose solutions are satisfied by (9.6.1) and (9.6.2).

For simplicity, assume that the environment is invariant during the study period. That is, $f$ and $g$ are independent of time. Obviously, economists care little about equation (9.6.1), while philosophers invent all sorts of reasons to say that economists are not "right" only to study (9.6.2) without devoting any attention to (9.6.1).

As the parameter $s$ is very small, we see that the discussions about the adjustment speeds and the time scale in the previous sections hold similarly for (9.6.1) and (9.6.2). In a short-term analysis, it is sufficient to investigate the behavior of (9.6.2) and neglect (9.6.1), while in a long-term analysis, "behavioral variables" $y$ are governed by "attitude" variables $x$ and it is thus sufficient to examine the dynamics of $x$. If the system is stable, we see that the short-term analysis of the economist and the long-term study of the philosopher are both valid since there is a deterministic relation between long-term and short-term analyses of human behavior. That is, in the short-term analyses, it is acceptable to treat "attitude" variables as fixed parameters, while in the long-term study we can neglect "behavior" variables.

If the system is unstable, the problem becomes very subtle. For instance, even for a short-term analysis we cannot effectively treat the variables $x$ as constant since even $s$ is sufficiently small that the dynamic behavior when $s$ is not zero may be dramatically different from that when $s$ is zero.

I would now like to quote Hume (1748) to end this section: "Ambition, avarice, self-love, vanity, friendship, generosity, public spirit; these passions, mixed in various degrees, and distributed through society, have been, from the beginning of the world, and still are, the source of all the actions and enterprises, which have even been observed among mankind."

## Appendix: The Slaving Principle for Stochastic Differential Equations

In this appendix, we discuss how the slaving principle can be applied to stochastic differential equations. The general method is given in Haken (1983) and Gardiner (1983). As the problem is very complicated, we should like to use two simple examples to show the main points of the method. These examples are provided by Gardiner (1983, Chap. 6).

As discussed above, it is often the case that a dynamic system can be described by (stochastic) differential equations which have widely differing response times and in which the behavior on a very short time scale is not of interest. Now, we consider how the slaving principle can be applied to this type of equation.

Consider Langevin's equation which describes motion of the "Brownian particle" as follows

$$\frac{dx}{dt} = v \, ,$$

$$m\frac{dv}{dt} = -bv + (2kbT)^{1/2}h(t) \, , \qquad (9.A.1)$$

where $T$ is absolute temperature, $k$ the Boltzmann's constant, $m$ the mass, and $h(t)$ represents exogenous shocks with zero mean. We now consider the situation in which the friction coefficient $b$ is not small but the mass $m$ is very small.

The corresponding Fokker-Planck equation for the distribution function $p(x, v, t)$ is given by

$$\frac{\partial p}{\partial t} = -\frac{\partial(vp)}{\partial x} + \frac{\partial(vpb/m)}{\partial v} + \frac{kbT}{m^2}\frac{\partial^2 p}{\partial v^2} \, . \qquad (9.A.2)$$

Introducing the position distribution function $p^*(x, t)$ by

$$p^*(x,t) = \int_{-\infty}^{\infty} p(x, v, t) dv \, ,$$

and letting $m \to 0$, we obtain the Fokker-Planck equation for $p^*(x, t)$

$$\frac{\partial p^*}{\partial t} = \frac{kT}{b}\frac{\partial^2 p^*}{\partial x^2} \, . \qquad (9.A.3)$$

This is a standard partial differential equation which can easily be solved under appropriate initial and boundary conditions.

We have thus eliminated the fast variable $v$, which is assumed to relax very rapidly to the value given by $v(t) = (2kT/b)^{1/2}h(t)$. We see that large $b$ forces the variable $v$ governed by the equation involving large $b$ to relax to a value given by assuming the slow variable $x$ to be constant. The fast variable is then effectively slaved by the slow variable.

As another example, consider the deterministic equations in Sect. 9.1:

$$\frac{dx}{dt} = -r_1 x - \alpha xy \, , \qquad (9.1.1)$$

$$\frac{dy}{dt} = -r_2 y + \beta x^2 \, . \qquad (9.1.2)$$

We have shown how the slaving principle can be applied to this system. The stochastic version of the system is given by

$$dx = -(r_1 x + \alpha xy)\, dt + C dW_1(t) \, , \qquad (9.A.4)$$

$$dy = \left(-r_2 y + \beta x^2\right) dt + D dW_2(t) \, , \qquad (9.A.5)$$

where $C$ and $D$ are constants, and $W_1(t)$ and $W_2(t)$ are independent of each other. If $r_2$ is sufficiently large, we may replace $y$ by the stationary solution (9.A.5) in terms of $x$ to obtain

$$y = \frac{\beta x^2}{k} ,$$

$$\frac{dx}{dt} = -r_1 x - \frac{\alpha \beta x^3}{k} .$$

The Fokker-Planck equation for (9.A.4) and (9.A.5) is

$$\frac{\partial p}{\partial t} = \left[ \frac{\partial (r_1 x + \alpha x y)}{\partial x} + \frac{1}{2} C^2 \frac{\partial^2}{\partial x^2} + \frac{\partial (r_2 y - \beta x^2)}{\partial y} + \frac{1}{2} D^2 \frac{\partial^2}{\partial y^2} \right] p , \tag{9.A.6}$$

where $p$ is defined by

$$p = \text{prob} \left( x, y, t | x_0, y_0, t \right) ,$$

$$p \left( x, y, t_0 | x_0, y_0, t_0 \right) = \delta \left( x - x_0, y - y_0 \right) . \tag{9.A.7}$$

As in Sect. 9.1, we wish to eliminate $y$. We introduce

$$z = y - \frac{\beta x^2}{r_2} . \tag{9.A.8}$$

For fixed $x$, the quantity $z$ has zero mean. In terms of $z$, we can write a Fokker-Planck equation as

$$\frac{\partial p}{\partial t} = \left( L_1^0 + L_2^0 + L_3^0 \right) p , \tag{9.A.9}$$

where

$$L_1^0 = \frac{\partial (r_2 z)}{\partial z} + \frac{1}{2} \frac{D^2 \partial^2}{\partial z^2} ,$$

$$\begin{aligned} L_2^0 = & \frac{\partial (\alpha z x)}{\partial z} + \frac{2 \beta^2 x^2 C^2}{r_2^2} \frac{\partial^2}{\partial z^2} \\ & - \frac{2 \beta x}{r_2} \frac{\partial (r_1 x + \alpha \beta x^3 / r_2 + x z)}{\partial z} \\ & - C^2 \frac{\beta x}{r_2} \frac{\partial^2}{\partial x \partial z} - C^2 \frac{\partial^2 (\beta x / r_2)}{\partial x \partial z} , \end{aligned}$$

$$L_3^0 = \frac{\partial (r_1 x + \alpha \beta x^3 / r_2)}{\partial x} + \frac{C^2}{2} \frac{\partial^2}{\partial x^2} .$$

In order for the limit $r_1 \to 0$ $dx/dt = -r_1 x - \alpha \beta x^3 / k$ to be a valid limiting form, there must exist an $A$ such that $\alpha \beta / r_2 = r_1 A$. For this limit to be recognizable, it must not be swamped by noise so we must also have $C^2/2 = r_1 B$ as $r_1 \to 0$. So

$$L_3^0 \to r_1 \left[ \frac{\partial (x + A x^3)}{\partial x} + B \frac{\partial^2}{\partial x^2} \right] , \quad \text{as } r_1 \to 0 .$$

In order for $L_1^0$ to be independent of $r_1$, we require $r_2$ to be independent of $r_1$. Hence, $\alpha \beta / r_2 = r_1 A$ means that $\alpha \beta$ must be proportional to $r_1$. Gardiner considers various possiblities.

First, consider the case of the silent slave (the term is due to Gardiner): $\alpha = ar_1$. In this case, $L_1^0$ is independent of $r_1$ while $L_2^0$ and $L_3^0$ are proportional to $r_1$. It can be shown that the usual elimination procedure yields

$$\frac{\partial p^*}{\partial t} = \left[\frac{\partial(x + Ax^3)}{\partial x} + B\frac{\partial^2}{\partial x^2}\right] p^* ,$$

where $p^*(x,t)$ is the position distribution. This corresponds to eliminating $y$ adiabatically, ignoring the fluctuations in $y$ and simply setting the deterministic value in the $x$ equation. Gardiner called it the "silent slave" since $y$ is slaved by $x$ and makes no contribution to the noise in the $x$ equation.

For the case of the noisy slave where $a$ and $b$ are proportional to $r_1^{1/2}$,

$$\alpha = ar_1^{1/2} , \quad \beta = br_1^{1/2} , \quad \alpha\beta = r_2 A ,$$

the proposition distribution is given by

$$\frac{\partial p^*}{\partial t} = \left[\frac{\partial f(x)}{\partial x} + \frac{\partial^2 g(x)}{\partial x^2}\right] p^* ,$$

where $f(x) = x\{1 - (aD/r_2)^2/2 + Ax^3\}$, $g(x) = B + (axD/r_2)^2/2$. This is termed the "noisy slave" since the slave makes its presence felt in the final equation by adding noise.

# 10. Implications of Synergetic Economics

> The existence of analogies between central features of various theories implies the existence of a general theory which underlies the particular theories and unifies them with respect to those central features.
>
> *P.A. Samuelson* (1947)

We have examined unstable behavior in different nonlinear economic dynamic systems. It has been argued that linearization and stability are not universal but limited cases in economic evolutionary processes. These emphases are different from what traditional economics focuses on. For instance, in *Foundations of Economic Analysis* Samuelson tried to find central features of economic phenomena in optimality, linearization and stability. With traditional comparative statics analysis and the correspondence principle we can only deal with systems in which small perturbations of parameters result in small changes of variables. In contrast to the traditional dynamics approach, this book studies those properties of dissipative systems in which small shifts of parameters cause qualitative changes of the dynamic behavior. We have shown that when the system becomes unstable, for instance, via perturbations in parameters, nonlinear terms become very important in determining the behavior of the system. In this chapter, we examine the economic implications of these observations.

## 10.1 Synergetic Economics and Its Relations to Synergetics

> ... we are not simply looking for truth, we are after interesting and enlightening truth, after theories which offer solutions to interesting problems. If at all possible, we are after deep theories.
>
> *Karl R. Popper* (1972)

Synergetic economics belongs to the field of theoretical economics. It deals with temporal and spatial economic evolution processes. In particular, synergetic economics deals with unstable nonlinear systems and focuses on nonlinear phenomena such as structural changes, bifurcations and chaos in economic evolution.

Many candidates for the name of this book such as "New Foundations of Economic Analysis", "New Evolutionary Economics", "Chaotic Economics" and "Synergetic Economics" were considered. We have selected "Synergetic Economics" to express our new approach to economic dynamics. The choice is influenced by Haken's synergetics.

Haken proposed synergetics as a general theory of the dynamic behavior of systems with particular characteristics. It deals with the cooperative interaction of many subsystems which thus engender macroscopic behavior of a self-organized nature. The focus of synergetics is on critical points where the system changes its macroscopic behavior and may undergo non-equilibrium phase transitions, including oscillations, spatial structures and chaos. The interest of synergetics is not merely restricted to transitions between equilibria and equilibrium-like attractors as limit cycles. It also tries to capture other transitions without a specific final form. Thus we can also consider synergetic economics as a field of synergetics.

It must be emphasized that although we develop synergetic economics from synergetics, we have also been strongly influenced by the works of Prigogine and others (e.g., Nicolis and Prigogine 1977, Prigogine 1980, Prigogine and Stengers 1984, Jantsch 1980) in forming the fundamental ideas about economic evolution presented in this book.

## 10.2 Relations to Traditional Dynamic Economics

> The growth of knowledge proceeds from old problems to new problems, by means of conjectures and refutations.
>
> *Karl. R. Popper* (1972)

Before examining the implications of synergetic economics for various economic problems, we discuss the relationship between synergetic economics and traditional economics. As synergetic economics is concerned with economic evolution, it is a field of dynamic economics. There are many theories such as business cycle theory and economic growth theory and many analytic methods such as the correspondence principle in dynamic economics. All of these theories and methods constitute the contents of traditional dynamic economics. Synergetic economics is an extension of traditional dynamic economics in the sense that the results of traditional dynamic economics can be explained within the framework of this new theory, and moreover it tries to explain other economic phenomena which have been neglected by traditional dynamic economics. The theories in traditional dynamic economics are not universal, but special cases from the point of view of synergetic economics. Although we cannot say that synergetic economics solves all of the problems in economic evolution, we can conclude that this new theory makes it possible for dynamic economics to explain/predict some dynamic economic processes which cannot be explained by the traditional theories and methods. Synergetic economics suggests a rather promising new direction for explaining the complexity of economic phenomena.

It is argued that some approaches in traditional economics, for example, the Arrow–Debreu competitive equilibrium system, are quite appropriate starting points for understanding economic phenomena. Traditional economics has suggested some fundamental economic mechanisms such as competition, co-operation and rational behavior of economic reality.

We have based synergetic economics upon some concepts. The concepts of rational behavior, stability and equilibrium, which play a fundamental role in development of traditional economics, are also important in synergetic economics. However, synergetic economics emphasizes the significance of concepts such as instabilities which are neglected in traditional economics. Synergetic economics finds sources of complexity of economic evolution in instability and nonlinearity, rather than stability and linearity (or linearization) as does traditional economics.

Business cycle theory is a main subject of traditional dynamic economics. This theory is also a main concern of synergetic economics. However, it is more than a mere renewed interest in formal theories of endogenous business cycles that have kept growing in recent years. We show that many economic mechanisms may produce oscillations. Business cycles can result from nonlinear interactions between different economic and political variables. They can occur not only in competitive economies, but also in planned economies.

Traditional business cycle theory is mainly concerned with regular (periodic) motion of economic variables. In traditional dynamic economics there is no theory which successfully interprets observed irregular motion of economic data with some endogenous mechanisms. Chaos has remained mysterious until the development of modern nonlinear dynamic theory. The concept of chaos is quite new in dynamic economic theory. Synergetic economics suggests some analytical methods to investigate exogenous chaos in economic systems. It is argued that chaos is the nature of any evolutionary economic system. The existence of chaos means that precise economic forecasting is almost impossible.

Synergetic economics provides some new insights into effects of stochastic processes upon economic evolution. It is argued that if a dynamic system is stable, the effects of noise with zero average values can be neglected in economic analysis. Such omission does not affect qualitative results of the analysis. Thus the point of view of the main streams in traditional economics about small fluctuations is valid under presumed stability. However, if systems are unstable, analysis of effects of noise becomes very complicated. Small fluctuations may cause dramatic changes in the behavior of dynamic systems.

It should be noted that the emphasis on instability can also be found in the works by Karl Marx, Keynes, Schumpeter and other economists, although these economists found different sources of instabilities. The "vision" of synergetic economics is very similar to the Schumpeterian vision about economic development. The Schumpeterian innovation shocks may be viewed as "energy implementations" leading to a qualitative change of the system. An economy without innovative forces is forced to remain in stable equilibrium, while innovation shocks may lead to chaotic behavior. However, this does not imply that what synergetic economics suggests has already been included in the works of Schumpeter. Even if people have the same vision about the same problem, reasoning processes may be different and such difference may bring about differences in the "degree of understanding". Moreover, the results of synergetic economics are testable using actual economic data. Mathematics plays a significant role in synergetic economics. Mathematics helps us express precisely what we mean by instability, cyclical development, chaos and so on. None of these can be found in the classical works just mentioned.

Imperfect information and irrationality are often emphasized by different authors in economic analysis. For instance, given the difficulty of following chaotic paths, Simon defined bounded rationality or satisfying outcome. He has argued that due to the difficulties of calculating an optimal strategy, economic agents will not behave in an optimising way, but will instead aim for a satisfactory outcome. The possibility of chaotic behavior may suggest another direction for interpreting Simon's bounded rationality.

Synergetic economics emphasizes interactions among various variables and among different levels of systems. Although the importance of such interactions has been recognized by "systems analysis", little deep new insight into social evolutionary processes has been obtained from this approach. Systems analysis is carried out under presumed stability. In this respect, it is still within the framework of traditional economics.

The introduction of nonlinearity and instability into economics may cause some new debates. For instance, the debate over which theory in economics is more realistic becomes more subtle. The existence of chaos affects the way in which an economic theory can be verified. The classical approach to verifying a theory is to make predictions and test them against experimental data. If the phenomena are chaotic, long-term predictions are intrinsically impossible. The process of verifying a theory thus becomes a much more delicate operation. Furthermore, synergetic economics may have especially negative implications on econometrics. As exact forecasting is proved to be impossible by the theory, the more sophisticated model structures and parameter estimations in econometrics may be viewed as superfluous. The impact of the existence of chaos seems to be rather negative not only in econometrics, but also on the development of theoretical economics. If the task of economics is not only to describe and explain economic phenomena in a historical fashion, but also to provide a basis for arguments in forecasting an actual economy, then the existence of chaos may lead to a failure of economics in telling precisely what may happen in the future.

In order to explain how to evaluate a new economic theory such as synergetic economics, we would like to illustrate some modern theories about growth of knowledge at the end of this section. This illustration only provides some hints on the problem from a rather broad point of view.

Just as material capital accumulation is a main concern of economics, the growth of knowledge of an individual and/or a society is one of the most important subjects in philosophy. There are different theories in the literature about the growth of knowledge. In what follows, we revise some points of view proposed by Popper, Kuhn and Lakatos, respectively.

An important approach to philosophy of science is suggested by Popper (1972). Central to Popper's approach is the concept of falsification. His argument is that empirical observation can never establish that a scientific generalization is true, for, however much evidence we can obtain in support of a theory, we can never be sure that the next observation will not turn out to be inconsistent with the theory. All that successful testing of a theory can do is fail to refute the theory. Such successful testing of a theory may be regarded as "confirming" the theory, in the sense that it increases our confidence in it, though this is not the same as proving the theory to be true. Although empirical observation cannot be used to verify a theory, it can be used to refute it. For Popper, falsifiability is the criterion with which to distinguish between science and nonscience. Scientific statements are in principle falsifiable. This emphasis on falsification leads Popper to stress the growth of scientific knowledge. Scientific knowledge, according to Popper, is not knowledge that has been established as true, but simply a body of generalizations which have, so far, survived attempts to refute them. Science progresses by progressively eliminating false hypotheses. Non-scientific statements, on the other hand, are unfalsifiable, in that they do not rule out the occurrence of anything.

Another approach in philosophy of science is proposed by Thomas Kuhn (1962). He uses normal science as the fundamental concept to explain the growth of scientific knowledge. Here, "normal science" means "research based upon one or more past scientific achievements that some particular scientific community acknowledges for a time as supplying the foundation for further practice" (Kuhn 1962). Aristotle's *Physica,* Newton's *Principles* and Lavoisier's *Chemistry* are cited as examples of such paradigms. Normal science stimulated by a new paradigm must be sufficiently unprecedented to attract an enduring group of adherents and sufficiently open-ended to leave all sorts of problems for scientists to solve. Normal science has several important characteristics, the main one being the abandonment of critical discourse in the sense that there is a set of assumptions which are not questioned, and a set of procedures which are followed. This is the disciplinary matrix within which normal science is carried on. In undertaking normal science, scientists are not following a series of explicit rules, but they are following an example. According to Kuhn, it is only such an uncritical attitude which permits the application of the theory to a large number of problems, enabling a large number of detailed aspects of the world to be investigated. This uncritical attitude is obviously different from Popper's theory. Moreover, according to Kuhn, if scientists spent all their time arguing over fundamentals, they would never manage to investigate "small" phenomena. It is due to the acceptance of a particular form of normal science which leads to a more rigid definition of a field of research and forms a scientific community. Those who do not accept its basic assumptions should be excluded from the community. For much of the time, development of normal science may be rather "stable". However, from time to time crises arise because of the discovery of anomalies, or facts which cannot be explained in terms of the paradigm. For most of the time anomalies can be ignored: they are simply facts that the theory cannot yet explain. A crisis owing to an anomaly may be produced either when it contains something that is fundamental to the paradigm, or when it is particularly important for external reasons. The failure of a paradigm may, Kuhn argues, produce bewilderment for the scientists concerned, since they do not know how to put it right. The scientists can no longer be guided by the paradigm. On the other hand, according to Kuhn, a crisis may also arise because the modifications required for the theory render it transparently unsatisfactory. In this case, more and more complicated systems are introduced into the theory, but the complexity of the system increases much more rapidly than the accuracy of its predictions. Finally, it becomes evident that something was fundamentally wrong with the whole system, and a crisis may result from an anomaly after it has persisted for a sufficiently long time.

A crisis usually causes a large number of ad hoc modifications to the paradigm concerned, and of divergent articulations of the paradigm. Scientists search randomly for answers, even turning to philosophy, something for which there is little place in normal science. Eventually, from these new articulations of the paradigm, a new exemplar emerges. For Kuhn, it is only in such period of revolutionary science when the fundamentals of the science are questioned, that the Popperian idea of theories being tested through confrontation with empirical evidence is applicable.

A rather different approach from that of Popper and Kuhn has been proposed by Lakatos (e.g., 1978). One of his modifications to Popper's scheme is to argue that the unit of appraisal should be the research program rather than an individual theory, or even a succession of theories. A research program consists mainly of two parts: a hard core and a positive heuristic. The hard core consists of provisionally accepted assumptions which, as long as the research program is continued, are treated as irrefutable. The positive heuristic is defined as a powerful piece of problem-solving machinery which defines problems, foresees anomalies, and turns them victoriously into examples according to a preconceived plan. A research program will be rejected if there is a better one (with excess empirical content over its rival) to replace it. The new one has to explain everything that its rival can explain, as well as predict some novel facts that the rival cannot. As this demarcation criterion for science allows ad hoc modification to a theory and leaves room for minor inconsistencies, it is more tolerant than Popper's criterion of falsification, though the criterion of better prediction is stricter. According to Lakatos' approach, the order in which Kuhnian puzzles are solved is determined not by current anomalies, but by theoretical considerations. It is mathematical problems, rather than anomalies, which determine the path pursued by science.

## 10.3 Competitive and Planned Economies and Synergetic Economics

> The Chinese philosophers saw reality, whose essence they called Tao, as a process of continuous flow and change. ... The Chinese gave this idea of cyclical patterns a definite structure by introducing the polar opposites yin and yang, the two poles that set the limits for the cycles of change .... All natural phenomena are manifestations of a continuous oscillation between the two poles. .... The natural order is one of dynamic balance between yin and yang.
>
> *F. Capra* (1983)

In traditional Chinese philosophy yin and yang had never been associated with moral values. What is good is not yin or yang but the dynamic balance between the two; what is bad or harmful is imbalance. In Confucianism the balance between yin and yang of society is assumed to be carried out by a good governor. In Taoism such balance is assumed to be determined by natural forces. In a sense we find some essential ideas of Confucianism in central planning theory, and some of Taoism in the competitive equilibrium approach, although modern economics is much more profound than these classical documents. If we interpret socialist and capitalist, respectively, as "yin" and "yang" in the Chinese tradition, we see that in practice, how to determine balance between yin and yang is still an unsolvable problem in modern economics.

However, intellectual and political battles between socialists and capitalists seem to have become less serious than before. The main reason is due to the fact that the "capitalist countries" have introduced many government interventions into their economic systems and the "socialist countries" have introduced competitive mechanisms. Mixture-type economies dominate the world. Since there are advantages and disadvantages in both competitive and planning systems, there are no definitive conclusions about which mechanism is preferable to the other.

We will discuss competitive and planning mechanisms from the point of view of synergetic economics. Economists view capitalist society differently. For instance, Karl Marx, Schumpeter and Keynes treated the competitive economy as an unstable system. In modern monetary economics it is argued that whether or not competitive economic systems are stable depends upon actual situations. On the other hand, in modern neoclassical growth theory, capitalist society is presumed to be stable. Even when economists such as Marx, Schumpeter and Keynes consider capitalist society to be unstable, sources of instability and their attitudes towards society are completely different.

From the point of view of synergetic economics there is no evolutionary economic system which can be kept stable forever. An evolutionary system is always subjected to changes due to external and internal forces. When systems cross some critical values of external parameters, sudden changes and chaos may occur. Although there are many beneficial aspects of pure competition, as capitalist society is potentially unstable, the conditions required for perfect competition can hardly be satisfied in the long term. The instability of competitive systems may bring inequality among people. As the economic system is characterized by chaos, some people can always make fortunes by chance. Fortune and misfortune must fall on someone sooner or later. Therefore, according to the unstable characteristics of the capitalist economy, some people can live very richly without working even if they are fortunate only once, while others may live very poorly with constant hard work. Inequality cannot be prevented by pure competitive mechanisms because of the existence of chaos. However, in such economic systems it is just because people know that there are chances for them to obtain positive profits that they make efforts to innovate and improve production efficiency. This point of view of synergetic economics is rather similar to views held by Schumpeter. All of the efforts to obtain positive profits make the whole system very unstable. It is impossible to distinguish a "causal" relation between instability and the efforts. It is just because of the existence of complicated interactions between instability and persistent efforts that capitalist systems are always in motion. In such societies it is not only ability, but ability with accumulated chances which make inequalities larger and larger. As purely competitive mechanisms cannot prevent inequality, a perfect capitalist society cannot last forever, since purely competitive mechanisms cannot be supported in a society with large inequalities among firms and individuals.

Synergetic economics argues that instability may result in unforecastable structural changes in dynamic economic systems, such as the Depression. To prevent such a depression, some stabilizer has to be introduced into competitive systems. Government can prevent the capitalist society from being destroyed by following stabilizing policies. It can prevent the occurrence of dramatic decreases in production such as in the Depression by undertaking appropriate activities. Keynes attempted to introduce the government to stabilize the economy. It is through certain activities of the government that social problems can be solved. For Keynes the complexity of economic evolution can be solved by appropriate interventions. This optimism is held in Keynesian analysis since it is presumed that stability can always be guaranteed through interventions by the government.

Since a capitalist economy is unstable, some followers of Marx tried to replace it by centrally planning systems to solve the problems of capitalist society. It is imagined that a chaotic economic life can be replaced by a peaceful and happy one guided by central planning. In such a society, the people are the governor of the country and every person is equal in almost every respect. Democracy is thought to be one of the main characteristics of socialist society. Since competition is replaced by central planning, it is argued that the growth of a socialist economy should be faster than that of a capitalist economy. The most important characteristic of a socialist country is that the greatest degree of happiness of the whole society can be easily achieved. Although the socialist society may prevent the economic system from being unstable by central planning, there is no theory to prove that efficiency and stability can be achieved simultaneously. If efficiency cannot be guaranteed, stability seems to be meaningless in the long term. If instability occurs, how the goals of socialists can be achieved is not clear.

## 10.4 Implications for Developed and Developing Economies

> After a time of decay comes the turning point. The powerful light that has been banished returns. There is movement, but it is not brought about by force ... . The movement is natural; arising spontaneously. For this reason the transformation of the old becomes easy. The old is discarded and the new is introduced. Both measures accord with time; therefore no harm results.
>
> *I Ching*

Economic development is a very complicated dynamic process which often contains elements of both success and failure. Different authors emphasize different aspects of economic development. According to the levels of industrialization, average income and other factors, the present economic systems are classified as developed and developing countries. From an economic point of view, it is difficult to classify an economy as developed or developing. One reason is that "development" is a comparative concept. Furthermore, it is argued that economic mechanisms in the developing countries do not work as "perfectly" as those in the developed countries.

The main economic problems confronted by the developed and developing countries are different. Developing countries are usually confronted with low efficiency, low income, corruption of officials, pollution and so on. There are problems of unemployment and high inflation in developed countries. These problems reflect different troubles we may meet in different stages of economic development. We will apply synergetic economics to provide some hints on the complexity of economic development.

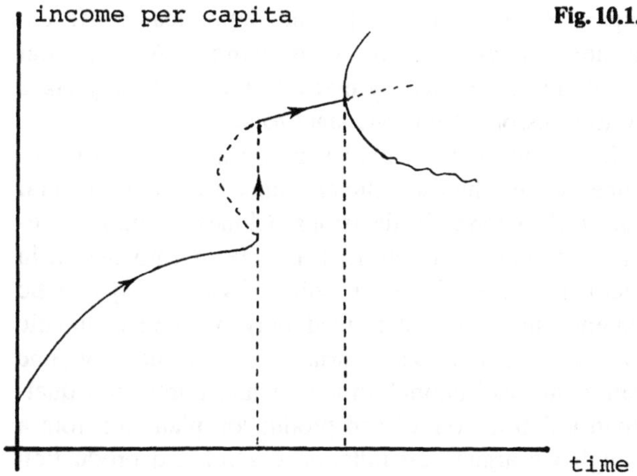


Fig. 10.1. An economic take-off

The American historian W.W. Rostow (1960) has introduced the idea of an economic "take-off" to describe structural changes in economic development. An economic take-off occurs when old inhibiting factors are overcome, production expands rapidly, and the economy moves into a long period of sustained growth. In terms of synergetic economics we interpret an economic "take-off" as a catastrophe phenomenon as illustrated in Fig. 10.1.

The crucial phase of such a "take-off" may be compressed into a relatively short period, and this structural change depends upon the whole structure of the system. A single factor shift cannot cause structural changes if the system is not located at a critical state. There may be many factors which can generate structural changes when the system is near the critical state. For example, in Britain, Rostow concludes, the economy "took off" in the two decades following 1783, a time span not greatly exceeding that of the American or French revolutions. This relative abruptness of the take-off contrasted with the centuries of preparation that proceeded it. The preparatorx stage has been traced all the way back to the early Middle Ages, which, by the insistence which prevailed then on discipline and on the performance of specified tasks at specified times, have been likened to a model for the organization of the labor force in a mine or mill. Among the factors that prepared Europe and America for industrialism are, for example, the capitalism of medieval and Renaissance bankers and merchants, the colonialism and merchatilism of the sixteenth and subsequent centuries, the rise of the competitive state system, the Protestant stress on hard work, and, especially, the brushing aside of tradition by the scientists of the seventeenth century and by the philosophers of the eighteenth.

Many people think that economic take-offs can occur in developing countries if they have sufficient financial support and other external conditions. However, in synergetic economics it is argued that structural changes occur if the system is near critical points. On the other hand, if the system is stable, small shifts of parameters can only result in small changes of economic states. As critical points are determined by the whole structure of the system, change in a single policy can hardly cause structural changes in economic development if many aspects of the society are not suitable for such sudden changes. No external change can dramatically affect a society if the whole society is not well prepared for such changes. As structural changes in economic development are determined by many factors, it usually takes a rather long time for a society to transform from one state to another.

Some politicians of developing countries only emphasize certain factors of economic development. For instance, some Chinese officials think that if the central government is able to make the right economic decisions, Chinese economic reforms can be successfully completed. Economic reforms have been emphasized in the documents of the government. However, from the point of view of synergetic economics, economic development cannot be determined only by pure economic factors. Interactions of various economic and social variables determine economic structures. Hence, for the government, institutional structures and population quality are much more important than inflation control and production planning from a long-term point of view. Although we argue that inflation control and production planning are short-term variables in comparison to institutional structures and population quality, this does not mean that we should not care about them; we argue that emphasis only on inflation control and production planning cannot provide a suitable environment for an economic "take-off".

## 10.5 Chance and Necessity in Economic Life

> All that exists in the universe is the fruit of chance and necessity. Tout ce qui existe dans l'univers est le fruit du hasard et de la nécessité.
>
> *Democritus* (from J. Mond, 1970)

It is obvious that almost every aspect of economic life, and life in general, is influenced by uncertainty. Life is full of misfortunes and opportunities. How, then, is the presence of misfortune and opportunity determined? Is there a God who controls one's life? Do misfortune and opportunity depend on one's effort? Is there any order in a chaotic life? These are the most essential questions in a human life. We give some explanations of these questions from the point of view of synergetic economics.

Synergetic economics argues that even if we can find deterministic mechanisms which control the behavior of human life, it is impossible to forecast all of the possible "trajectories" of the behavior. Life is full of chaos because of the existence of nonlinear interactions among the elementary variables under influences from the environment (see Sect. 9.5). Such chaotic behavior is not proposed to be determined by a God, but by the complexity of the structure which determines the way in which the individual will react to different external impacts. Moreover, the structure is also subjected to change, although its speed of change may be very slow.

Although there are many uncertain factors which may influence a human life, human behavior is not so randomly determined as it appears. There are deterministic mechanisms which provide one's capacity to prevent misfortunes and find opportunities. It is argued that such "capacity" can be increased by one's effort to learn. In this sense necessity and chance are strongly related to one's effort.

The argument of increasing "capacity" by effort similarly holds for a society. There is necessity of historical development, although chances may have a significant role in changing actual development paths. For instance, there are some societies which have become very rich by chance. However, if these societies cannot use these chances to improve potentials for further development, the chances cannot have great effects on economic development in the long term. How chances can change a society strongly depends on the social structures.

According to synergetic economics, nonlinear dynamic cooperation and competition among participants may result in chaotic phenomena which are actually beyond anyone's ability to forecast. It is during chaotic periods that people can have opportunities. On the other hand, necessity results from stability. In a permanently stable society, people can rarely hope for opportunities. There is no change in the society. This means that the social development evolves in a deterministic way. It is argued that in such a society there are few people who have hopes, which is an important factor in inspiring people to work.

## 10.6 Policy Decision in a Chaotic World

> The ideas of economists and political philosophers both when they are right and when they are wrong, are more powerful than is commonly understood. ... Practical men, who believe themselves to be quite exempt from any intellectual influences, are usually the slaves of some defunct economists.
>
> *J.M. Keynes* (1936)

In any modern economic system, government plays a significant role in directing economic development. It is important to investigate the effects of various policies upon different aspects of the economy such as economic growth, and income distribution both in the short term and long term. Evaluation of economic policies is a very complicated subject. It may be said that in practice there are almost no policies which benefit every group of the population both in the short term and in the long term. How to make a balanced combination of policies to make the population feel happy is difficult for any government.

Any policy decision should be based upon forecasting the events of the future. A policy may either stabilize the system or make behavior more chaotic. If the future is forecastable, we can evaluate the effects of a policy. However, if the economic system is chaotic, it is impossible to precisely foretell the effects of the policy. We now provide some examples to illustrate the complexity of policy decisions in an unstable system.

First, consider an example of the economic growth model suggested by Andersson and Zhang (1990). There are three variables – the capital for production, the capital for R&D activities and the knowledge stock – in the growth model. The dynamic equations of the change rates of capital stocks are similar to those of neoclassical growth models. Knowledge growth depends on learning by doing and the efficiency of R&D activities. The efficiency is, in turn, related to the knowledge policy of the government. As shown by Andersson and Zhang (1990), with respect

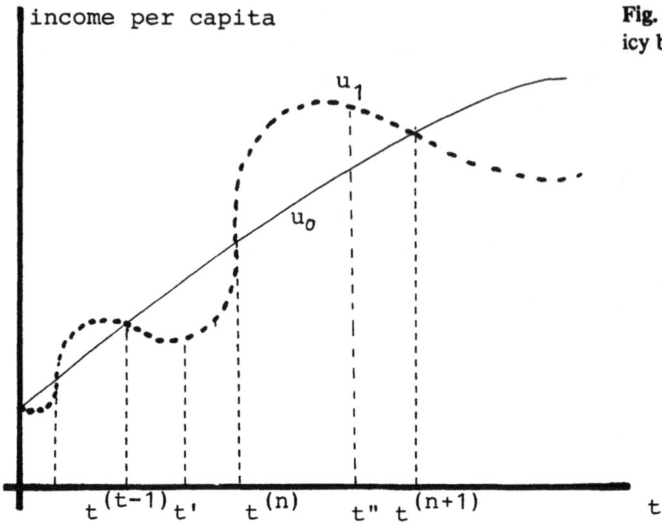


**Fig. 10.2.** How will knowledge policy be decided?

to the knowledge parameter chosen by the government there are different curves of income over time $t$. We describe curves in a $t$-$w$ plane ($t$ : time, $w$ : income) for the knowledge parameter $u$. For $u = u_0$, the system is stable and is at equilibrium; the real income per capita grows at a constant rate $n$. It is assumed that the government may choose another value $u_1$ of $u$, where the equilibrium loses its stability and Hopf bifurcation occurs. The cycle is shown in Fig. 10.2.

It should be mentioned that it is possible to choose an interval of time $[t^{(n-1)}, t^{(n)}]$ during which the sum of income along the cyclical economy is greater than the sum of income in equilibrium.

We assume that the effect of policy making is judged only according to its effects upon the change of income. We may ask the following question: for the households, which of the policy decisions, $u_0$ or $u_1$, is more "desirable"? Obviously, from the figure we can see that the total sum of income for $u_1$ is larger than that for $u_0$ for a very long period. Hence $u_1$ is more desirable than $u_0$ in the long term. However, if the judgement is limited to a certain period or point, the results may differ. For the interval $[t^{(n-1)}, t^{(n)}]$ the policy decision $u_1$ is less desirable than $u_0$; for the interval $[t^{(n)}, t^{(n+1)}]$ $u_0$ is less desirable than $u_1$. At time $t'$, $u_0$ is more desirable than $u_1$; at time $t''$, $u_1$ is more desirable than $u_0$. Thus the judgement of the effects of the knowledge policy is dependent upon the time scale under consideration. It is not difficult to see that there are many tricks which the politician can play on the households.

We may consider a more subtle question: in practice which of the policies, $u_0$ or $u_1$, may make the households feel happier in the short and long term? In the short term this depends upon the interval of time in which the effects of policy decisions are felt. In the long term, even if the total sum of income for $u_1$ is larger than that for $u_0$, it may intuitively be agreed that in practice the policy $u_0$ may make the households happier because for $u_0$ there is no period for which income relatively decreases.

From this example we see that in an unstable economic system it may be very complicated for the analyst to judge relationships between the happiness of people and policy decisions. It must be emphasized that how to analyse the implications of various policies for morals, justice, fairness and so on in an unstable economic system is much more difficult than it first appears.

## 10.7 Relations Between Microeconomics and Macroeconomics

The study of relationships among the whole and the components of a system is one of the most important subjects in sciences. It is generally agreed that the sum of the parts is not equal to the whole. However, there should exist relationships between them because the whole consists of the parts.

Economists have made great efforts to discover relationships between microeconomic behavior and macroeconomic variables, although most of these studies are limited to static analysis. The most elegant system which economists have discovered is the general equilibrium model (see Arrow and Hahn 1971). In this system the macroscopic variables – prices – are determined by microscopic competitive behavior of households and firms. Under some conditions there is a unique correspondence between the microscopic behavior and the macroscopic variables. However, this approach does not deal with dynamic behavior of households and firms, although various price adjustment processes have been suggested. In what follows, we discuss how to simultaneously take the dynamics of microscopic behavior and macroeconomic variables into account within the same framework.

It is argued that economic systems cannot be "reduced" to a single scheme. It is necessary to clearly define various levels of description and to find conditions that permit us to pass from one level to another. In synergetics, the process of describing real systems can be carried out hierarchically (Haken 1983). For instance, we may generally distinguish several levels of description as shown in Fig. 10.3.

The main mechanism of this description can be described as follows: close to the instability point we may distinguish between stable and unstable collective motions (modes). The stable modes are slaved by the unstable modes and can be eliminated. The remaining unstable modes serve as order parameters determining the macroscopic behavior of the system. The resulting equations for the order parameters can be grouped into a few universality classes which describe the dynamics of the order parameters. As the dimension of the reduced system may be very low, we can explain dynamic behavior which may appear to be unsolvable. The reduction process shown in Fig. 10.3 has been applied to economic dynamics by Weidlich and Haag (e.g., 1983). We illustrated this approach in Sect. 7.4. The model in Sect. 7.4 was derived upon the basis of individual behavior. However, the problem in this approach is whether it is appropriate to describe behavior of entrepreneurs by the stochastic approach. We have more knowledge about human behavior than about elementary particles in physics. As discussed in Sect. 9.5, microscopic behavior may follow some deterministic equations subject to influences from a random environment.

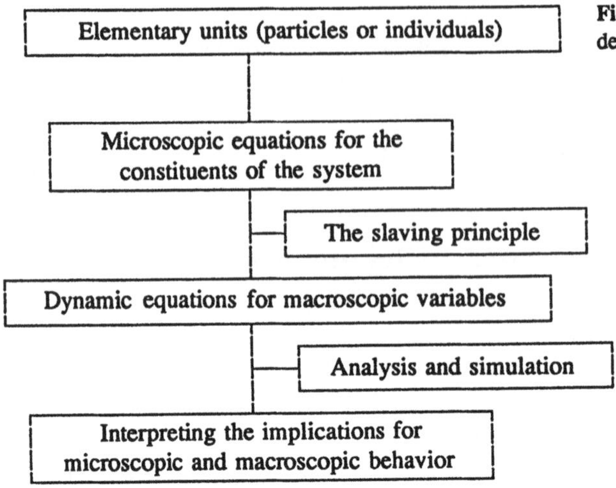


**Fig. 10.3.** Hierarchy for the levels of description of dynamic systems

We suggest the following procedure for analyzing the evolution of an economic system. For simplicity we neglect any random effects upon the system and limit our discussion to pure economic aspects.

We assume that the system consists of $n$ entrepreneurs (including firms and households). The economic behavior of each entrepreneur is characterized by an $m$-dimensional vector, $x_i$ (= $(x_{1i}, \dots x^i_m)$) ($i = 1, \dots, n$). For instance, the variables $x_{ij}$ may represent the actual consumption or production of entrepreneur $i$. There are $q$ macroeconomic variables $y_k$ ($k = 1, \dots, q$) which may represent prices, wages, interest rates and so on. The variables $y$ [= $(y_1, \dots, y_q)$] and $x_i$ are time dependent.

First, consider dynamics of the microscopic variables. It is assumed that each entrepreneur has perfect information about the macroscopic variables at each point of time. It is assumed that each entrepreneur makes a decision on consumption (production) upon the basis of current consumption (production) and the values of the macroscopic variables. We omit "direct interactions" among entrepreneurs. The dynamic behavior of the $i$th entrepreneur is proposed to be generally described as follows

$$\frac{dx_{ij}}{dt} = sf_{ij}(x_i, y) , \qquad (10.7.1)$$

where the parameter $s$ represents adjustment speed of the microscopic variables. Various approaches in microeconomics have different points of view about how to specify the functional forms $f_{ij}$.

The dynamics of the macroscopic variables are proposed to be generally described by

$$\frac{dy_k}{dt} = g_k(x_1, \dots, x_n, y) , \qquad (10.7.2)$$

where $g_k$ are assumed to continuously differentiable.

The dynamics of the whole system consist of (10.7.1) and (10.7.2). It is not difficult to see that some of the adjustment processes suggested in the equilibrium approach may be considered as a special case of our general system. For instance, if we assume that $s$ is sufficiently large and $y_k$ is the price of the $i$th good, then under appropriate conditions the system may be reduced to

$$f_{ij}(x_i, y) = 0 ,$$

$$\frac{dy_k}{dt} = g_k(x_1, \ldots , x_n, y) .$$

Thus, the dynamic system only consists of the motion of the prices. It should be noted that, in general, equilibrium approach exchanges occur only when prices arrive at equilibrium. However, in synergetic economics an "adaptive" point of view is accepted. That is, exchanges may occur even when prices are not at equilibrium.

The dimension of the general system is usually very high. As the system is potentially unstable, from synergetic economics we know that very complicated behavior may occur in the system. However, applying the analytical methods developed in this book (such as the slaving principle and the center manifold theorem), we can reduce this high-dimensional problem to a relatively low one. Thus, it is still possible for us to understand some properties of the dynamic system.

Finally, it must be emphasized that this section only provides some quite general ideas. An actual analysis of the proposed procedure may be rather complicated.

# 11. Conclusions and Prospects for Further Research

> Many people have a passionate hatred of abstraction, chiefly, I think, because of its intellectual difficulty; but as they do not wish to give this reason they invent all sorts of others that sound grand. ... Those who argue in this way are in fact concerned with matters quite other than those that concern science.
>
> *Bertrand Russell*

Like economic phenomena, economics is itself unstable in its development process. Synergetic economics is obviously located in a series of evolutionary orders of economic analysis. In his *Foundations of Economic Anylsis,* Paul A. Samuelson broadly classified the development of analytical economics into five steps. First, in Walras we have the final culmination of the notation of determinateness of equilibrium on the statical level. Pareto and others took a second step which laid the basis of a theory of comparative statics. The third step, which is characterized by maximizing action within an economic unit, was carried out by Johnson, Slutsky, Hicks, Allen and other economists. The fourth advance is due to the discovery of the correspondence principle. "A natural fifth step to take after we have investigated the response of a system to change in given parameters is to investigate its behavior as a result of the passage of time". Furthermore, Samuelson emphasized that "The usefulness of any theoretical structure lies in the light which it throws upon the way economic variables will change when there is a change in some datum or parameter. This commonplace holds as well in the realm of dynamics as in statics. It is a logical next step, therefore, to begin to create a theory of comparative dynamics. This will include the theory of comparative statics as a special case, and indeed all of the earlier five subjects, but it will cover a much richer terrain". Obviously, this fifth step has been cultivated in this book. This step took such a long time because it is only recently that the development of mathematics has supplied us with the powerful analytical methods which are necessary in understanding genuine dynamical behavior. Behind any scientific discovery we can always find historically accumulated knowledge which supports the work. Even a small man can see further than a giant if he sits on the giant's head.

We have argued that economic systems may run through a hierarchy of instabilities in which more and more structured patterns evolve. Such instabilities are caused by changes of external parameters and can lead to a new spatio-temporal pattern of the system. We have presented a number of explicit examples to show how such behavior can occur and have presented different analytical methods to cope with these problems. In particular, we have been concerned with sudden (structural) changes, the existence of limit cycles and chaos, the role of stochastic processes in economic evolution, and the effects of the time scale and the adjustment speeds in economic analysis.

The existence of economic chaos has significant implications for economic forecasting, methodologies and so on. It should be noted that the discovery of chaos has been based upon more fundamental, testable and acceptable concepts. It has not started from the concept of chaos. Chaos resulted from orders and certain rational mechanisms. Chaos is an observed phenomenon, not a mechanism. That is, fundamental mechanisms which create chaos are not random at all.

In a sense, chaos is not solely negative. It is not just the destruction of what exists. Chaos is also potentially positive. The hope for our future is that out of crises of our time may emerge what may be called a "great turning point" or a new, more positive direction for economic evolution. Although the existence of economic chaos means limitations for economic predictability during chaotic or maximally transitional states, this discovery also suggests new possibilities for improving forecasting within these limits by identifying patterns that foreshadow either impending chaos or potential order out of chaos. This capacity suggests how actual problems may be alleviated by using such theory to develop more effective "early warning systems".

We have also argued that from a psychological point of view human nature consists of only a few basic aspects: aspiration for knowledge, egoism, altruism, love, and so on. It is the combination of these aspects under different environments that creates the complexity of economic behavior.

This study is only a starting point of synergetic economics. We are confronted with more difficult analytic problems. Synergetic economic systems are described by unstable nonlinear dynamic equations of high dimensions with different adjustment speeds. Analytically, we cannot hope to completely understand the behavior of such systems. To illustrate the difficulty, we quote Arnold: "Nonintegrable problems of dynamics appeared inaccessible to tools of modern mathematics".

In this book we have examined and suggested many ideas. However, "new ideas, unless carefully elaborated, painstakingly defended, and 'pushed', simply will not tell" (Schumpeter 1934).

# References


Aftalion, A. (1913): *Les Crises Périodiques de Surproduction*, two vols. (Rivière, Paris)

Alonso, W. (1964): *Location and Land Use* (Harvard University Press, Cambridge, MA)

Anderson, P.W., Arrow, J.K., Pines, D. (eds.) (1988): *The Economy as an Evolving Complex System* (Addison-Wesley, New York)

Andersson, Å.E. (1986): "The Four Logistical Revolutions", Papers of Regional Science Association **59**, 1–12

Andersson, Å.E., Batten, D.F. (1988): "Creative Nodes, Logistical Networks, and the Future of the Metropolis", Transportation **14**, 281–293

Andersson, Å.E., Zhang, W.B. (1988): "The Two Dimensional Continuous Spatial Input-Output System", Ricerche Economiche **XLII**, 2

Andersson, Å.E., Zhang, W.B. (1988a): "Decision Centralization and Decentralization in a Dynamic Economic System", J. Comp. Appl. Math. **22**, 317–337

Andersson, Å.E., Zhang, W.B. (1990): "Endogenous Technological Changes and Economic Growth" in M. Chatterji, R. Kuenne (eds.) *Dynamics and Conflict in Regional Structural Change* (The Macmillan Press, London)

Andronov, A.A., Pontrjagin, L. (1937): "Systems Grossiers", Dokl. Akad. Kauk., SSSR **14**, 247–251

Araujo, A.P., Scheinkman, J.A. (1977): "Smoothness, Comparative Dynamics and Turnpike Property", Econometrica **45**, 601–620

Arrow, K.J. (1962): "The Economic Implications of Learning by Doing", Review of Economic Studies **24**, 155–173

Arrow, K.J., Hahn, F.H. (1971): *General Competitive Analysis* (Holden-Day, San Francisco)

Arrow, K.J., Hurwicz, L. (1958): "On the Stability of the Competitive Equilibrium I", Econometrica **XXVI**, 522–552

Arrowsmith, D.K., Place, C.M. (1982): *Ordinary Differential Equations* (Chapman and Hall, London)

Barro, R.J. (ed.) (1989): *Modern Business Cycle Theory* (Blackwell, Oxford)

Beckmann, M.J. (1952): "A Continuous Model of Transportation", Econometrica **20**

Beckmann, M.J., Puu, T. (1985): *Spatial Economics: Density, Potential, and Flow* (North-Holland, Amsterdam)

Benhabib, J., Day, R.H. (1981): "Rational Choice and Erratic Behaviour", Rev. Economic Studies **48**, 459–471

Benhabib, J., Day, R.H. (1982): "A Characterization of Erratic Dynamics in the Overlapping Generations Model", J. Economic Dynamics and Control **4**, 37–55

Benhabib, J., Miyao, T. (1981): "Some New Results of the Dynamics of the Generalized Tobin Model", International Economic Review **22**, 589–596

Benhabib, J., Nishimura, K. (1979): "The Hopf Bifurcation and the Existence and Stability of Closed Orbits in Multisector Models of Optimal Economic Growth", J. Economic Theory **21**, 421–444

Benhabin, J., Nishimura, K. (1986): *Endogenous Fluctuations in the Barro-Becker Theory of Fertility* (New York University)

Boldrin, M., Montrucchio, L. (1986): "On the Indeterminacy of Capital Accumulation Paths", J. Economic Theory **40**, 26–39

Britton, N.F. (1986): *Reaction-Diffusions and Their Applications to Biology* (Academic, New York)

Brock, W.A., Scheinkman, J.A. (1976): "Global Asymptotic Stability of Optimal Control Systems With Applications to the Theory of Optimal Economic Growth", J. Economic Theory **12**, 164–190

Burmeister, E., Dobell, A.R. (1970): *Mathematical Theories of Economic Growth* (Macmillan, New York)

Capra, F. (1983): *The Turning Point – Science, Society, and the Rising Culture* (Fontana, London)
Carr, J. (1981): *Applications of Centre Manifold Theory* (Springer, Berlin, Heidelberg)
Carr, J., Muncaster, R.G. (1983): "The Application of Centre Manifolds to Amplitude Expansions", J. Differential Equations **50**, 260–288
Cass, D., Shell, K. (1976): "The Structure and Stability of Competitive Dynamical Systems", J. Economic Theory **12**, 31–70
Casti, J.L., (1985): *Nonlinear System Theory* (Academic, New York)
Chang, W.W., Smyth, D.J. (1971): "The Existence and Persistence of Cycles in a Non-linear Model: Kaldor's 1940 Model Re-examined", Rev. Economic Studies **38**, 37–44
Chow, S.N., Hale, J.K. (1982): *Methods of Bifurcation Theory* (Springer, Berlin, Heidelberg)
Clark, J.M., (1917): "Business Acceleration and the Law of Demand: A Theoretical Factor in Economic Cycles", J. Political Economy **25**, 217–235
Clark, C.W. (1976): *Mathematical Bioeconomics* (Wiley, New York)
Coddington, E.A., Levinson, N. (1955): *Theory of Ordinary Differential Equations* (McGraw-Hill, New York)
Coullet, P., Tresser, J. (1978): "Iterations d'Endomorphismes et Groupe de Renormalisation", J. Phys. (Paris) Coll. **39**, C 5–25
Cournot, A. (1838): *Recherches sur les Principes Mathematiques de la Theorie des Richesses* (Rivière, Paris)

Dana, R.A., Montrucchio, L. (1986): "Dynamic Complexity in Duopoly Games", J. Economic Theory, 40–56
Day, R.H. (1983): "The Emergence of Chaos From Classical Economic Growth", Quarterly Journal of Economics **98**, 201–213
Day, R.H., Schafer, W. (1985): "Keynesian Chaos", J. Macroeconomics **7**, 277–295
Debreu, G. (1959): *Theory of Value* (Yale University Press)
Dendrinos, D.S., Mullally, H. (1983): *Urban Evolution Studies in the Mathematical Ecology of Cities* (Oxford University Press, Oxford)
Deneckere, R., Pelikan, S. (1986): "Competitive Chaos", J. Economic Theory **40**, 13–45
Deneubourg, J.L., de Palma, A., Kahn, D. (1979): "Dynamic Models of Competition Between Transportation Modes", Environment and Planning **2**, 665–673
Desai, M. (1973): "Growth Cycles and Inflation in a Model of the Class Struggle", J. Economic Theory **6**, 527–545
Dominguez, K.M., Fair, R.C., Shapiro, M.D. (1988): "Forecasting the Depression: Harvard Versus Yale", Am. Economic Rev. **78**, 595–612

Eckalbar, J.C. (1985): "Inventory Fluctuations in a Disequilibrium Macro Model", Economic Journal **95**, 976–991

Feigenbaum, M.J. (1978): "Quantitative University for a Class of Nonlinear Transformations", J. Stat. Phys. **19**, 25
Flaschel, P. (1984): "Some Stability Properties of Goodwin's Growth Model: A Critical Elaboration", J. Economics **44**, 63–69
Freedman, H.I., Waltman, P. (1975): "Perturbation to Two-Dimensional Predator–Prey Equations", SIAM J. Appl. Math. **28**, 1–10
Friedman, M. (1953): "The Methodology of Positive Economics", in *Essays in Positive Economics.* Reprinted in *Readings in Microeconomics*, ed. by W. Breit, H.W. Hochman (1986) (Holt, Rinehart and Winston, London)

Gabisch, G., Lorenz, H.-W. (1986): *Business Cycle Theory* (Springer, Berlin, Heidelberg)
Gale, D. (1973): "Pure Exchange Equilibrium of Dynamic Economic Model", J. Economic Theory **6**, 12–36
Gardiner, C.W. (1983): *Handbook of Stochastic Methods* (Springer, Berlin, Heidelberg)
Gilmore, R. (1981): *Catastrophe Theory for Scientists and Engineers* (Wiley, New York)
Golubitsky, M., Schaeffer, F.G. (1984): *Singularities and Groups in Bifurcation Theory*, Vol. I. (Springer, Berlin, Heidelberg)

Golubitsky, M., Schaeffer, F.G. (1988): *Singularities and Groups in Bifurcation Theory*, Vol. II (Springer, Berlin, Heidelberg)
Goodwin, R.M. (1951): "The Nonlinear Accelerator and the Persistence of Business Cycles", Econometrica **XIX**, 1–18
Goodwin, R.M. (1967): "A Growth Model", in *Socialism and Growth*, ed. by C.H. Feinstein (Cambridge University Press, Cambridge)
Grandmont, J.M. (1985): "On Endogenous Competitive Business Cycles", Econometrica **53**, 535–572
Grandmont, J.M. (1986): "Periodic and Aperiodic Behavior in Discrete One-Dimensional Dynamical System", in *Contributions to Mathematical Economics*, ed. by W. Hildenbrand, A. Mac-Colell (North-Holland, Amsterdam)
Grossmann, S., Thomae, S. (1977): "Invariant Distributions and Stationary Correlation Functions of One-Dimensional Discrete Processes", Z. Naturforsch. **32A**, 1353
Gu, Z.M., Nefedov, N.N., O'Malley, R.E. (1989): "On Singular Singularity Perturbed Initial Value Problems", SIAM J. Appl. Math. **49**, 1–25
Guckenheimer, J., Holmes, P. (1983): *Nonlinear Oscillations, Dynamical Systems, and Bifurcations of Vector Fields* (Springer, New York)
Guel, O., Rössler, O.E. (eds.) (1979): *Bifurcation Theory and Applications in Scientific Disciplines* (The New York Academy of Sciences)

Haavelmo, T. (1954): *A Study in the Theory of Economic Evolution* (North-Holland, Amsterdam)
Hahn, F. (1982): "Stability" in *A Handbook of Mathematical Economics*, Vol. 2, ed. by K.J. Arrow, M.D. Intriligator (North-Holland, Amsterdam)
Haken, H. (1975): "Analogy Between Higher Instabilities in Fluids and Lasers", Phys. Lett. **53A**, 77–78
Haken, H. (1977): *Synergetics: An Introduction* 3rd ed. (1983) (Springer Berlin, Heidelberg)
Haken, H. (1983): *Advanced Synergetics* (Springer, Berlin, Heidelberg)
Hao, B.L. (1984): *Chaos* (World Scientific, Singapore)
Harrod, R.F. (1936): *The Trade Cycle* (Oxford University Press, Oxford)
Hassard, B.D., Kazarinoff, N.D., Wan, Y.H. (1981): *Theory and Applications of Hopf Bifurcation* (Cambridge University Press, Cambridge)
Hawtrey, R.G. (1913): *Good and Bad Trade: An Inquiry into the Causes of Trade Fluctuations* (Constable, London)
Hayek, F.A. von (1933): *Monetary Theory and the Trade Cycle* (Harcourt, Brace & Co., New York)
Hicks, J.R. (1939): *Value and Capital.* (Clarendon, Oxford)
Hicks, J.R. (1950): *A Contribution to the Theory of the Trade Cycle* (Clarendon, Oxford)
Hicks, J.R. (1965): *Capital and Growth* (Oxford University Press, Oxford)
Hopf, E. (1942): *Abzweigung einer Periodischen Lösung von einer Stationären Lösung eines Differential-Systems* Ber. Math. Phys. (Sächsische Akad. der Wissenschaften, Leipzig) Translated in Marsden and McCracken (1976)
Hume, D. (1748): *An Enquiry Concerning Human Understanding*
Huseyin, K. (1986): *Multiple Parameter Stability Theory and Its Applications – Bifurcations, Catastrophes, Instabilities* (Clarendon, Oxford)

Iooss, G., Joseph, D.D. (1980): *Elementary Stability and Bifurcation Theory* (Springer, New York)
Isard, W. (1977): "Strategic Elements of a Theory of Major Structure Change", Papers, Regional Science Association **38**, 1–14

Jantsch, E. (1980): *The Self-Organizing Universe – Scientific and Human Implications of the Emerging Paradigm of Evolution* (Pergamon Press)
Jullien, B. (1988): "Competitive Business Cycles in an Overlapping Generations Economy with Productive Investment", J. Economic Theory **46**, 45–65

Kaldor, N. (1940): "A Model of the Trade Cycle", Economic Journal **50**, 78–92
Kalecki, M. (1935): "A Macro-dynamic Theory of Business Cycles", Econometrica **3**, 327–344
Kalecki, M. (1937): "A Theory of the Business Cycle", Rev. Economic Studies **4**, 77–97

Kevorkian, J., Cole, J.D. (1981): *Perturbation Methods in Applied Mathematics* (Springer, New York)
Keynes, J.M. (1936): *The General Theory of Employment, Interest and Money* (Harcourt, Brace, New York)
Knobloch, E. (1981): "Chaos in a Segmented Disc Dynamo", Phys. Lett. **82A**, 439–440
Koopmans, T. (1965): "On the Concept of Optimal Growth" in *The Econometric Approach to Development Planning* (Rand McNally, Chicago) pp. 225–287
Kuhn, T.S. (1962): *The Structure of Scientific Revolutions* (Chicago University Press, Chicago)

Lakatos, I. (1978): *The Methodology of Scientific Research Programme: Philosophical Papers*, Vol. 1 (Cambridge University Press, Cambridge)
Landau, L.D. (1944): C.R. Dokl. Acad. Sci. USSR **44**, 311
Landau, L.D., Lifshitz, E.M. (1959): *Fluid Mechanics* (Pergamon, Oxford)
Leijonhwfrud, A. (1973): "Effective Demand Failure", Swedish Journal of Economics **75**, 27–48
Levinson, N., Smith, O.K. (1942): "A General Equations for Relaxation Oscillations", Duke Math. J. **9**, 382–403
Lewis, W.A. (1955): *The Theory of Economic Growth* (George Allen & Unwin Ltd., London)
Li, T.Y., Yorke, Y. (1975): "Period Three Implies Chaos", American Mathematical Monthly **82**, 985-992
Lin, J., Kahn, P.B. (1977): "Limit Cycles in Random Environments", SIAM J. Appl. Math. **32**, 260–291
Long, N.V., Siebert, H. (1985): "Lay-off Restraints, Employment Subsidies, and the Demand for Labour", in *Optimal Control Theory and Economic Analysis* 2, ed. by G. Feichtinger (North-Holland, Amsterdam)
Lorenz, E. (1963): "Deterministic Nonperiodic Flow", J. Atmos. Sci. **20**, 130–141
Lorenz, H.W. (1986): "On the Uniqueness of Limit Cycles in Business Cycle Theory", Metroeconomica **XXXVIII**, 281–293
Lorenz, H.W. (1987): "International Trade and the Possible Occurrence of Chaos", Economics Lett. **23**, 135–138
Lucas, R.E. (1975): "An Equilibrium Model of the Business Cycle", J. Political Economy **83**, 1113–1144

Mangel, M. (1980): "Small Fluctuations in Systems With Multiple Cycles", SIAM J. Appl. Math. **38**, 120–138
Manneville, P., Pomeau, Y. (1979): "Intermittency and the Lorenz Model", Phys. Lett. **75A**, 1
Marsden, J.E., McCracken, M. (1976): *The Hopf Bifurcation and Its Applications* (Springer, New York)
Marx, K. (1867–1894): *Capital*, 3 vols., translated by S. Moore and E. Aveling. (Lawrence and Wishart, London)
Mas-Colell, A. (1985): *The Theory of General Economic Equilibrium: A Differentiable Approach* (Cambridge University Press, Cambridge)
Mees, A. (1975): "The Revival of Cities in Medieval Europe: An Application of Catastrophe Theory", Regional Science and Urban Economics **5**, 403–425
Metzler, L.A. (1941): "The Nature and Stability of Inventory Cycles", Rev. Econ. Stat. **23**, 113–129
Mill, J.S. (1848): *Principles of Political Economy* (Penguin, Harmondsworth, 1970)
Mitchell, W.C. (1913): *Business Cycles* (University of California Press, Berkeley)
Mond, J. (1970): *Le Hasard et la Nécessité* (Editions Du Seuil, Paris)

Nelson, R.R., Winter, S.G. (1982): *An Evolutionary Theory of Economic Change* (Harvard University Press, Cambridge, MA)
Newhouse, S., Ruelle, D., Takens, F. (1978): "Occurrence of Strange Axiom – An Attractor near Quasiperiodic Flow on $T^m, m \geqslant 3$", Commun. Math. Phys. **64**, 35
Nicolis, G., Prigogine, I. (1977): *Self-organization in Nonequilibrium Systems* (Wiley, New York)
Nikaido, H. (1968): *Convex Structures and Economic Theory* (Academic, New York)

O'Malley, R.E. (1988): "On Nonlinear Singularly Perturbed Initial Value Problems", SIAM Rev. **30**, 193–212

Pareto, V. (1908): *Manual of Political Economy*

Pedlosky, J., Frenzen, C. (1980): "Chaotic and Periodic Behavior in Finite Amplitude Baroclinic Waves", J. Atmos. Sci. **37**, 1177–1196

Peixoto, M.M. (1977): "Generic Properties of Ordinary Differential Equations", in *Studies in Ordinary Differential Equations*, ed. by J. Hale, Studies in Mathematics: The Mathematical Association of America, **14**, 52

Piremnne, H. (1925): *Medieval Cities*, translated into English by F.D. Halsey (1952) (Princeton University Press, Princeton, New Jersey)

Pigou, A.C. (1927): *Industrial Fluctuations* (Macmillan, London)

v.d. Ploeg, F. (1983): "Predator-prey and Neo-classical Models of Cyclical Growth", J. Economics **43**, 235–256

v.d. Ploeg, F. (1987): "Growth Cycles, Induced Technical Change and Perpetual Conflict over the Distribution of Income", J. Macroeconomics **9**, 1–12

Popper, K.R. (1972): *Objective Knowledge – An Evolutionary Approach* (Oxford University Press, Oxford)

Prigogine, I. (1980): *From Being into Becoming: Time and Complexity in the Physical Sciences* (Freeman, San Francisco)

Prigogine, I., Stengers, I. (1984): *Order out of Chaos: Man's Dialogue with Nature* (New Science Library, Boulder)

Puu, T. (1981): " Structural Stability and Change in Geographical Space", Environment and Planning A **13**

Puu, T. (1986): "Multiplier-Accelerator Models Revisited", Regional Science and Urban Economics **16**

Puu, T. (1987): "Complex Dynamics in Continuous Models of the Business Cycle", in *Economic Evolution and Structural Adjustment*, ed. by. D. Batten, J. Casti, B. Johansson (Springer, Berlin, Heidelberg)

Ricardo, D. (1817): *Principles of Political Economy and Taxation* (Penguin, Harmondsworth, 1971)

Robertson, D.H. (1915): *A Study of Industrial Fluctuations* (Aldwych, London, 1948)

Rostow, W.W. (1960): *The Stages of Economic Growth* (Cambridge University Press, Cambridge)

Ruelle, D., Takens, F. (1971): "On the Nature of Turbulence", Commun. Math. Phys. **20**, 167

Samuelson, P.A. (1939): "Interactions Between the Multiplier and the Principle of Acceleration", Rev. Economics and Statistics **21**, 75–78

Samuelson, P.A. (1947): *Foundations of Economic Analysis* (Harvard University Press, Cambridge, MA)

Samuelson, P.A. (1972): *The Collected Scientific Papers of Paul A. Samuelson*, Vol. III (MIT Press, Cambridge, MA)

Sargent, T.J. (1979): *Macroeconomic Theory* (Academic, New York)

Sattinger, D.H. (1973): *Topics in Stability and Bifurcation Theory* (Springer, Berlin, Heidelberg)

Schinasi, G.J. (1982): "Fluctuations in a Dynamic Intermediate-Run IS-LM Model: Applications of the Poincaré-Bendixon Theorem", J. Economic Theory **28**, 369–375

Schumpeter, J.A. (1934): *The Theory of Economic Development* (Harvard University Press, Cambridge, MA)

Schumpeter, J.A. (1939): *Business Cycles: A Theoretical, Historical, and Statistical Analysis of the Capitalist Process*, 2 vols. (McGraw-Hill, New York)

Schumpeter, J.A. (1975): *Capitalism, Socialism and Democracy* (Harper & Row, New York)

Schuster, H.G. (1988): *Deterministic Chaos – An Introduction*, 2nd revised ed. (VCH, Weinheim, F.R.G.)

Semmler, W. (ed.) (1985): *Competition, Instability and Nonlinear Cycles* (Springer, Berlin, Heidelberg)

Semmler, W. (1986): "On Nonlinear Theories of Economic Cycles and the Persistence of Business Cycle", Mathematical Social Sciences **12**, 47–76

Shah, A., Desai, M. (1981): "Growth Cycles With Induced Technical Change", The Economic Journal **91**, 1006–1010

Smale, S. (1967): "Differentiable Dynamical Systems", Bull. Amer. Math. Soc. **73**, 747–817

Smith, A. (1776): *An Inquiry into the Nature and Causes of the Wealth of Nations* (University of Chicago Press, Chicago, 1976)

Sparrow, C. (1982): *The Lorenz Equations, Bifurcations, Chaos, and Strange Attractors* (Springer, Berlin, Heidelberg)

Solow, R. (1956): "A Contribution to the Theory of Growth", Quarterly Journal of Economics **70**, 65–94
Solow, R. (1988): "Growth Theory and After", Am. Economic Review **78**, 307–317 (This is the lecture delievered in Stockholm when he received the Nobel Prize in Economic Science)
Spiethoff, A. (1953): "Business Cycles", Int. Econ. Pap. **3**, 75–171 (first published as "Krisen" in *Handwörterbuch der Staatswissenschaften*, Vol. 6, 1925, 8–91)
Steindal, A., Feichtinger, G., Hartl, R.F., Sorger, G. (1986): "On the Optimality of Cyclical Employment Policies – A Numerical Investigation", J. Economic Dynamics and Control **10**, 457–466
Stutzer, M. (1980): "Chaotic Dynamics and Bifurcation in a Macro Model", J. Economic Dynamics and Control **2**, 253–276
Swan, T.W. (1956): "Economic Growth and Capital Accumulation", Economic Record **XXXII**, 334–361

Takayama, A. (1985): *Mathematical Economics*, 2nd ed. (Cambridge University Press, Cambridge)
Tobin, J. (1965): "Money and Economic Growth", Econometrica **33**, 671–684
Tobin, J. (1969): "A General Equilibrium Approach to Monetary Theory", J. Money, Credit, Banking **1**, 15–29
Torre, V. (1977): "Existence of Limit Cycles and Control in Complete Keynesian System by Theory of Bifurcations", Econometrica **45**, 1457–1466
Tugan-Baranowskii, M.I. (1894): *Les Crises Industrielles en Angleterre* (Giard & Brière, Paris)

Varian, H.R. (1979): "Catastrophe Theory and the Business Cycle", Economic Enquiry **17**, 14–28
Velupillai, K. (1978): "Some Stability Properties of Goodwin's Growth Cycle Model", J. Economics **39**, 245–257

Walras, L. (1874): *Elements of Pure Economics*, translated by W. Jaffe (1954) (Geoge Allen and Unwin, London)
Weidlich, W., Haag, G. (1983): *Quantitative Sociology* (Springer, Berlin, Heidelberg)
Wicksell, K. (1898): *Interest and Prices: A Study of the Causes Regulating the Value of Money*, translated by R.F. Kahn (1936, Macmillan, London)
Wicksell, K. (1901): *Lecture on Political Economy*, Vol. 1, *General Theory*, translated by E. Classen (Routledge and Kegan, London, 1934)
Wiggins, S. (1988): *Global Bifurcations and Chaos – Analytical Methods* (Springer, Berlin, Heidelberg)
Wilson, A.G. (1981): *Catastrophe Theory and Bifurcation: Application to Urban and Regional Systems* (Choom Helm, London)

Ye, Y.Q. et al. (1986): *Theory of Limit Cycles*, translations of Mathematical Monographs (American Mathematical Society)
Yorke, E.D., Yorke, A. (1978): "Metastable Chaos Transition to Substained Chaotic Behavior in the Lorenz Model", J. Stat. Phys. **31**, 263–277

Zarnowitz, V. (1985): "Recent Work on Business Cycles in Historical Perspective: A Review of Theories and Evidence", J. Economic Literature **23**, 523–580
Zeeman, E.C. (1977): *Catastrophe Theory: Selected Papers: 1972–1977* (Addison-Wesley, Reading, MA)
Zhang, W.B. (1988a): "Limit Cycles in van der Ploeg's Model of Economic Growth and Conflict Over the Distribution of Income", J. Economics **48**, 159–173
Zhang, W.B. (1988b): "Hopf Bifurcations in Multisector Models of Optimal Economic Growth", Economics Lett. **26**, 329–334
Zhang, W.B. (1988c): "The Pattern Formation of an Urban System", Geographical Analysis **20**, 75–84
Zhang, W.B. (1988d): "Brain Drain and Economic Cycles with International Migration: A Case of Minimum Wage in the Unskilled Sector", J. Development Economics, **2**
Zhang, W.B. (1988e): "Urbanizing Processes With Moving Boundaries", Geographical Analysis **20**, 328–339

Zhang, W.B. (1988f): *Limit Cycles in an Optimal Employment Policy Model*, Umeå University Studies No. 183 (University of Umeå)
Zhang, W.B. (1989): *Economic Development – Nonlinearity, Instability and Non-Equilibrium* Doctoral Dissertation presented to Department of Economics, Umeå Economic Studies No. 198, University of Umeå
Zhang, W.B. (1989a): *Economic Growth and Technological Change*. International Journal of Systems Science (to appear)
Zhang, W.B. (1989b): "Oscillations in Rodriguez's Model of Entry and Price Dynamics", J. Economic Dynamics and Control **13**, 485–497
Zhang, W.B. (1989c): *Coexistence and Separation of Two Residential Groups – An Interactional Spatial Dynamic Approach*. Geographical Analysis **21**, 91–102
Zhang, W.B. (1989d): "Existence of Aperiodic Time-Dependent Solutions in Optimal Growth Economy With Three Sectors", International Journal of Systems Science **20**, No. 10, 1943–1953
Zhang, W.B. (1989e): *Urban Structural Changes in Continuous Time and Space* Umeå University Studies No. 191 (University of Umeå)
Zhang, W.B. (1989f): *Short-Run Inventory Oscillations in the Eckalbar's Disequilibrium Macro Model*. Applied Mathematics and Computation (to appear)
Zhang, W.B. (1990): "Stability Versus Instablity in Urban Pattern Formation", Occasional Paper Series on Socio-Spatial Dynamics **1**, 41–56
Zhang, W.B. (1990a): "The Complexity of Nonlinear Dynamic Economic Systems – The Kaldorian Model With Government Policy of Bond Finance", J. Mathematical Sociology **15** (4)
Zhang, W.B. (1990b): *Economic Dynamics – Growth and Development* (Springer, Berlin, Heidelberg)

# Subject Index

MIX
Papier aus verantwortungsvollen Quellen
Paper from responsible sources
FSC® C105338

If you have any concerns about our products,
you can contact us on
**ProductSafety@springernature.com**

In case Publisher is established outside the EU,
the EU authorized representative is:
**Springer Nature Customer Service Center GmbH**
**Europaplatz 3, 69115 Heidelberg, Germany**

Printed by Libri Plureos GmbH
in Hamburg, Germany